Foundation Engineering Mathematics

Mathematics plays a central role in modern culture, and a basic understanding of the nature of mathematics is required for scientific literacy. This new textbook will prepare readers to continue to develop analytical and numerical skills through the study of a variety of mathematical techniques. The statistical element of this textbook enhances readers' ability to organise and interpret data. Most of the topics covered in this textbook are widely used in various areas of engineering, including industrial engineering, to analyse complex systems, optimise processes and make informed decisions to improve efficiency, productivity and reliability in various industrial settings.

From the complexities of double integration and ordinary differential equations to the complexities of linear systems of differential equations, Fourier series and Laplace transform, *Foundation Engineering Mathematics* unfolds with careful attention to detail, offering readers a structured approach to mastering these fundamental topics. Each chapter of this book is carefully presented to provide a balance between theoretical foundations and practical applications, ensuring that readers not only grasp the underlying principles but also appreciate their relevance in real-world engineering scenarios. Each chapter is accompanied by practical examples, illustrative diagrams and engineering applications to reinforce understanding and demonstrate the relevance of mathematical concepts in engineering practice.

Whether you're a student embarking on your journey into the world of mathematics or an experienced engineer seeking to deepen your understanding of mathematical concepts, this book serves as an invaluable resource, guiding you through the complexities of mathematical theory and its engineering applications.

A solutions manual and a set of PowerPoint slides are available for qualified textbook adoptions.

Engineering Mathematics and Operations Research

Tools, Techniques, Theory, and Applications Used in Manufacturing and Management Science

Series Editor: Aliakbar Montazer Haghighi

The main aim of this new book series is to publish books by qualified authors in the areas related to mathematics and its applications in engineering and operations research areas. Thus, the series will publish rigorous and quality books related to a variety of areas of mathematics and science related to engineering. All books will include some applications, tools, and theories as well as the latest research and development of mathematics and its applications in engineering, manufacturing, and management sciences areas. Such topics that will be covered in this series are: vector analysis, linear algebra, fuzzy mathematics, sequences, series, probability, statistics, reliability, numerical analysis, approximation, teaching of mathematics for engineers, complex analysis, ordinary differential equations, differential equations, partial differential equations, stochastic processes, random walk, point process, Brownian motions, birth and death processes, and queueing theory.

Books in this series will be of interest to undergraduate and graduate students as course textbooks as well as reference books.

Optimization Techniques and Associated Applications
Edited by Sandeep Singh and Sandeep Dalal

Foundation Engineering Mathematics
By Faridon Amdjadi and Dharminder Singh

For more information about this series, please visit: www.routledge.com/Engineering-Mathematics-and-Operations-Research/book-series/CRCEMORTTTA

Foundation Engineering Mathematics

Faridon Amdjadi and Dharminder Singh

CRC Press is an imprint of the
Taylor & Francis Group, an **informa** business

Designed cover image: Shutterstock – Fouad A. Saad

First edition published 2025
by CRC Press
2385 NW Executive Center Drive, Suite 320, Boca Raton FL 33431

and by CRC Press
4 Park Square, Milton Park, Abingdon, Oxon, OX14 4RN

CRC Press is an imprint of Taylor & Francis Group, LLC

ISBN: 978-1-032-62740-3 (hbk)
ISBN: 978-1-032-63068-7 (pbk)
ISBN: 978-1-032-63069-4 (ebk)

DOI: 10.1201/9781032630694

Typeset in Times
by Apex CoVantage, LLC

Access the Instructor and Student Resources: www.Routledge.com/9781032627403

Contents

Preface ix
About the Authors xi

Chapter 1 Double Integration 1

1.1 Introduction 1
1.2 Double Integral over Rectangular Regions 3
1.2.1 Exercise 7
1.3 Double Integrals over Non-Rectangular Regions 8
1.4 Reversing the Order of Integration 15
1.4.1 Exercises 18
1.5 Double Integration over Polar Rectangular Regions 19
1.6 Double Integration over General Polar Regions 25
1.6.1 Exercises 29
1.7 Engineering Applications 30
1.7.1 Density Distributions of a Two-Dimensional Thin Layer 30
1.7.2 Centre of Mass of a Two-Dimensional Thin Layer (Lamina) 33
1.7.3 Moment of Inertia of a Two-Dimensional Thin Layer (Lamina) 36
1.7.4 Exercises 37

Chapter 2 Ordinary Differential Equations 39

2.1 Introduction 39
2.2 Vibrating Spring 39
2.3 ODEs – The Basics 42
2.4 Separable ODEs 44
2.4.1 Exercises 48
2.5 Linear Equations 49
2.6 First-Order Linear Equations 50
2.6.1 Exercises 55
2.7 Second-Order Linear ODEs with Constant Coefficients 56
2.7.1 Homogeneous Equations with Constant Coefficients 57
2.7.2 Exercises 63
2.8 Homogeneous Equations, the Initial Value Problems 63
2.8.1 Exercise 66
2.9 Non-Homogeneous Equations 67
2.9.1 The Method of Undetermined Coefficients 67
2.9.2 Exercises 73
2.10 Numerical Solutions of Differential Equations 75
2.10.1 Euler Method 75

2.10.2 Implementation of the Euler Method Using Excel 78
2.10.3 Exercises 81
2.11 Engineering Applications 82
2.11.1 Falling Objects 82
2.11.2 Mixture of Solutions 85
2.11.3 Vibration of a Spring-Mass System 87
2.11.4 Fluid Flow Streamlines 93
2.11.5 Heat Transfer – Fourier's Law of Heat Conduction 93
2.11.6 Electrical Circuit 95
2.11.7 Exercises 98

Chapter 3 Laplace Transform 101

3.1 Introduction 101
3.2 Definition of the Laplace Transform 102
3.3 Linear Property of the Laplace Transform 102
3.4 Inverse Laplace Transform 104
3.5 The First Shifting Theorem 106
3.5.1 Exercises 107
3.6 Inverse Laplace Transform Using Completing the Square 108
3.6.1 Exercises 109
3.7 Derivatives and the Laplace Transform 110
3.7.1 Using Laplace Transform to Solve ODEs 111
3.7.2 Exercises 114
3.8 The Unit Step Function 114
3.8.1 Products Involving Unit Step Functions 116
3.8.2 Laplace Transform of the Unit Step Function 119
3.9 The Second Shifting Theorem 120
3.9.1 Exercises 123
3.10 Dirac Delta Function 129
3.10.1 Exercises 134
3.11 Convolution 137
3.11.1 Visual Explanation 138
3.11.2 Some Useful Property of Convolution 144
3.11.3 Laplace Transform of $f * g$ 145
3.11.4 Exercises 145
3.12 Application in Control Engineering 148
3.13 Table of Laplace Transforms 150

Chapter 4 Linear Systems of Differential Equations 153

4.1 Introduction 153
4.2 Eigenvalues 154
4.3 Eigenvectors 155
4.3.1 Exercises 160
4.4 Repeated Eigenvalues 161

4.5 Complex Eigenvalues and Associated Eigenvectors 162
4.5.1 Exercises 168
4.6 Diagonalisation of a Matrix 169
4.6.1 Exercises 172
4.7 Homogeneous Linear Systems 173
4.7.1 Solving 2 × 2 Systems (Distinct Real Eigenvalues) 174
4.7.2 Solving 2 × 2 Systems (Algebraic Multiplicity 2, Geometric Multiplicity 1) 177
4.7.3 Solving 2 × 2 Second-Order Systems 181
4.7.4 Solving 3 × 3 Systems (3 Distinct Real Eigenvalues) 184
4.7.5 Solving 3 × 3 Systems (Algebraic Multiplicity 2 and Geometric Multiplicity 2) 185
4.7.6 Solving 3 × 3 Systems (Algebraic Multiplicity 2 and Geometric Multiplicity 1) 187
4.7.7 Solving 3 × 3 Systems (Algebraic Multiplicity 3 and Geometric Multiplicity 2) 190
4.7.8 Solving 3 × 3 Systems (Algebraic Multiplicity 3 and Geometric Multiplicity 1) 196
4.7.9 Exercises 199
4.8 Systems with Complex Eigenvalues 203
4.8.1 Exercises 208
4.9 Engineering Applications 208
4.9.1 A Two-Tank Mixing Problem 209
4.9.2 A Three-Tank Mixing Problem 210
4.9.3 Modelling 2 Degrees-of-Freedom Spring-Mass-Damper System 215

Chapter 5 Fourier Series 217

5.1 Introduction 217
5.2 Periodic Functions 217
5.3 Adding Periodic Signals with Unequal Frequencies 219
5.4 Adding Signals When Frequencies Are Integer Multiple of Smallest Frequency 225
5.5 The Beat Phenomenon 226
5.6 Frequency Domain (Frequency Spectrum) 229
5.6.1 Exercises 231
5.7 Fourier Series of the Square Wave 233
5.8 Fourier Series of Periodic Functions 236
5.9 Frequency Spectrum of the Fourier Series 238
5.9.1 Exercises 249
5.10 Engineering Applications 255
5.10.1 Vibration Analysis 255
5.10.2 Voltage Output of a Rectifier 257
5.10.3 Exercises 258

Chapter 6 Statistics 260

6.1 Introduction 260
6.2 Population 261
6.3 Sample 261
6.4 Random Variable 261
6.5 Qualitative Variables 261
6.6 Quantitative Variables 261
6.7 Descriptive Statistics 262
6.7.1 Summarising Data 262
6.7.2 Frequency Distribution (Discrete Variable) 262
6.8 The Mean, Standard Deviation and Median 266
6.8.1 Exercises 267
6.9 Continuous Quantitative Variables and Probability Distribution 271
6.10 Normal Distribution 274
6.10.1 Standard Normal Distribution 276
6.10.2 Exercises 279
6.10.3 Standardising the Random Variable X 279
6.10.4 Exercises 281
6.10.5 Calculating the X-Value by Knowing the Probability 282
6.10.6 Exercises 285
6.11 Estimation and Confidence Intervals 286
6.11.1 Point Estimation 286
6.11.2 Distribution of the Sample Mean 287
6.11.3 Exercises 289
6.11.4 Interval Estimates 289
6.11.5 Exercises 292
6.11.6 The t-Distribution 293
6.11.7 Exercises 296
6.12 Hypothesis Testing 297
6.12.1 Exercises 300
6.13 Correlation and Regression 301
6.13.1 Introduction 301
6.13.2 Correlation 303
6.13.3 Regression 304
6.13.4 Exercises 306

Index 311

Preface

Welcome to *Engineering Mathematics*, a comprehensive handbook designed to provide a solid foundation in mathematical techniques essential to engineering applications.

The principles of mathematics are based on creativity and logic and are essential for learning many scientific subjects. For many people, the beauty of mathematical concepts lies in their intellectual challenges, but for engineers, mathematics is a tool without which a simple task cannot be done. Engineering students cannot deal with a very simple electrical or mechanical system without the underlying mathematical knowledge. For engineering students, it is essential to understand the beauty of mathematical concepts and how to apply them in engineering.

Due to the underlying importance of mathematics in scientific practices and for its need in the advancement of practical techniques, a basic understanding of mathematics is essential for all students in particular for those who are training to be an engineer. This book is crafted with the aim of equipping engineering students and professionals with the necessary mathematical tools to tackle real-world problems across a wide range of disciplines and analyse complex problems, break them down into mathematical models and derive solutions. For engineering students who are interested in improving their mathematics skills, we provide systematic and logical approaches to problem-solving.

This textbook provides basic knowledge to students completing a degree programme in engineering at the university level. The aim is to meet the mathematical needs of students of varying abilities.

The audience of this book is Level 2 engineering students, and it provides useful coverage of mathematical topics for any field in engineering, including students who study mechanical or electrical engineering.

In writing this book, we have adopted a simple structure, and in some cases, we have avoided or minimised long proofs and theorems. We have given precise and short explanations about most of the topics, and we have introduced numerous examples for a very simple explanation.

Although there is literature on engineering mathematics, our approach is innovative, and the book brings new initiatives in some areas. This book is structured to cover a diverse array of mathematical topics, each presented in a clear and concise manner.

In the chapter on double integration, we have introduced a new approach to defining double integrals in Cartesian, as well as polar coordinate, systems. This approach relates the volume of a solid to a double integral rather than linking it to a double summation, which is used in the readily available literature. We explore the principles of double integration and their applications in calculating volumes and moments of inertia, essential for solving problems in mechanics, fluid dynamics and structural analysis.

The ordinary differential equations chapter describes the theory and techniques for solving ordinary differential equations, a cornerstone in modelling dynamic systems and predicting their behaviour over time. A comprehensive application of differential equations is presented focusing on falling objects, the mixture of solutions and the vibration analysis of spring-mass-damper mechanical systems, where we also discuss beat and resonance phenomena.

The chapter on Laplace transform focuses on understanding the concept of convolution using visual explanation and its use in linear time-invariant systems. The Laplace transforms covered in this book provide a convenient way for solving linear differential equations and constructing transfer functions. They also have important applications in reliability engineering, where engineers can model dynamic systems, estimate the mean time between failures and design maintenance strategies.

We have included a chapter on solving homogeneous linear systems of ordinary differential equations and have focused on systems associated with "defective matrices", as well as systems associated with complex eigenvalues and eigenvectors. We have described how to analyse and solve systems of linear differential equations, which is critical to the understanding of coupled dynamic systems and control theory applications.

The chapter on Fourier series covers the combination of waves with unequal frequencies and prepares students to understand the reasoning behind the construction of Fourier series. We explore the fascinating phenomena that arise from combining waves of different frequencies, including beat phenomena and resonance. The chapter also covers engineering applications in the field of vibration analysis and electrical circuits.

The statistics chapter includes descriptive as well as inferential statistics, short and precise, and the approach is more practical. We cover statistical methods and their engineering applications, including data analysis, continuous probability distribution (normal distribution), estimation of a population parameter, hypothesis testing and simple linear regression.

Each chapter is carefully presented to provide a balance between theoretical foundations and practical applications, ensuring that readers not only grasp the underlying principles but also appreciate their relevance in real-world engineering scenarios. Each chapter is accompanied by practical examples, illustrative diagrams and engineering applications to reinforce understanding and demonstrate the relevance of mathematical concepts in engineering practice. In each chapter, we have dedicated a section, solely, to the engineering applications, which includes vibration analysis, calculating the centre of mass and calculating the moment of inertia, free fall and mixing problems, among others. In addition, exercises and problems are provided at the end of each section to facilitate learning and assess understanding.

About the Authors

Dr. Faridon Amdjadi is a lecturer and professor in the Department of Mechanical Engineering at Glasgow Caledonian University. He is an applied mathematician whose research areas of interest include symmetric bifurcation theory, mode interactions and chaos and reaction-diffusion problems. He has published many papers in prestigious international journals. Dr. Amdjadi is a dedicated and accomplished lecturer with extensive experience in teaching engineering students. With a strong academic background and a passion for education, he has spent over a decade shaping the minds of future engineers. Dr. Amdjadi taught various mathematics modules at undergraduate and postgraduate levels to mathematics and engineering students and continues to inspire and empower his students, preparing them to excel in the ever-evolving field of engineering.

Dr. Dharminder Singh completed his PhD in 2017 at Glasgow Caledonian University (GCU), which focused on the use of the discrete element method to model the breakage of particles. Before joining GCU as a lecturer, Dharminder worked as a research and development engineer on a knowledge-transfer partnership between the University of Huddersfield and Trillium Flow Technologies, where he conducted research on the condition monitoring of control valves and the development of Smart valves. Dr. Singh is a lecturer in the Department of Mechanical Engineering at GCU and has published at conferences and in reputed international journals. He has taught various engineering modules which are underpinned by mathematics.

1 Double Integration

1.1 INTRODUCTION

The concept of a double integral is very similar to that of a single integral, except that double integrals are usually defined over a two-dimensional region. This makes the evaluation of integrals slightly different from single integrals that are evaluated over one-dimensional intervals.

The single integral of the function $y = f(x)$ over the interval $[a,b]$ (definite integral) defines the area under the curve $y = f(x)$ and above the x-axis, which can be denoted as follows:

$$\int_a^b f(x)dx.$$

The single integral can be linked to the total area of infinitesimal rectangular elements that also cover the area under the curve $y = f(x)$ and above the x-axis. To elaborate this, we divide $[a,b]$ into n small sub-intervals $[x_{i-1}, x_i]$, $1 \le i \le n$, of width $\delta x = \frac{b-a}{n}$. In each sub-interval, we construct a rectangle (see Figure 1.1a). The ith rectangle with the area $f(x_i)\delta x$ is shown in Figure 1.1b, where $f(x_i)$ is the height (length) of the ith rectangle whose width is δx.

The area, S, under the curve $y = f(x)$ and above the x-axis can be approximated by adding the area of n rectangles:

$$S \approx \sum_{i=1}^{n} f(x_i)\delta x.$$

The exact area is obtained as $n \to \infty$:

$$S = \sum_{i=1}^{\infty} f(x_i)\delta x = \int_a^b f(x)dx \tag{1.1}$$

Therefore, Equation 1.1 shows the relationship between the single integral and the total area of infinitesimal rectangular elements.

In this chapter, we introduce a "new approach" to define double integrals over a two-dimensional region R in Cartesian coordinates. The idea is to use the single integral of one-variable functions to obtain an exact definition for the double integral of two-variable functions. The same idea will be extended to the double integral in circular regions, where polar coordinates are used.

The applications of double integration in engineering are extensive. In Section 1.7, we consider the application of double integrals for calculating density

DOI: 10.1201/9781032630694-1

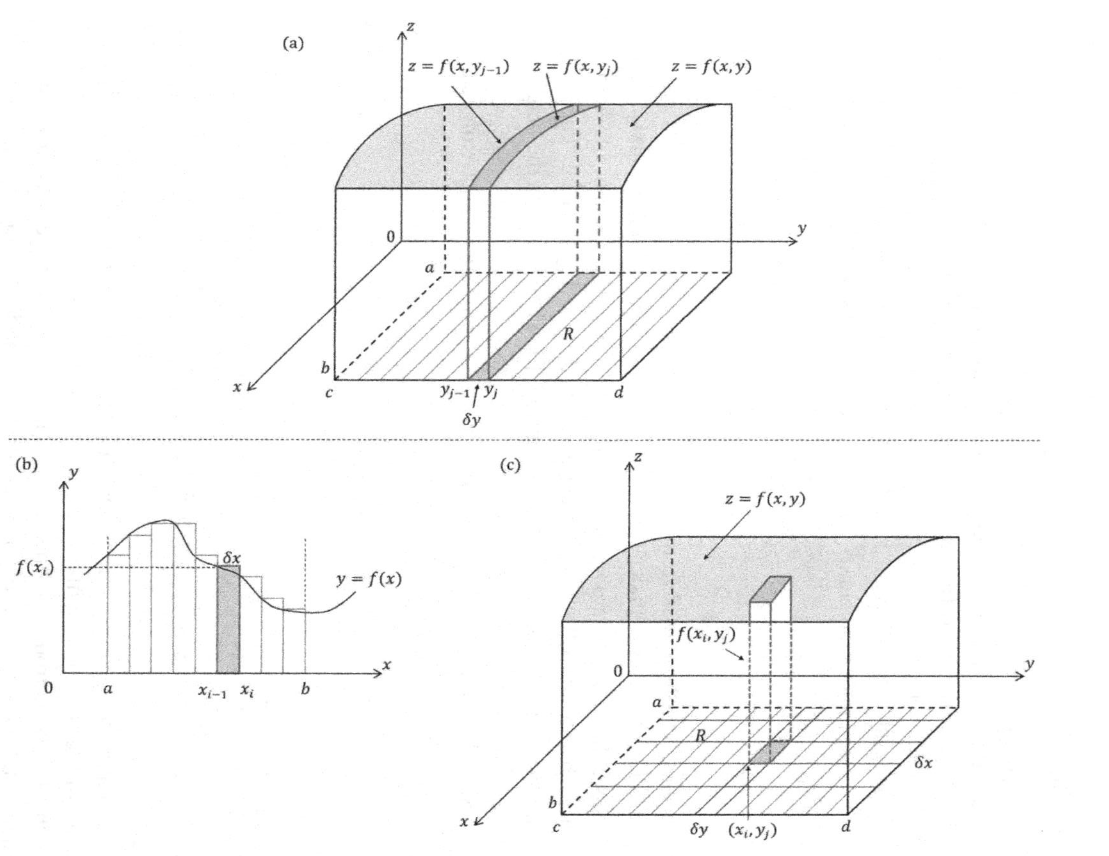

FIGURE 1.1 (a) A solid defined by the function $z = f(x, y)$ over the rectangular region R. The volume of the solid is approximated by adding the volume of n cuboidal elements with the volume of $v_j, 1 \le j \le n$. The jth cuboidal element is highlighted and $\delta y = \frac{d-c}{n}$. (b) The area under $y = f(x)$ is approximated by adding the area of n rectangles of height $f(x_i), 1 \le x_i \le n$ and width δx. The ith rectanglular element with the area $f(x_i)\delta x$ is highlighted. (c) The

distributions. In that section, by integrating the mass or density function over the object's region, we show how to employ double integration to determine the coordinates of the centre of mass of a two-dimensional solid. In Section 1.7, we also consider the application of using double integration to find the moment of inertia of a thin layer about the coordinate axes. All these topics are accompanied by numerical examples.

1.2 DOUBLE INTEGRAL OVER RECTANGULAR REGIONS

We begin this section by considering a surface defined over a rectangular region. Suppose $z = f(x,y) > 0$ is a two-variable function defined over the region $R = \{(x,y) \in R^2 : a \le x \le b,\ c \le y \le d\}$

The region R is a rectangle consisting of points (x, y), and the graph of the function f defines the surface that is obtained using the height $z = f(x,y)$ at these points. The graph of the function f along with the region R is shown in Figure 1.1a.

The notation

$$\iint_R f(x,y)\,dR$$

denotes the double integral of the function $z = f(x,y)$ over the region R. We are going to explain what we mean by the double integral of the function $z = f(x,y)$ of two variables x and y over the region R.

Let us calculate the volume of the solid under the surface $z = f(x,y)$ and above the region R. We divide the interval $[c,d]$ into n sub-intervals, $[y_{j-1}, y_j]$, $1 \le j \le n$, of width $\delta y = \frac{d-c}{n}$. We use these sub-intervals to divide R into n sub-rectangles and slice the solid into n small cuboidal elements, with the volume $v_j, 1 \le j \le n$, The jth element (highlighted in Figure 1.1a) is bounded at the bottom by the jth sub-rectangle of width δy and at the top by a small surface area between the two curves $z = f(x, y_{j-1})$ and $z = f(x, y_j)$.

The volume v_j can be approximated in two steps:

Step 1: Evaluate the area under the curve $z = f(x, y_j)$ in the range $a \le x \le b$ using single integral and denote it by $A(y_j)$, that is

$$A(y_j) = \int_a^b f(x, y_j)\,dx, \quad c \le y_j \le d. \tag{1.2}$$

Step 2: Multiply $A(y_j)$ by δy:

$$v_j \approx A(y_j)\delta y.$$

The volume, V, of the solid can be approximated by the total volume of such small cuboidal slices in the range between c and d:

$$V \approx \sum_{j=1}^{n} v_j = \sum_{j=1}^{n} A(y_j)\delta y.$$

Taking the limit as $n \to \infty$ and using the definition of the single integral gives

$$V = \sum_{j=1}^{\infty} A(y_j)\delta y = \int_{c}^{d} A(y)\,dy. \tag{1.3}$$

Now, we use Equation 1.2 and replace $A(y)$ in Equation 1.3:

$$V = \int_{y=c}^{y=d}\left[\int_{x=a}^{x=b} f(x,y)\,dx\right]dy = \int_{y=c}^{y=d}\int_{x=a}^{x=b} f(x,y)\,dx\,dy.$$

Thus, we have shown that the double integral evaluates the volume of a solid bounded from the top by the surface defined by $z = f(x,y)$ and from the bottom by the region R.

If we divide the interval $[a,b]$ into m sub-intervals of width $\delta x = \frac{b-a}{m}$ instead of dividing $[c,d]$ and use them to slice the solid into m small pieces in the y-direction we can get

$$V = \int_{a}^{b}\int_{c}^{d} f(x,y)\,dy\,dx.$$

Hence,

$$V = \int_{a}^{b}\int_{c}^{d} f(x,y)\,dydx = \int_{c}^{d}\int_{a}^{b} f(x,y)\,dx\,dy.$$

Key Concepts: We can obtain a relation between the double integral and the double summation, similar to the relation between the single integral and the summation presented in Equation 1.1:

- Express $A(y_j)$ defined in Equation 1.2 using a summation.

$$A(y_j) = \int_{a}^{b} f(x,y_j)\,dx = \sum_{i=1}^{\infty} f(x_i,y_j)\delta x, \ \delta x = \frac{b-a}{m}, \ m \to \infty$$

Substituting it in Equation 1.3, we obtain

$$V = \sum_{j=1}^{\infty}\sum_{i=1}^{\infty} f(x_i,y_j)\delta x\delta y.$$

Hence, double summation and double integration express the solid volume shown in Figure 1.1, which is

$$V=\sum_{j=1}^{\infty}\sum_{i=1}^{\infty}f\left(x_i,y_j\right)\delta x\delta y=\int_{y=c}^{y=d}\int_{x=a}^{x=b}f\left(x,y\right)dxdy=\iint_R f\left(x,y\right)dR.$$

- The relationship between the volume of a solid and the double summation can also be described as follows: divide $\left[a,b\right]$ into m sub-intervals of width $\delta x=\frac{b-a}{m}$ and divide $\left[c,d\right]$ into n sub-intervals of width $\delta y=\frac{d-c}{n}$. In each sub-interval, draw lines parallel to the x-axis and y-axes. In this way, the region R is divided into $N=m\times n$ small sub-rectangles with an area element of $\delta R=\delta x\times\delta y$ (see Figure 1.1c). Now in each area element δR construct a cuboid with a volume of $V_{i,j}=f\left(x_i,y_j\right)\times\delta x\times\delta y, 1\le i\le m,$ and $1\le j\le n$ where $\left(x_i,y_j\right)$ is a sample point in a typical area element. Summing up these volume elements and letting $n\to\infty$ and $m\to\infty$ gives:

$$V=\sum_{j=1}^{\infty}\sum_{i=1}^{\infty}V_{i,j}=\sum_{j=1}^{\infty}\sum_{i=1}^{\infty}f\left(x_i,y_j\right)\delta x\delta y=\iint_R f\left(x,y\right)dR.$$

The relationship between double integrals and double summation is also used in Section 1.7, "Engineering Applications".

Now, we will explain how to integrate the iterated integral

$$\int_{x=a}^{x=b}\int_{y=c}^{y=d}f\left(x,y\right)dydx.$$

The integral is evaluated inside out in two steps:

Step 1: Integrate $\int_{y=c}^{y=d}f\left(x,y\right)dy$ with respect to y regarding x as constant.

Step 2: Integrate the result of Step 1 with respect to x between the limits $x=a$ and $x=b$.

Example 1.1 Assume the two-variable function $z=f\left(x,y\right)=4+x-y$ is defined over $R=\left\{\left(x,y\right)\in R^2:0\le x\le 1,\ 0\le y\le 2\right\}$. Sketch the graph of the function f over the region R and evaluate its volume.

Solution

To draw the graph, we first draw the rectangular region R and then determine the values of the function f at the corners of the rectangle, which gives four points above the xy plane. The graph can be obtained by joining the 4 points together as shown in Figure 1.2.

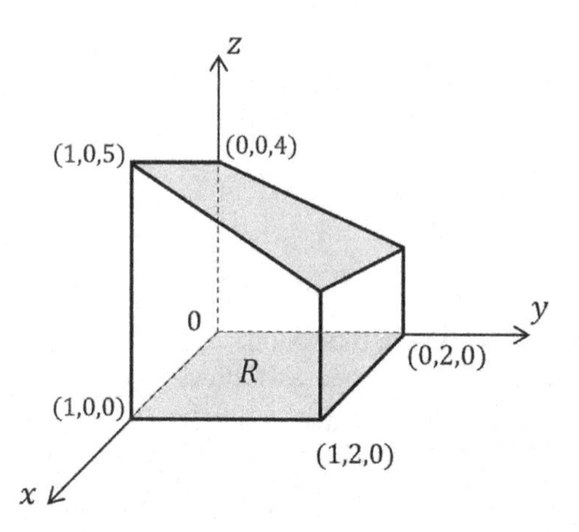

FIGURE 1.2 A solid defined by the function $f(x,y) = 4 + x - y$ over the rectangular region $R = \{(x,y) \in R^2 : 0 \le x \le 1, 0 \le y \le 2\}$.

To determine the volume of the solid, we must evaluate the following double integral:

$$V = \iint_R (4 + x - y)\,dR.$$

This integral can be integrated in two ways:

1. Replace dR with $dydx$ (write the integral in y-direction):

$$V = \int_{x=0}^{x=1}\int_{y=0}^{y=2} (4 + x - y)\,dydx.$$

Step 1: Evaluate the inner integral (treat x as constant)

$$\text{inner integral} = \int_{y=0}^{y=2} (4 + x - y)\,dy \;= \left[4y + xy - \frac{1}{2}y^2\right]_{y=0}^{y=2}$$

$$= (8 + 2x - 2) - 0 = 6 + 2x.$$

Step 2: Evaluate the outer integral using the result of Step 1:

$$\text{outer integral} = \int_{x=0}^{x=1} (6 + 2x)\,dx = \left[6x + x^2\right]_{x=0}^{x=1} = (6 + 1) - 0 = 7.$$

2. Replace dR with $dxdy$ (write the integral in x-direction):

$$V = \int_{y=0}^{y=2}\int_{x=0}^{x=1} (4 + x - y)\,dxdy.$$

Step 1: Evaluate the inner integral (treat y as constant):

$$\text{inner integral} = \int_{x=0}^{x=1} (4+x-y)\,dx = \left[4x+\frac{1}{2}x^2 - yx\right]_{x=0}^{x=1}$$
$$= \left(4+\frac{1}{2}-y\right) - 0 = \frac{9}{2} - y.$$

Step 2: Evaluate the outer integral using the result of Step 1:

$$\text{outer integral} = \int_{y=0}^{y=2} \left(\frac{9}{2}-y\right) dy = \left[\frac{9}{2}y - \frac{1}{2}y^2\right]_{y=0}^{y=2}$$
$$= (9-2) - 0 = 7.$$

Notice that

$$\int_{x=0}^{x=1}\int_{y=0}^{y=2} (4+x-y)\,dy\,dx = \int_{y=0}^{y=2}\int_{x=0}^{x=1} (4+x-y)\,dx\,dy = 7.$$

This means that the order of the integral over the rectangular region can easily be reversed without any complications, and both integrals give the same value.

NOTE: For the continuous function $z = f(x,y)$, the preceding result can be extended to any general region; that is $\iint_R f(x,y)\,dy\,dx = \iint_R f(x,y)\,dx\,dy$ is true for rectangular and non-rectangular regions.

1.2.1 Exercise

Evaluate the following integrals:

(a) $\int_0^2\int_0^1 xy^2\,dy\,dx$ (b) $\int_{-1}^2\int_1^3 x^2y^3\,dx\,dy$

(c) $\int_1^3\int_3^4 \left(2y^2+5x\right)dy\,dx$ (d) $\int_3^4\int_1^3 \left(2y^2+5x\right)dx\,dy$

(e) $\int_{-1}^2\int_1^2 y\,dx\,dy$ (f) $\int_1^2\int_{-1}^2 x\,dy\,dx$

(g) $\int_0^{\pi/2}\int_1^2 x\sin(xy)\,dy\,dx$ (h) $\int_1^2\int_0^{\pi/2} x\sin(xy)\,dx\,dy$

Answers

(a) $\frac{2}{3}$ (b) $\frac{65}{2}$ (c) $\frac{208}{3}$

(d) $\frac{208}{3}$ (e) $\frac{3}{2}$ (f) $\frac{9}{2}$

(g) 1 (h) 1

1.3 DOUBLE INTEGRALS OVER NON-RECTANGULAR REGIONS

In Section 1.2, we described double integrals over regions that are rectangular, but not all regions are rectangular.

The double integral $\iint_R f(x,y)dR$ over general regions can be explained in a similar way to the earlier description for rectangular regions. However, by omitting the details, we only explain the calculation of double integrals defined over non-rectangular regions. To be able to calculate the double integral of the function $z = f(x,y)$ over regions that are non-rectangular, we need to understand the shape of the general forms of the integration regions.

Although the region of integration is a projection of the surface defined by a function $z = f(x,y)$ in the xy-plane, however the process of integration will be easier if we can identify the type of integration regions. In Cartesian coordinates, unlike rectangular regions, non-rectangular regions are divided into two regions: **type 1 and type 2** regions. In this section, we explain these regions and give examples to clarify this topic.

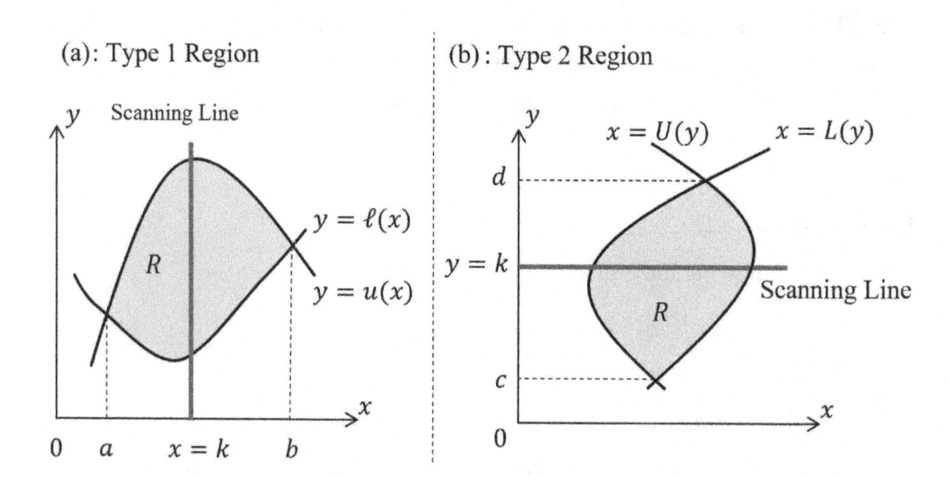

FIGURE 1.3 (a) A type 1 region bounded by the curves $y = \ell(x)$, $y = u(x)$ and the lines $x = a$, $x = b$ with a vertical scanning line at $x = k$. (b) A type 2 region bounded by the curves $x = L(y)$, $x = U(y)$ and the lines $y = c$, $y = d$ with a horizontal scanning line at $y = k$.

DEFINITION: TYPE 1 REGION

Suppose the region R is bounded at the bottom by $y=\ell(x)$, at the top by $y=u(x)$, on the left by the line $x=a$ and on the right by $x=b$, then R is said to be a type 1 region (see Figure 1.3a). The functions $y=\ell(x)$ and $y=u(x)$ are the lower and upper limits, respectively. For a type 1 region, the vertical line $x=k$, for $a<k<b$, that intersects R at the curves $y=\ell(x)$ and $y=u(x)$ is called the **vertical scanning line**.

For type 1 regions, each vertical scanning line that passes through R intersects first the lower and then the upper bounds of the integration region, defining the limits of the inner integral. Therefore, when R is considered as a type 1 region, the double integral $\iint_R f(x,y)dR$ can be described as the following iterated integral:

$$\int_{x=a}^{x=b}\left\{\int_{y=\ell(x)}^{y=u(x)} f(x,y)dy\right\}dx \quad \text{or} \quad \int_a^b\int_{\ell(x)}^{u(x)} f(x,y)dydx.$$

This iterated integral (in which dR is replaced by $dy\,dx$) is called **integration in the *y*-direction**.

KEY CONCEPTS

- When a double integral is written in the y-direction, then dR is replaced by $dydx$.
- The limits of the inner integral are functions of the x-variable, but the limits of the outer integral are constant numbers.

DEFINITION: TYPE 2 REGION

Suppose R is bounded on the left by $x=L(y)$, on the right by $x=U(y)$ at the bottom by the horizontal line $y=c$ and at the top by $y=d$, then R is a type 2 region (see Figure 1.3b). The functions $x=L(y)$ and $x=U(y)$ are called the left and right limits, respectively. For a type 2 region, the line $y=k$, for $c<k<d$, that intersects R at the curves $x=L(y)$ and $x=U(y)$ is called the **horizontal scanning line**.

For type 2 regions, each horizontal scanning line that passes through R intersects first the left and then the right boundaries of the integration region, defining the limits of the inner integral. Therefore, when R is considered as a type 2 region, the double integral $\iint_R f(x,y)dR$ can be described as the following iterated integral:

$$\int_{y=c}^{y=d}\left\{\int_{x=L(y)}^{x=U(y)} f(x,y)\,dx\right\}dy \text{ or } \int_{c}^{d}\int_{L(y)}^{U(y)} f(x,y)\,dxdy.$$

This iterated integral (in which dR is replaced by $dxdy$) is called **integration in the *x*-direction**.

KEY CONCEPTS

- When a double integral is set out in the x-direction, dR is replaced by $dxdy$.
- The limits of the inner integral are functions of the y-variable, but the limits of the outer integral are constant numbers.

Example 1.2 The double integral $\iint_R (3-x-y)dR$ is defined over the region R, where R is bounded by the following three lines:

$$y = x,\ x = 0 \text{ and } y = 1.$$

(a) Sketch the graph of the region R.
(b) Write the iterated integral $\iint_R (-x-y)\ dR$ in y-direction and evaluate it.
(c) Write the iterated integral $\iint_R (3-x-y)dR$ in x-direction and evaluate it.

Solution

(a) The graph of the region R is shown in Figure 1.4.

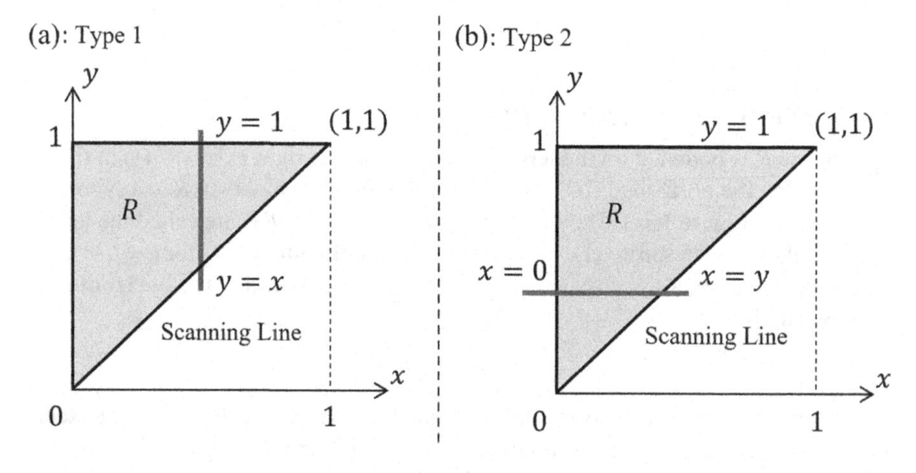

FIGURE 1.4 (a) A type 1 region bounded by the lines $y = x$, $y = 1$ and $x = 0$; the scanning line is vertical. (b) The same region but presented as type 2; the scanning line is horizontal.

(b) We consider the region R as type 1 and write the integral in the y-direction; that is the scanning lines are vertical, and dR is replaced by $dydx$.

The vertical scanning lines, one of which is shown in Figure 1.4a, starts at $y = x$ and finishes at $y = 1$. Thus, the limits of integration in the order of integrating first with respect to y and then x can be described as follows:

$$\begin{aligned} y = x \quad &\to \quad y = 1 \quad \text{(limits of the inner integral)} \\ x = 0 \quad &\to \quad x = 1 \quad \text{(limits of the outer integral).} \end{aligned}$$

Hence, the iterated integral over the region R can be written as follows:

$$\iint_R (3 - x - y)\, dR = \int_{x=0}^{x=1} \left\{ \int_{y=x}^{y=1} (3 - x - y)\, dy \right\} dx.$$

First, we evaluate the inner integral treating x as constant:

$$\begin{aligned} \text{inner integral} = \int_{y=x}^{y=1} (3 - x - y)\, dy &= \left[3y - xy - \frac{1}{2}y^2 \right]_{y=x}^{y=1} \\ &= \left(3 - x - \frac{1}{2} \right) - \left(3x - x^2 - \frac{1}{2}x^2 \right) \\ &= \frac{5}{2} - 4x + \frac{3}{2}x^2. \end{aligned}$$

Using the result obtained, we evaluate the outer integral:

$$\begin{aligned} \text{outer integral} = \int_{x=0}^{x=1} \left(\frac{5}{2} - 4x + \frac{3}{2}x^2 \right) dx &= \left[\frac{5}{2}x - 2x^2 + \frac{1}{2}x^3 \right]_{x=0}^{x=1} \\ &= \left(\frac{5}{2} - 2 + \frac{1}{2} \right) - 0 \\ &= 1. \end{aligned}$$

(c) The region R can be considered as a type 2 region and the integral can be written in the x-direction. In this case, the scanning lines are horizontal and dR is replaced by $dxdy$.

The horizontal scanning lines, one of which is shown in Figure 1.4b, starts at $x = 0$ and finishes at $x = y$. Thus, the limits of integration in the order of first x and then y can be described as follows:

$$x = 0 \quad \rightarrow \quad x = y \quad \text{(limits of the inner integral)}$$
$$y = 0 \quad \rightarrow \quad y = 1 \quad \text{(limits of the outer integral)}.$$

Hence, the iterated integral over the region R can be written as follows:

$$\iint_R (3-x-y)\,dR = \int_{y=0}^{y=1} \left\{ \int_{x=0}^{x=y} (3-x-y)\,dx \right\} dy.$$

The inner integral is evaluated first (regarding the y constant):

$$\begin{aligned}\text{inner integral} &= \int_{x=0}^{x=y} (3-x-y)\,dx = \left[3x - \frac{1}{2}x^2 - yx \right]_{x=0}^{x=y} \\ &= \left(3y - \frac{1}{2}y^2 - y^2 \right) - 0 \\ &= 3y - \frac{3}{2}y^2.\end{aligned}$$

Using the result of the inner integral, we obtain the outer integral:

$$\begin{aligned}\text{outer integral} &= \int_{y=0}^{y=1} \left(3y - \frac{3}{2}y^2 \right) dy \\ &= \left[\frac{3}{2}y^2 - \frac{1}{2}y^3 \right]_{y=0}^{y=1} = \left(\frac{3}{2} - \frac{1}{2} \right) - 0 = 1.\end{aligned}$$

We must emphasise that the result of part (c) is consistent with the answer obtained earlier in part (b).

KEY CONCEPTS

- To obtain the limits of integration, the region R must be plotted, and the scanning lines used to obtain the limits of the inner integral.
- The following is true:

$$\iint_R f(x,y)\,dR = \int_a^b \int_{y=\ell(x)}^{y=u(x)} f(x,y)\,dydx = \int_c^d \int_{x=L(y)}^{x=U(y)} f(x,y)\,dxdy.$$

Example 1.3 Assume the region R is bounded by the line $y = x$, the line $y = 2x$ and the line $x = 1$.

(a) Sketch the graph of the region R.

(b) Consider R as type 1 region. Write the iterated integral of $\iint_R (x+y)dR$ in y-direction and evaluate it.
(c) Consider R as type 2 region. Write the iterated integral of $\iint_R (x+y)dR$ in x-direction and evaluate it.

Solution

(a) The graph of the region R is shown in Figure 1.5.
(b) Vertical scanning lines, one of which is shown in Figure 1.5a, intersect the boundary of R first on the line $y = x$ and then on the line $y = 2x$. Thus, the limits of integration in the order of integrating first with respect to y and then x can be described as follows:

$$y = x \quad \to \quad y = 2x \quad \text{(limits of the inner integral)}$$
$$x = 0 \quad \to \quad x = 1 \quad \text{(limits of the outer integral)}.$$

Hence, the double integral over R can be written as

$$\iint_R (x+y)dR = \int_{x=0}^{x=1} \left\{ \int_{y=x}^{y=2x} (x+y)dy \right\} dx.$$

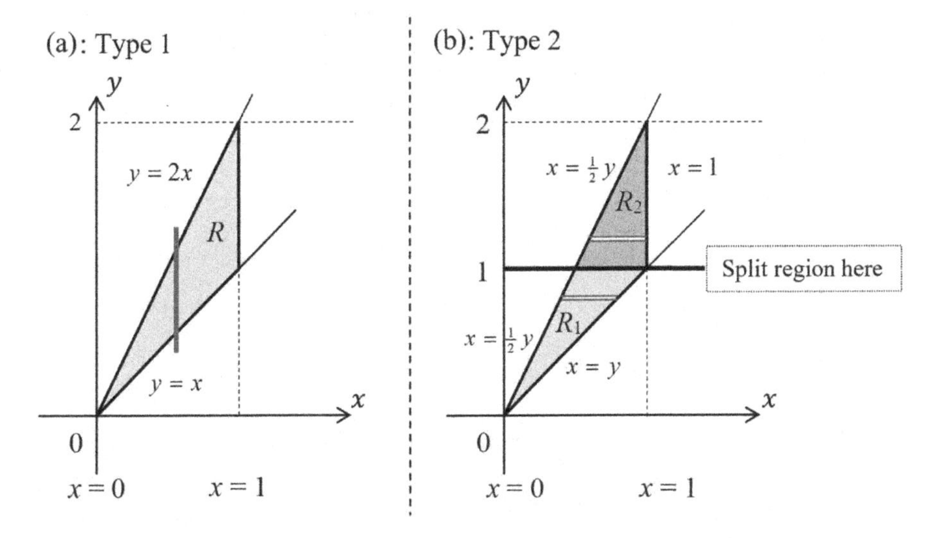

FIGURE 1.5 (a) A region bounded by the lines $y = x$, $y = 2x$ and $x = 1$. It is presented as a type 1 region with a vertical scanning line. (b) The same region but presented as type 2 region. The horizontal scanning lines for $0 \le y \le 2$ intersect the right boundary of R on different lines.

We evaluate the inner integral first, regarding x constant:

$$\text{inner integral} = \int_{y=x}^{y=2x} (x+y)\,dy$$

$$= \left[xy + \frac{1}{2}y^2 \right]_{y=x}^{y=2x}$$

$$= \left[x\times(2x) + \frac{1}{2}(2x)^2 \right] - \left[x\times x + \frac{1}{2}x^2 \right]$$

$$= 4x^2 - \frac{3}{2}x^2 = \frac{5}{2}x^2.$$

Then we evaluate the outer integral:

$$\text{outer integral} = \int_0^1 \frac{5}{2}x^2\,dx$$

$$= \left[\frac{5}{2}\times\frac{1}{3}x^3 \right]_{x=0}^{x=1}$$

$$= \frac{5}{6}\times 1^3 - \frac{5}{6}\times 0^3 = \frac{5}{6}.$$

(c) To write the iterated integrals in the x-direction, the horizontal scanning lines, for $0 \le y \le 2$ intersect the left boundary of the region R on $x = \frac{1}{2}y$. But for the intersection with the right boundary of the region R, for $0 \le y \le 1$, intersect it on $x = y$ and, for $1 \le y \le 2$, intersect it on $x = 1$ (see Figure 1.5b). Because the scanning lines for $0 \le y \le 2$ intersect the right boundary of R on different lines, thus, to set up the iterated integral in the x-direction requires dividing the region of integration R into two regions, R_1 and R_2, as shown in Figure 1.5b.

In the region R_1 $(0 \le y \le 1)$, the scanning lines intersect the region boundary first on $x = \frac{1}{2}y$ and then on $x = y$. Thus, in R_1,

$$x = \frac{1}{2}y \quad \rightarrow \quad x = y \quad (\text{limits of the inner integral})$$

$$y = 0 \quad \rightarrow \quad y = 1 \quad (\text{limits of the outer integral}).$$

In the region R_2 $(1 \le y \le 2)$, the scanning lines intersect the region boundary first on $x = \frac{1}{2}y$ and then on $x = 1$. Thus, in R_2,

$$x = \frac{1}{2}y \quad \rightarrow \quad x = 1 \quad (\text{limits of the inner integral})$$

$$y = 1 \quad \rightarrow \quad y = 2 \quad (\text{limits of the outer integral}).$$

Therefore, to set up the integral in the order of dx first and then dy, we can write

$$\iint_R (x+y)\,dR = \int_0^1 \int_{x=\frac{y}{2}}^{x=y} (x+y)\,dxdy + \int_1^2 \int_{x=\frac{y}{2}}^{x=1} (x+y)\,dxdy$$

$$= \left[\frac{7}{24} + \frac{13}{24}\right] = \frac{5}{6}.$$

Note: Clearly, for this example, integration in the y-direction is much simpler as it requires only "one" integration. Nevertheless, integration in the x-direction involves considerably more effort because it requires evaluating "two" iterated integrals.

1.4 REVERSING THE ORDER OF INTEGRATION

In Example 1.3, we evaluated a double integral in two ways by considering the region of integration first as a type 1 and then as a type 2. In this example, the calculation becomes more difficult when R is treated as a type 2 region, and the integral is first evaluated in the order of dx and then dy. The process of writing the double integral in a different order is called "reversing the order of integration".

In some cases, reversing the integration order helps us evaluate complex or even impossible integration. In this section, we provide examples to illustrate such complex integrations.

Example 1.4 Evaluate the following double integral:

$$\int_{x=0}^{x=1} \int_{y=x}^{y=1} \frac{\sin(y)}{y}\,dydx.$$

Solution

The integral is written in the y-direction. In the given order, the inner integral $\int \frac{\sin(y)}{y}\,dy$ cannot be integrated using readily available methods, but by reversing the order of integration and writing it as $\iint_R \frac{\sin(y)}{y}\,dx\,dy$, the function $\frac{\sin(y)}{y}$ can be considered as a constant term.

To reverse the order and write the integral in the x-direction, first, we identify the region R using the limits of integration. These limits are considered in Example 1.2, and its graph is shown in Figure 1.4; hence the integral in reverse order can be written as follows (see Figure 1.4b):

$$\int_{y=0}^{y=1} \int_{x=0}^{x=y} \frac{\sin(y)}{y}\,dxdy.$$

This integral is integrated as follows:

$$\text{inner integral} = \int_{x=0}^{x=y} \frac{\sin(y)}{y} dx \ (\text{y is constant})$$

$$= \left[\frac{\sin(y)}{y} x\right]_{x=0}^{x=y} = \frac{\sin(y)}{y} \times y - \frac{\sin(y)}{y} \times 0 = \sin(y)$$

$$\text{outer integral} = \int_{y=0}^{y=1} \sin(y) dy = \left[-\cos(y)\right]_{y=0}^{y=1}$$

$$= -\cos(1) + \cos(0) = 1 - \cos(1).$$

NOTE: The process of reversing the order of integration is very difficult until you understand the topic. First, you have to draw the integration region and then use the scanning lines to get the new limits in reverse order.

Example 1.5 Evaluate the following integral

$$\int_{x=0}^{x=1} \int_{y=\sqrt{x}}^{y=1} \sqrt{y^3 + 1}\, dy dx.$$

Solution

With the current order of integration, it is impossible to evaluate the inner integral using the readily available methods of integration. With the presented order, the limits of integration are as follows:

$$y = \sqrt{x} \quad \rightarrow \quad y = 1 \quad (\text{limits of the inner integral})$$

$$x = 0 \quad \rightarrow \quad x = 1 \quad (\text{limits of the outer integral}).$$

Hence, the region of integration can be easily plotted and is shown in Figure 1.6. To reverse the order of integration we draw the horizontal scanning lines to identify the limits of inner integral (see Figure 1.6b). Note that $y = \sqrt{x}$ and $y = 1$ intersect at the point $(1,1)$ and $y = \sqrt{x} \Rightarrow x = y^2$.

The horizontal scanning lines intersect the boundary of R first at $x = 0$ and then at $x = y^2$. Hence, the limits of integration in the order of integrating first with respect to x and then y are as follows:

$$x = 0 \quad \rightarrow \quad x = y^2 \quad (\text{limits of the inner integral})$$

$$y = 0 \quad \rightarrow \quad y = 1 \quad (\text{limits of the outer integral}).$$

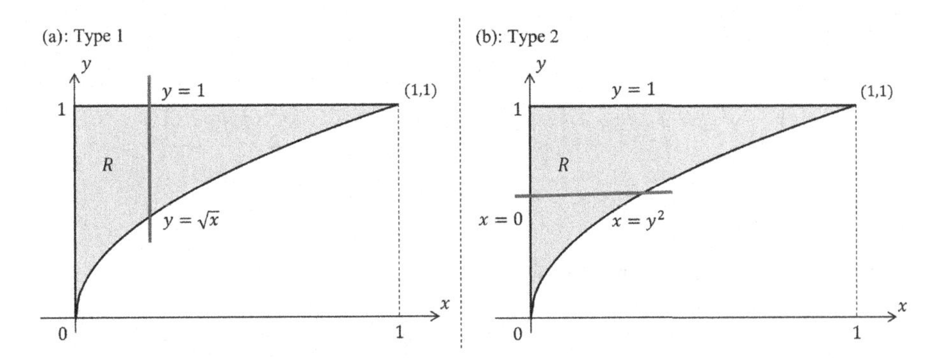

FIGURE 1.6 (a) The region R is bounded by $x=0,\ y=1$, and $y=\sqrt{x} \Rightarrow x=y^2$, a vertical scanning line is presented. (b) Same region but presented as type 2. The horizontal scanning line intersects the left boundary at $x=0$ and right boundary at $x=y^2$, defining the limits of the inner integral.

Thus, in x-direction, the integral is written as

$$\int_{y=0}^{y=1}\int_{x=0}^{x=y^2} \sqrt{y^3+1}\,dx\,dy.$$

We integrate the inner integral first, regarding y as constant:

$$\text{inner integral} = \int_{x=0}^{x=y^2} \sqrt{y^3+1}\,dx = \sqrt{y^3+1}\left[x\right]_{x=0}^{x=y^2}$$
$$= y^2\sqrt{y^3+1} - 0 = y^2\sqrt{y^3+1}.$$

We use this result of the inner integral to find the outer integral:

$$\text{outer integral} = \int_{y=0}^{y=2} y^2\sqrt{y^3+1}dy.$$

We use the substitution rule to calculate this integral:

$$u = y^3+1 \Rightarrow du = 3y^2dy \ or\ \frac{1}{3}du = y^2dy.$$

We use the lower and upper limits of the outer integral to find the values for the variable u:

$$y=0 \Rightarrow u = 0^3+1 = 1$$
$$\text{y} = 1 \Rightarrow u = 1^3+1 = 2.$$

Thus,

$$\int_{y=0}^{y=1} y^2\sqrt{y^3+1}dy = \frac{1}{3}\int_{u=1}^{u=2}\sqrt{u}\,du = \frac{1}{3}\left[\frac{2}{3}u^{\frac{3}{2}}\right]_{u=1}^{u=2} = \frac{2}{9}\left(2\sqrt{2}-1\right).$$

1.4.1 Exercises

1. Evaluate the following integrals:

(a) $\int_0^2\int_0^x \left(x^2+y^2\right)dydx$ (b) $\int_0^1\int_0^{2y}(2x+3y)dxdy$

(c) $\int_{-1}^2\int_{x^2}^{x+2} dydx$ (d) $\int_0^a\int_0^{\sqrt{a^2-y^2}} x^3dxdy$

2. Evaluate the integral $\iint_R y\,dR$, where R is bounded by the parabola $x^2 = 4y$ and the line $y = 1$.
3. Reverse the order of integration for the integrals in Exercise 1.
4. Reverse the order and integrate the following integrals:

(a) $\int_0^1\int_{\sqrt{y}}^1 \sqrt{x^3+1}dxdy$

(b) $\int_0^1\int_{x^2}^1 x^3\sin\left(y^3\right)dydx$

(c) $\int_0^1\int_{\sqrt{x}}^1 \cos\left(\frac{y^3+1}{2}\right)dydx$

5. For each of the following integrals:

(a) $\int_0^1\int_0^{2y} y\,dxdy + \int_1^3\int_0^{3-y} ydxdy$

(b) $\int_0^1\int_0^{\sqrt{4x-x^2}} dydx + \int_1^2\int_0^{\sqrt{4-x^2}} dydx$

- draw the region of integration;
- reverse the order of integration and express it in the form of a single iterated integral;
- evaluate the integral.

Answers

1. (a) $\frac{16}{3}$ (b) $\frac{10}{3}$ (c) $\frac{9}{2}$ (d) $\frac{2a^5}{15}$
2. $\frac{8}{5}$

3. (a) $\int_0^2\int_0^x \left(x^2+y^2\right)dydx = \int_0^2\int_{x=y}^{x=2}\left(x^2+y^2\right)dxdy = \frac{16}{3}$

(b) $\int_0^1\int_0^{2y}(2x+3y)\,dxdy = \int_0^2\int_{y=\frac{x}{2}}^{y=1}(2x+3y)\,dydx = \frac{10}{3}$

(c) $\iint_R dR = \int_{y=0}^{y=1}\left\{\int_{x=-\sqrt{y}}^{x=\sqrt{y}}dx\right\}dy + \int_{y=1}^{y=4}\left\{\int_{x=y-2}^{x=\sqrt{y}}dx\right\}dy$

$=\left[4/3+19/6\right]=9/2$

(d) $\int_0^a\int_0^{\sqrt{a^2-y^2}} x^3 dxdy = \int_0^a\int_{y=0}^{y=\sqrt{a^2-x^2}} x^3 dydx = \frac{2}{15}a^5$

4. (a) $\frac{2}{9}\left(2^{3/2}-1\right)$ (b) $\frac{1}{12}\left(1-\cos(1)\right)$ (c) $\frac{2}{3}\left(\sin(1)-\sin\left(\frac{1}{2}\right)\right)$

5. (a) $\int_0^2\int_{y=\frac{x}{2}}^{y=3-x} ydydx = 4$ (b) $\int_0^{\sqrt{3}}\int_{2-\sqrt{4-y^2}}^{\sqrt{4-y^2}} dxdy = -\sqrt{3}+\frac{4\pi}{3}$

1.5 DOUBLE INTEGRATION OVER POLAR RECTANGULAR REGIONS

When dealing with single integrals, we often use a process called changing the integration variable. This process can also be applied to double integrals when their region of integration is circular. Often, in circular regions, it is essential to convert the Cartesian coordinates (x,y) to the polar coordinates (r,θ), where $x = r\cos(\theta)$ and $y = r\sin(\theta)$, allowing tedious integration to be easily computed.

For example, the projection of the hemisphere $f(x,y)=\sqrt{a^2-x^2-y^2}$ in the xy-plane is a disc of radius a and the integral $\iint_R f(x,y)dR$ in Cartesian coordinates can be written as

$$\iint_R \sqrt{a^2-x^2-y^2}\,dR = \int_{x=-a}^{x=+a}\int_{y=-\sqrt{a^2-x^2}}^{y=+\sqrt{a^2-x^2}}\sqrt{a^2-x^2-y^2}\,dydx. \tag{1.4}$$

The integral in Equation 1.4 is difficult to calculate in Cartesian coordinates (current form), but it can be easily evaluated when written in polar coordinates(r,θ). We will return to this integral in Example 1.7.

In Section 1.1, we considered double integration over a rectangular region and expressed its relationship with the volume of a solid. Now, we consider a polar rectangular region R bounded by the arcs $r=a$, $r=b$ and the rays $\theta=\alpha$, $\theta=\beta$ as shown in Figure 1.7. Thus,

$$R=\left\{(r,\theta)\in R^2 : a\le r\le b, \alpha\le\theta\le\beta\right\}.$$

Suppose a two-variable function defined over the polar rectangular region R is denoted by $z = f(r,\theta)$. Our aim is to explain the relationship between the double integral and the volume, V, of a solid below $z = f(r,\theta)$ and above R (shown in Figure 1.7) in polar coordinates.

Let us divide the interval $[\alpha,\beta]$ into n sub-intervals $[\theta_{j-1},\theta_j], 1 \le j \le n$, of width $\delta\theta = \frac{\beta-\alpha}{n}$ and use the rays $\theta = \theta_j$, $1 \le j \le n$ to divide the region R into n polar sub-rectangles (the jth sub-rectangle is highlighted in Figure 1.7a). Using these sub-rectangles, cut the solid into small pieces with the volume $v_j, 1 \le j \le n$. The jth element (highlighted in Figure 1.7a) is bounded at the bottom by the jth sub-rectangle and at the top by the surface area between the curves $z = f(r,\theta_{j-1})$ and $z = f(r,\theta_j)$.

The volume, v_j, of the jth element, which is clearly shown in Figure 1.7b, can be approximated in two steps:

Step 1: Evaluate the area under the curve $z = f(r,\theta_j)$, for $a \le r \le b$, using single integration and denote it by $A(\theta_j)$, that is

$$A(\theta_j) = \int_a^b f(r,\theta_j)\,dr. \tag{1.5}$$

Step 2: Multiply $A(\theta_j)$ by $\delta s = r \times \delta\theta$. Arc length δs is the length of an arc subtended at the angle $\delta\theta$ and can be drawn anywhere between $r = a$ and $r = b$ (see Figure 1.7b):

$$v_j \approx A(\theta_j) \times \delta s = A(\theta_j) \times r \times \delta\theta.$$

Note that $A(\theta_j) = \int_a^b f(r,\theta_j)\,dr$ is a function of the variable $\alpha \le \theta_j \le \beta$.

The total volume, V, of the solid can be approximated by the sum of the n elements with volume v_j:

$$V \approx \sum_{j=1}^{j=n} v_j = \sum_{j=1}^{j=n} A(\theta_j) \times r \times \delta\theta.$$

Taking the limit as $n \to \infty$ and using the definition of the single integral gives

$$V = \sum_{j=1}^{\infty} A(\theta_j) \times r\delta\theta = \int_{\theta=\alpha}^{\theta=\beta} A(\theta)\,r\,d\theta.$$

Now, we use Equation 1.5 and replace $A(\theta)$:

$$V = \int_{\theta=\alpha}^{\theta=\beta}\left[\int_{r=a}^{r=b} f(r,\theta)\,dr\right] \times r\,d\theta = \int_{\theta=\alpha}^{\theta=\beta}\int_{r=a}^{r=b} f(r,\theta)\,r\,dr\,d\theta.$$

(a)

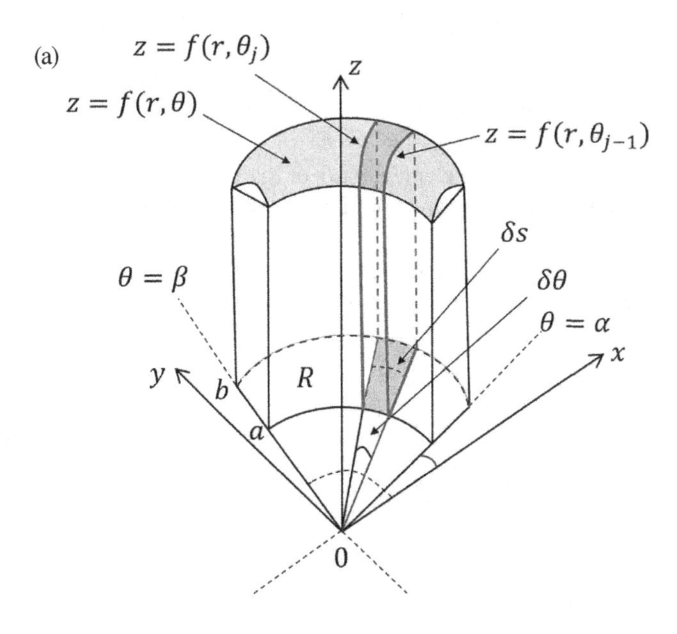

(b)

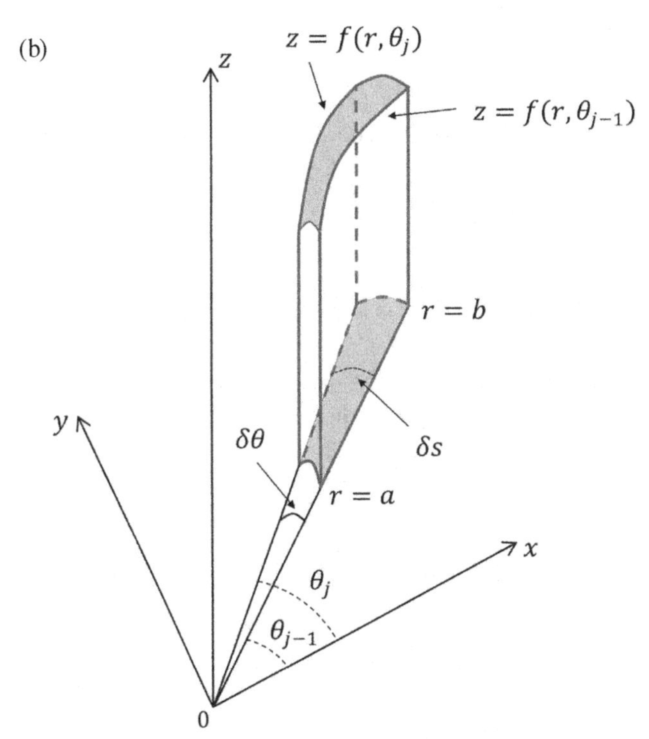

FIGURE 1.7 (a) The solid defined by $z = f(r,\theta)$ over a circular region R, which is bounded by the arcs $r = a$, $r = b$ and the rays $\theta = \alpha$, $\theta = \beta$. The solid is sliced into small pieces with the volume $v_j, 1 \leq j \leq n$. The jth element with the volume v_j is highlighted. (b) Close-up of the jth element with the volume v_j.

KEY CONCEPTS

- It is often necessary to write the double integral in the polar coordinate system for circular regions.
- In polar coordinates, the element *dxdy* or *dydx* changes to $rdrd\theta$.
- In polar coordinates, the limits of integration depend on the variables r and θ.
- The limits of the inner integral depend on the variation of r and can be a function of θ for general regions. However, the limits of the outer integral are always constant values.

Example 1.6 Evaluate the following integral by writing it in polar coordinates:

$$\iint_R e^{x^2+y^2}\,dR,$$

where

$$R=\left\{(x,y)\in R^2 : x^2+y^2\le 1, y\ge 0\right\}.$$

Solution

To convert Cartesian coordinates to polar coordinates, we use the following equations:

$$x=r\ \cos\theta \qquad \text{and} \qquad y=r\ \sin\theta.$$

Thus,

$$x^2+y^2=r^2\left(\cos^2\theta+\sin^2\theta\right)=r^2.$$

The graph of the region R is shown in Figure 1.8a, which is the part of the disc $x^2+y^2\le 1$ above the x-axis. The region R in polar coordinates can be described as follows:

$$R=\left\{(r,\theta)\in R^2 : 0\le r\le 1, 0\le\theta\le\pi\right\}.$$

The limits of the inner integral, the radial change, varies between $r=0$ and $r=1$, and the limits of the outer integral, the angular change, varies between $\theta=0$ and $\theta=\pi$. Hence,

$$\iint_R e^{x^2+y^2}\,dR=\int_0^{\pi}\int_0^{1} e^{r^2}\,rdrd\theta.$$

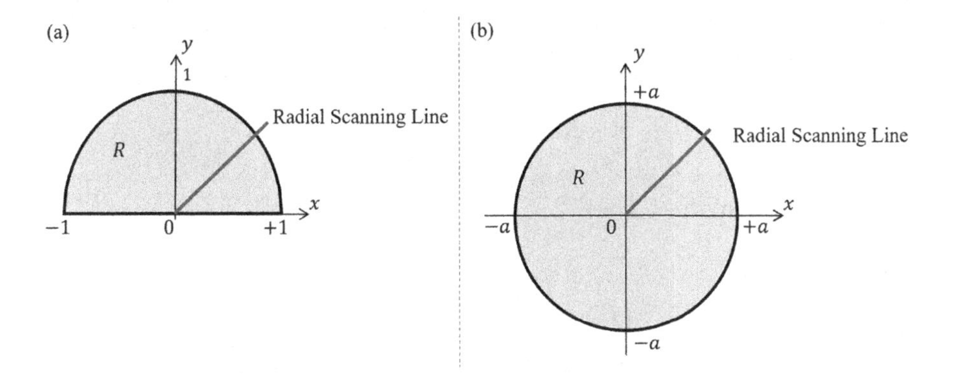

FIGURE 1.8 (a) A circular region R is bounded by $r = 1$, and the rays $\theta = 0$ and $\theta = \pi$. The radial scanning line starts at $r = 0$ and finishes at the boundary of the region at $r = 1$. (b) A circular disc R is bounded by $r = a$. The radial scanning line starts at $r = 0$ and finishes at the boundary of the region. The angle, varies in the interval $0 \le \theta \le 2\pi$.

The inner integral can be evaluated by the substitution $u = r^2$:

$$\text{inner integral} = \int_0^1 e^{r^2} r dr = \frac{1}{2}\int_0^1 e^u du = \frac{1}{2}(e-1),$$

which implies

$$\text{outer integral} = \frac{1}{2}(e-1)\int_0^{\pi} d\theta = \frac{\pi}{2}(e-1).$$

Example 1.7 Evaluate the volume of a hemisphere of radius a, which is bounded at the top by $z = +\sqrt{a^2 - x^2 - y^2}$ and at the bottom by the plane $z = 0$.

Solution

In Cartesian coordinates, the volume of the hemisphere can be calculated using the integral in Equation 1.4, where

$$R = \{(x, y) : x^2 + y^2 \le a^2\}$$

is the projection of the hemisphere of radius a in the xy-plane ($z = 0$).

We can write the region of integration in polar coordinates as

$$R = \{(r, \theta) : 0 \le r \le a, 0 \le \theta \le 2\pi\}.$$

The graph of the region R, which is a disc of radius a, is shown in Figure 1.8b. The two-variable function in polar coordinates is

$$z = \sqrt{a^2 - x^2 - y^2} = \sqrt{a^2 - r^2}$$

and dR changes to $rdrd\theta$. Thus,

$$V = \iint_R \sqrt{a^2 - r^2}\, rdrd\theta.$$

The limits of the inner integral, the radial change, varies between $r = 0$ and $r = a$, and the limits of the outer integral, the angular change, varies between $\theta = 0$ and $\theta = 2\pi$. Hence,

$$V = \iint_R \sqrt{a^2 - r^2}\, rdrd\theta = \int_{\theta=0}^{\theta=2\pi} \int_{r=0}^{r=a} \sqrt{a^2 - r^2}\, rdrd\theta.$$

First, we calculate the inner integral:

$$\text{inner integral} = \int_{r=0}^{r=a} \sqrt{a^2 - r^2}\, rdr$$

We use the substitution method:

$$u = a^2 - r^2 \quad du = -2rdr \text{ or } r\,dr = -\frac{1}{2}du$$

The lower and upper limits of integration for the variable u are

$$r = 0 \Rightarrow u = a^2$$
$$r = a \Rightarrow u = 0.$$

Thus,

$$\text{inner integral} = \int_{r=0}^{r=a} \sqrt{a^2 - r^2}\, rdr = \int_{u=a^2}^{u=0} \sqrt{u}\left(-\frac{1}{2}du\right)$$

$$= -\frac{1}{2}\int_{u=a^2}^{u=0} u^{\frac{1}{2}} du$$

$$= -\frac{1}{2}\left[\frac{2}{3}u^{\frac{3}{2}}\right]_{a^2}^{0}$$

$$= -\frac{1}{3}(0)^{\frac{3}{2}} + \frac{1}{3}(a^2)^{\frac{3}{2}} = \frac{1}{3}a^3$$

$$\text{outer integral} = \int_{\theta=0}^{\theta=2\pi} \frac{1}{3}a^3 d\theta = \frac{1}{3}a^3 \int_{\theta=0}^{\theta=2\pi} d\theta = \frac{1}{3}a^3 \left[\theta\right]_0^{2\pi} = \frac{2\pi a^3}{3}.$$

NOTE: We showed that the volume of a hemisphere of radius a is $V = \frac{2\pi a^3}{3}$, and hence, the volume of the sphere of radius a is $2 \times V = \frac{4\pi a^3}{3}$.

1.6 DOUBLE INTEGRATION OVER GENERAL POLAR REGIONS

In Section 1.5, we described double integration over polar rectangular regions. This process can be extended to the general polar regions shown in Figure 1.9. General regions in Cartesian coordinates are classified as type 1 and type 2. However, in polar coordinates, general polar regions R are defined as follows:

$$R = \{(r,\theta) : f_1(\theta) \le r \le f_2(\theta), \alpha \le \theta \le \beta\}.$$

A general polar region R shown in Figure 1.9 is bounded by the two curves $r = f_1(\theta)$, $r = f_2(\theta)$ and the rays $\theta = \alpha$, $\theta = \beta$. The theory for describing the relationship

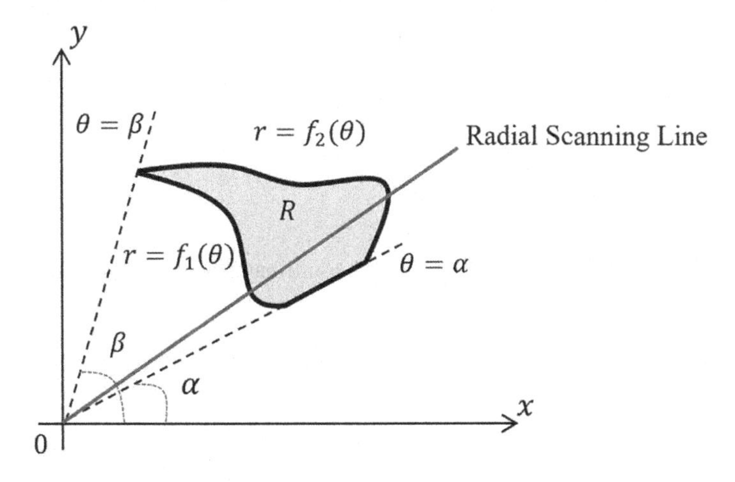

FIGURE 1.9 A polar region R is bounded by $r = f_1(\theta)$, $r = f_2(\theta)$, and rays $\theta = \alpha$, $\theta = \beta$. A radial scanning is also presented.

between the double integration and the volume of a solid for general polar regions is similar to that given in Section 1.5 for the polar rectangular region. However, we only explain computing integrals over general polar regions in this section.

For integrals over polar regions, the limits of the inner integral, the radial variation, are defined by the intersection of the radial scanning lines with the boundary of the region R. The radial scanning lines start at the origin O and continue towards the boundary of the region R, one of which is shown in Figure 1.9. The iterated integral, for general regions in polar coordinates, can be written as

$$\iint_R f(r,\theta)\,dR = \int_{\theta=\alpha}^{\theta=\beta}\int_{r=f_1(\theta)}^{r=f_2(\theta)} f(r,\theta)\,r\,dr\,d\theta.$$

Example 1.8 Find the solid volume bounded at the top by the paraboloid $z = 16 - x^2 - y^2$ and at the bottom by the circular disc $(x-2)^2 + y^2 = 2^2$.

Solution

A diagram of a cylinder intersecting a paraboloid with their projection in the xy-plane is shown in Figure 1.10. The required volume can be evaluated using the following double integral:

$$\iint_R \left(16 - x^2 - y^2\right) dR,$$

where the region R is a disc of radius 2 and centred at the point $(2,0)$ shown in Figure 1.10b.

Using $x = r\cos(\theta)$ and $y = r\sin(\theta)$ the equation of paraboloid is written as

$$z = 16 - \left(r^2\cos^2\theta + r^2\sin^2\theta\right) = 16 - r^2.$$

Similarly, the equation of the circle in polar coordinates is as follows:

$$(x-2)^2 + y^2 = 2^2 \Rightarrow$$
$$x^2 - 4x + 4 + y^2 = 4 \Rightarrow$$
$$r^2 = 4r\cos(\theta) \Rightarrow r = 4\cos(\theta) \text{ or } r = 0.$$

The radial scanning lines start at the origin and intersect the boundary of R at $r = 4\cos(\theta)$, and the angle θ varies from $\theta = -\frac{\pi}{2}$ to $\theta = \frac{\pi}{2}$ (see Figure 1.10b). Hence,

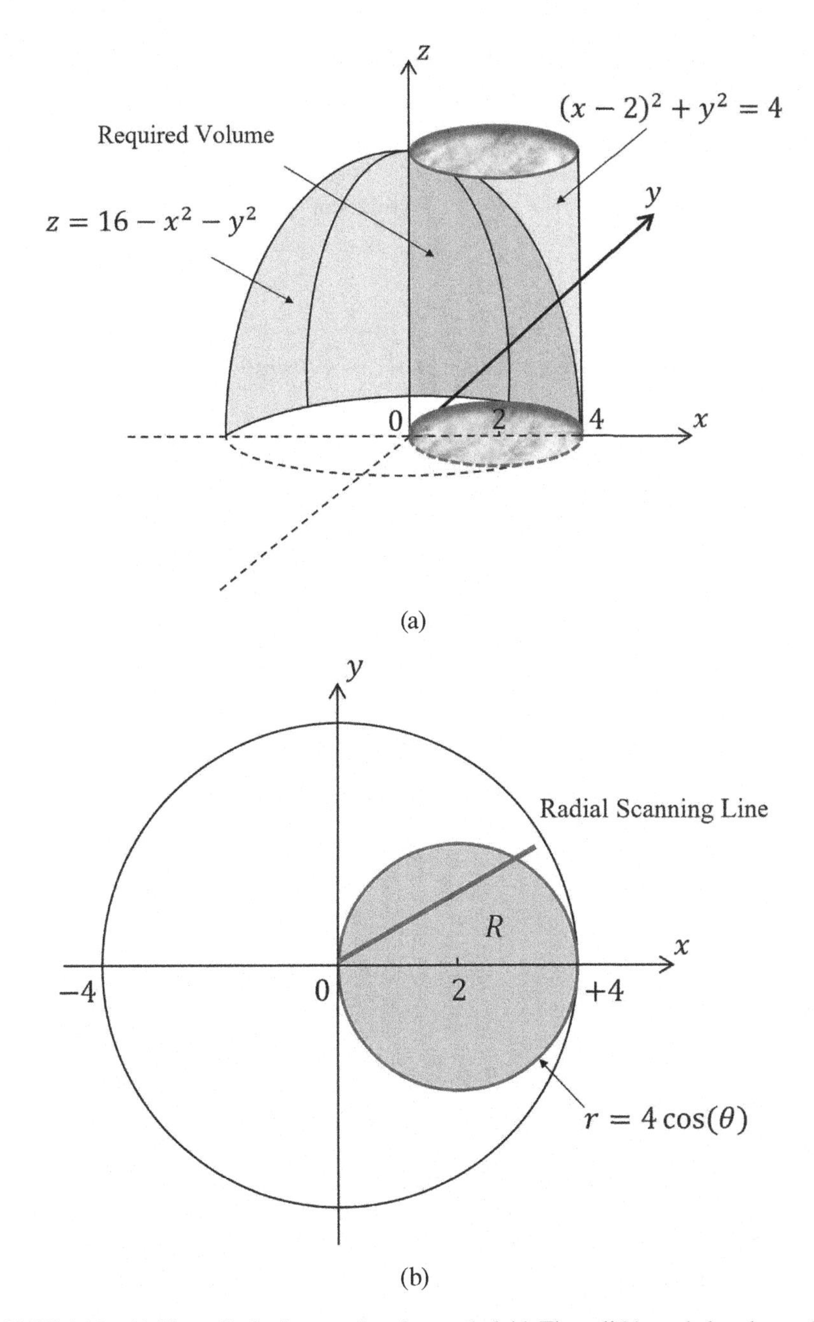

FIGURE 1.10 (a) The cylinder intersecting the paraboloid. The solid bounded at the top by a paraboloid and at the bottom by a disc of radius 2. (b) Projection of paraboloid and cylinder in the xy-plane. The outer circle is the projection of the paraboloid, and the inner circle is the projection of the cylinder in the xy-plane. The region R is a disc of radius 2 centred at the point (2, 0).

the limits of integration in the order of first dr and then $d\theta$ can be represented as follows:

$$r = 0 \quad \rightarrow \quad r = 4\cos(\theta) \quad (\text{limits of the inner integral})$$

$$\theta = -\frac{\pi}{2} \quad \rightarrow \quad \theta = +\frac{\pi}{2} \quad (\text{limits of the outer integral}).$$

Thus,

$$\iint_R \left(16 - x^2 - y^2\right) dR = \iint_R \left(16 - r^2\right) r dr d\theta$$

$$= \int_{\theta=-\frac{\pi}{2}}^{\theta=\frac{\pi}{2}} \int_{r=0}^{r=4\cos(\theta)} \left(16\mathrm{r} - \mathrm{r}^3\right) dr d\theta$$

$$= \int_{\theta=-\frac{\pi}{2}}^{\theta=\frac{\pi}{2}} \left[16\frac{r^2}{2} - \frac{r^4}{4}\right]_{r=0}^{r=4\cos(\theta)} d\theta$$

$$= 64 \int_{\theta=-\frac{\pi}{2}}^{\theta=\frac{\pi}{2}} \left[2\cos^2\theta - \cos^4\theta\right] d\theta$$

$$= 64 \int_{\theta=-\frac{\pi}{2}}^{\theta=\frac{\pi}{2}} \left[\cos^2\theta - \frac{1}{4}\sin^2 2\theta\right] d\theta$$

$$= 64 \int_{\theta=-\frac{\pi}{2}}^{\theta=\frac{\pi}{2}} \left[\frac{5}{8} + \frac{3}{8}\cos 2\theta\right] \mathrm{d}\theta = 40\pi.$$

NOTES

- To evaluate

$$\int_{\theta=-\frac{\pi}{2}}^{\theta=\frac{\pi}{2}} \left[2\cos^2\theta - \cos^4\theta\right] d\theta,$$

  we have used the trigonometric identity $\cos^2\theta = \frac{1+\cos(2\theta)}{2}$.

- In Cartesian coordinates, we obtain the following integral using the vertical scanning lines:

$$\iint_R \left(16 - x^2 - y^2\right) dy dx = \int_{x=0}^{x=4} \int_{y=-\sqrt{4x-x^2}}^{y=+\sqrt{4x-x^2}} \left(16 - x^2 - y^2\right) dy dx.$$

- Evaluating the integral in Cartesian coordinates should imply the same answer as 40π.

1.6.1 EXERCISES

1. For the following integrals, determine the region of the integration and then evaluate the integral.

   (a) $\int_0^{\pi/2}\int_0^3 re^{-r^2}\,dr\,d\theta$ (b) $\int_{-\pi/2}^{\pi/2}\int_0^4 r^2\cos\theta\,dr\,d\theta$

2. Write down the following integrals in polar coordinates and then evaluate the integral.

   (a) $\iint_R e^{-x^2-y^2}\,dR$ where $R=\{(x,y):1\le x^2+y^2\le 4\}$

   (b) $\iint_R 2(x+y)\,dR$ where $R=\{(x,y):x^2+y^2\le 9, x\ge 0\}$

3. Given the double integral $\iint_R x\,dR$, where R is the region in the first quadrant that lies between two circles $x^2+y^2=4$ and $x^2+y^2=2x$:

   (a) Sketch the graph of the region R.
   (b) Write the region R in polar coordinates.
   (c) Write the iterated integral in polar coordinates and evaluate it.

4. Write down the following integral in polar coordinates and evaluate it.

$$\int_{y=0}^{y=\frac{1}{\sqrt{2}}}\int_{x=y}^{x=\sqrt{1-y^2}} y\,dx\,dy$$

5. Draw the region R for the following integral:

$$I=\int_0^3\int_{\frac{3}{4}x}^{\frac{4}{3}x} xy^2\,dy\,dx+\int_3^4\int_{\frac{3}{4}x}^{\sqrt{25-x^2}} xy^2\,dy\,dx$$

   Change to polar coordinates to show that

   (a) $I=\int_{\theta_1}^{\theta_2}\int_0^5 r^4\sin^2\theta\cos\theta\,dr\,d\theta$, where $\sin\theta_1=\frac{3}{5}$ and $\sin\theta_2=\frac{4}{5}$.

   (b) $I=\frac{185}{3}$.

Answers

1. (a) Value $=\frac{\pi}{4}\left(1-e^{-9}\right)$, Graph: Quarter of a circle in the first quadrant, with radius 3.

   (b) Value $=\frac{128}{3}$, Graph: Half of a circle to the right side of the y-axis, with radius 4.

2. (a) $\pi\left(e^{-1}-e^{-4}\right)$ (b) 36
3. (a) See Figure 1.11a.
   (b) The radia l changes are from $r = 2\cos\theta$ to $r = 2$ and the angular changes are from $\theta = 0$ to $\theta = \frac{\pi}{2}$.
   (c) $\frac{8}{3}-\frac{\pi}{2}$
4. $\frac{1}{6}\left(2-\sqrt{2}\right)$
5. See Figure 1.11b for the graph of the region R.

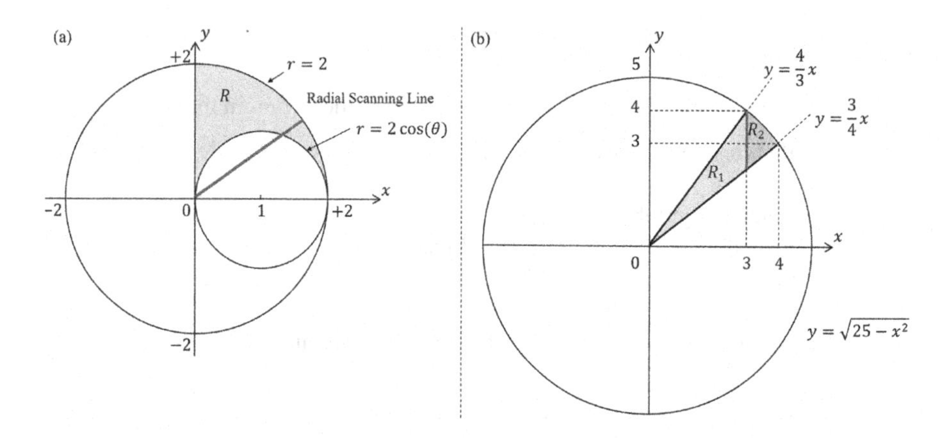

FIGURE 1.11 (a) The radial scanning lines start at $r = 2\cos\theta$ and finish at $r = 2$. (b) The region R_1 is associated with the first integral and R_2 with the second one. The angles θ_1 and θ_2 are the angles between $y = \frac{3}{4}x$ and $y = \frac{4}{3}x$, with the positive x-axis, respectively.

1.7 ENGINEERING APPLICATIONS

Double integration is a mathematical technique that plays a crucial role in calculating the centre of mass and moment of inertia for various objects. These concepts are fundamental in engineering, and double integration simplifies their computation in complex systems. By breaking down complex objects with non-uniform density and irregular shapes into infinitesimally small elements and linking them to the concept of double integral, we can determine their centre of mass and moment of inertia, which are crucial for understanding their mechanical properties and behaviour in engineering applications.

1.7.1 Density Distributions of a Two-Dimensional Thin Layer

Suppose we consider the density distribution of a thin layer (lamina) shown in Figure 1.12. Without loss of generality, we can assume the solid to be a two-dimensional object which occupies the region R in the xy-plane (see Figure 1.12b). Assume the

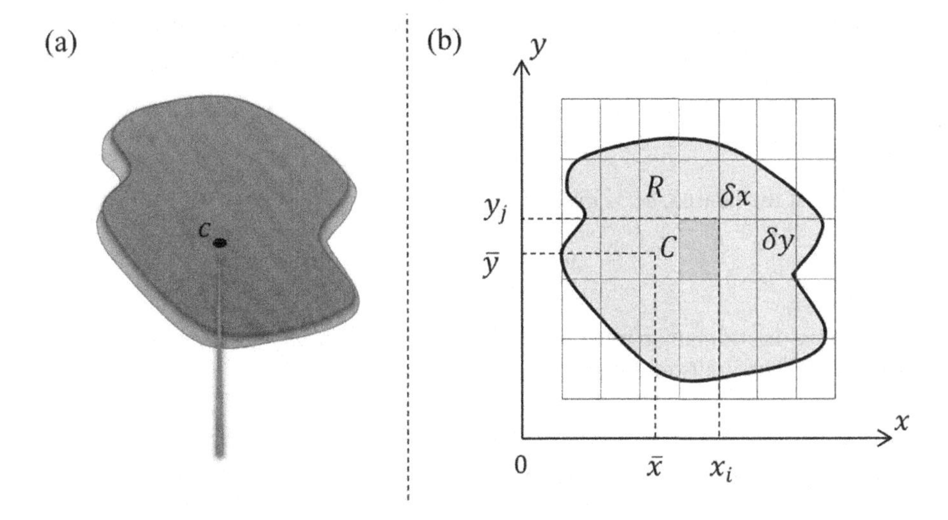

FIGURE 1.12 (a) A two-dimensional thin plate (lamina) with the centre of mass at the point C. (b) The projection of the lamina in xy-plane is presented and a small rectangular element $\delta R = \delta x \delta y$ is highlighted.

density of the lamina denoted by $\rho(x, y)$ (defined as mass per unit area) varies at each position of R, and hence,

$$\text{Total mass} = \rho(x, y) \times \text{ Total Area.}$$

To obtain the total mass of the lamina, we assume that the region R is covered by grid lines parallel to x and y axes. These lines divide the projected region R into $N = k \times \ell$ small rectangles (of width δx and length δy), where k and ℓ are the number of grid lines in the x and y directions required to cover the region R. Some of these rectangles lie entirely within the region R, some outside of the region R and some can be partly inside the region R. We shall consider all rectangles, which lie completely inside the region. A typical rectangle inside the region R with a sample point (x_i, y_j) and the area element $\delta R = \delta x \delta y$ is highlighted in Figure 1.12 (b). The mass of this rectangle can be approximated by $\rho(x_i, y_j)\delta R$. Hence, the total mass, M, of the lamina is approximated as

$$M \approx \sum_{i=1}^{i=k} \sum_{j=1}^{j=\ell} \rho(x_i, y_j) \delta x \delta y$$

and exactly as

$$M = \sum_{i=1}^{\infty} \sum_{j=1}^{\infty} \rho(x_i, y_j) \delta x \delta y = \iint_R \rho(x, y) dR.$$

Example 1.9 Suppose we have a lamina that occupies a triangular area in the xy-plane as shown in Figure 1.13. The density of the lamina varies as $\rho(x,y)=2x+3y$ g/cm^2. Determine the mass of the lamina.

Solution

The mass of the lamina, M, is given by

$$M=\iint_R \rho(x,y)\,dR,$$

where R is the triangular region in Figure 1.13. We set up the integration in the y-direction. The limits of integration are

$$y=0 \;\rightarrow\; y=\frac{4}{3}x \quad \text{(limits of the inner integral)}$$

$$x=0 \;\rightarrow\; x=3 \quad \text{(limits of the outer integral)}.$$

Therefore,

$$M=\int_{x=0}^{x=3}\int_{y=0}^{y=\frac{4}{3}x}(2x+3y)\,dydx$$

$$\text{Inner integral}=\int_{y=0}^{y=\frac{4}{3}x}(2x+3y)\,dy=\left[2xy+\frac{3}{2}y^2\right]_{y=0}^{y=\frac{4}{3}x}$$

$$=\frac{8}{3}x^2+\frac{8}{3}x^2=\frac{16}{3}x^2$$

$$\text{Outer integral}=\int_0^3\frac{16}{3}x^2dx=\left[\frac{16}{9}x^3\right]_0^3=48.$$

Therefore, the mass of the lamina is 48 g.

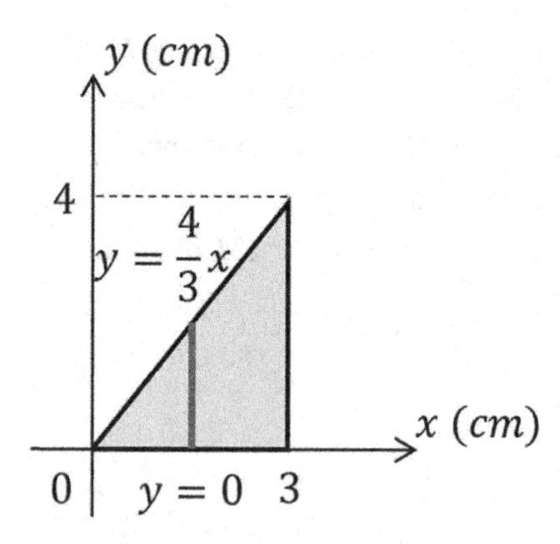

FIGURE 1.13 The projection of a triangular lamina in the xy-plane with density $\rho(x,y)=2x+3y$ g/cm^2.

1.7.2 Centre of Mass of a Two-Dimensional Thin Layer (Lamina)

Before considering the centre of mass, C in Figure 1.12, for a two-dimensional thin plate, let's look at the concept of centre of mass by considering a simple system consisting of two point masses m_1 and m_2. Suppose a thin rod connects these two masses and we want to find a point on the rod such that if we place a fulcrum (support) on this point, the system will be balanced (see Figure 1.14).

The rod will balance if

$$m_1 d_1 = m_2 d_2. \tag{1.6}$$

Suppose $\bar{x}$ is the centre of mass and we place the fulcrum at $\bar{x}$ distance away from the origin, the mass m_1 at x_1 distance away and the mass m_2 at x_2 distance away. Hence, $d_1 = x_1 - \bar{x}$ and $d_2 = \bar{x} - x_2$. Substituting d_1 and d_2 in equation (1.6) and rearranging the variables we obtain

$$\bar{x} = \frac{m_1 x_1 + m_2 x_2}{m_1 + m_2} = \frac{\text{Total moment about the origin}}{\text{Total mass}}.$$

If the system consists of n discrete point masses distributed in one dimension, then the preceding formula can be described as

$$\bar{x} = \frac{M_0}{M}$$

$$\text{where } M_0 = \sum_{i=1}^{i=n} m_i x_i = \text{Total moment of the system}$$

$$M = \sum_{i=1}^{i=n} m_i = \text{Total mass of the system.}$$

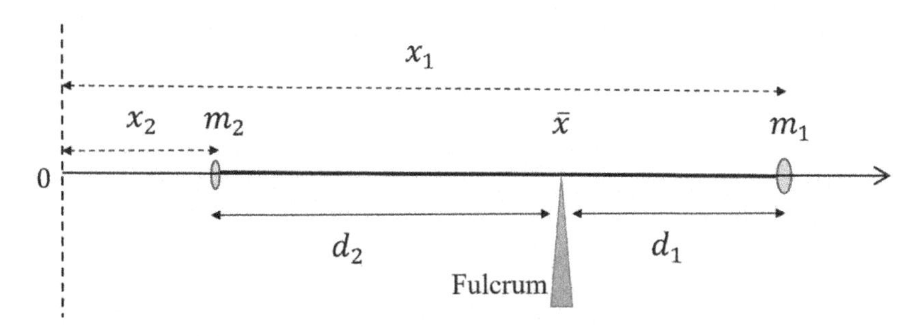

FIGURE 1.14 A simple system consisting of two point masses m_1 and m_2. The mass m_1 is at a distance x_1 from the origin O, and the mass m_2 at a dsitance x_2 away from the orgin.

This formula can be extended to a system consisting of n discrete point masses distributed in the xy-plane, in this case the coordinate of the centre of mass, $(\bar{x}, \bar{y})$, can be found using the following equations:

$$\bar{x} = \frac{M_y}{M} \quad \text{and} \quad \bar{y} = \frac{M_x}{M},$$

where M_x = total moment of the system about x-axis, M_y = total moment of the system about y-axis and M = total mass of the system.

Now, we explain how to determine the centre of mass of the lamina shown in Figure 1.12a. Its projection is shown in Figure 1.12b in the xy -plane and $\rho(x, y)$ is its variable density distribution at any point (x, y).

To find the coordinates of the centre of mass $C = (\bar{x}, \bar{y})$ of the layer, we are going to use a similar idea explained earlier for finding the centre of mass of point masses in two dimensions, except that we have a two-dimensional lamina with the projected region R shown in Figure 1.12b. The total mass of the lamina was found in Section 1.7a; hence, we only need to find its total moments about the x- and y-axes.

We divide the region R into small rectangles as described in Section 1.7a. The moments of a typical small rectangle in Figure 1.12b (considering it as a point mass) about the x- and y-axes are $\delta M_x = y_j \rho(x_i, y_j) \delta x \delta y$ and $\delta M_y = x_i \rho(x_i, y_j) \delta x \delta y$, respectively. Hence, the total moments about the axes can be approximated as

$$M_x \approx \sum_{i=1}^{i=k} \sum_{j=1}^{j=\ell} y_j \rho(x_i, y_j) \delta x \delta y$$

$$M_y \approx \sum_{i=1}^{i=k} \sum_{j=1}^{j=\ell} x_i \rho(x_i, y_j) \delta x \delta y$$

and exactly as

$$M_x = \sum_{i=1}^{\infty} \sum_{j=1}^{\infty} y_j \rho(x_i, y_j) \delta x \delta y = \iint_R y \rho(x, y) dR$$

$$M_y = \sum_{i=1}^{\infty} \sum_{j=1}^{\infty} x_i \rho(x_i, y_j) \delta x \delta y = \iint_R x \rho(x, y) dR.$$

Hence,

$$\bar{x} = \frac{M_y}{M} = \frac{\iint_R x \rho(x, y) dR}{\iint_R \rho(x, y) dR}$$

$$\bar{y} = \frac{M_x}{M} = \frac{\iint_R y \rho(x, y) dR}{\iint_R \rho(x, y) dR}.$$

Example 1.10 Determine the coordinates of the centre of mass of the lamina described in Example 1.9.

Solution

In Example 1.9, we determined that the total mass of the lamina is $M = 48$g. Therefore, we only need to evaluate the moments about the x- and y-axes.

Moment about the x-axis:

$$M_x = \iint_R y\rho(x,y)\,dR = \int_{x=0}^{x=3}\int_{y=0}^{y=\frac{4}{3}x} y(2x+3y)\,dy\,dx$$

$$\text{Inner integral} = \int_{y=0}^{y=\frac{4}{3}x} \left(2xy+3y^2\right)dy = \left[xy^2+y^3\right]_{y=0}^{y=\frac{4}{3}x}$$

$$= \frac{16}{9}x^3 + \frac{64}{27}x^3 = \frac{112}{27}x^3$$

$$\text{Outer integral} = \int_0^3 \frac{112}{27}x^3\,dx = \left[\frac{28}{27}x^4\right]_0^3 = 84.$$

Therefore, $M_x = 84$ g-cm.

Moment about the y-axis:

$$M_y = \iint_R x\rho(x,y)\,dR = \int_{x=0}^{x=3}\int_{y=0}^{y=\frac{4}{3}x} x(2x+3y)\,dy\,dx$$

$$\text{Inner integral} = \int_{y=0}^{y=\frac{4}{3}x} \left(2x^2+3xy\right)dy = \left[2x^2y+\frac{3}{2}xy^2\right]_{y=0}^{y=\frac{4}{3}x}$$

$$= \frac{8}{3}x^3 + \frac{8}{3}x^3 = \frac{16}{3}x^3$$

$$\text{Outer integral} = \int_0^3 \frac{16}{3}x^3\,dx = \left[\frac{4}{3}x^4\right]_0^3 = 108.$$

Therefore, $M_y = 108$ g-cm.

Now, we can determine the coordinates of the centre of mass:

$$\overline{x} = \frac{M_y}{M} = \frac{108}{48} = 2.25\text{cm}$$

and

$$\overline{y} = \frac{M_x}{M} = \frac{84}{48} = 1.75\text{cm}.$$

Therefore, the coordinates of the centre of mass of the lamina is $(2.25, 1.75)$.

1.7.3 Moment of Inertia of a Two-Dimensional Thin Layer (Lamina)

Suppose, we want to determine the moment of inertia (defined as the second moment of mass about an axis of rotation) of the lamina shown in Figure 1.12 about the x- and y-axes. We divide the region R into small rectangles as described in Section 1.7a. The moment of inertia of a typical rectangle (with a sample point (x_i, y_j), represented in Figure 1.12b) about the x- and y-axes can be approximated by $\delta I_x = y_j^2 \rho(x_i, y_j)\delta x \delta y$ and $\delta I_y = x_i^2 \rho(x_i, y_j)\delta x \delta y$, respectively.

For the region R, the total moment of inertia about the x- and y-axes are approximated as

$$I_x \approx \sum_{i=1}^{k}\sum_{j=1}^{\ell} y_j^2 \rho(x_i, y_j)\delta x \delta y \quad I_y \approx \sum_{i=1}^{k}\sum_{j=1}^{\ell} x_i^2 \rho(x_i, y_j)\delta x \delta y$$

and exactly as

$$I_x = \sum_{i=1}^{\infty}\sum_{j=1}^{\infty} y_j^2 \rho(x_i, y_j)\delta x \delta y = \iint_R y^2 \rho(x, y)\, dR$$

$$I_y = \sum_{i=1}^{\infty}\sum_{j=1}^{\infty} x_i^2 \rho(x_i, y_j)\delta x \delta y = \iint_R x^2 \rho(x, y)\, dR.$$

Example 1.11 Determine the moment of inertia of the lamina, described in Example 1.9, about both the x-axis and the y-axis.

Solution

Moment of inertia about the x-axis:

$$I_x = \iint_R y^2 \rho(x, y)\, dR = \int_{x=0}^{x=3}\int_{y=0}^{y=\frac{4}{3}x} y^2 (2x + 3y)\, dy\, dx$$

$$\text{Inner integral} = \int_{y=0}^{y=\frac{4}{3}x} \left(2xy^2 + 3y^3\right) dy$$

$$= \left[\frac{2}{3}xy^3 + \frac{3}{4}y^4\right]_{y=0}^{y=\frac{4}{3}x} = \frac{128}{81}x^4 + \frac{64}{27}x^4 = \frac{320}{81}x^4$$

$$\text{Outer integral} = \int_0^3 \frac{320}{81}x^4\, dx = \left[\frac{64}{81}x^5\right]_0^3 = 192.$$

Therefore, $I_x = 192$ g-cm^2.

Moment of inertia about the y-axis:

$$I_y = \iint_R x^2 \rho(x,y)\,dR = \int_{x=0}^{x=3}\int_{y=0}^{y=\frac{4}{3}x} x^2(2x+3y)\,dy\,dx$$

$$\text{Inner integral} = \int_{y=0}^{y=\frac{4}{3}x} \left(2x^3+3x^2y\right)dy$$

$$= \left[2x^3y + \frac{3}{2}x^2y^2\right]_{y=0}^{y=\frac{4}{3}x} = \frac{8}{3}x^4 + \frac{8}{3}x^4 = \frac{16}{3}x^4$$

$$\text{Outer integral} = \int_0^3 \frac{16}{3}x^4\,dx = \left[\frac{16}{15}x^5\right]_0^3 = 259.2.$$

Therefore, $I_y = 259.2$ g-cm².

1.7.4 Exercises

1. For the lamina projection in the xy-plane described in parts a–d, determine the mass of the lamina.
   (a) The projection of the lamina is bounded by $y = x^2$, $y = 0$ and $x = 3$ with the density function $\rho(x,y) = x^2 + 1$.
   (b) The projection of the lamina is a triangle bounded by $y = x$, $y = 2x$ and $x = 2$ with the density function $\rho(x,y) = xy$.
   (c) The projection of the lamina is a square bounded by $y = 0$, $y = 3$, $x = 0$ and $x = 3$ with the density function $\rho(x,y) = x^2 + xy$.
   (d) The projection of the lamina is a circle with the density function $\rho(x,y) = x^2 + y^2$. The centre of the circle is at the origin, and its radius is 5.
2. Determine the coordinates of the centre of mass of each lamina described in parts a–d of Exercise 1.
3. Determine the moment of inertia, about the x- and y-axes, of each lamina described in parts (a)–(d) of Exercise 1.
4. The moment of inertia I of the solid with uniform density ρ_0 about the origin is given by

$$I = \iint_R \rho_0\left(x^2 + y^2\right)dR,$$

   where $R = \left\{(x,y) : 10 \le x^2 + y^2 \le 20, x \ge 0\right\}$. Evaluate the integral and find the moment of inertia I.
5. A thin plate of uniform thickness and unform density, ρ_0, occupies the region of xy-plane bounded by the parabola $y = x^2$ and the line $y = x + 2$. Evaluate its moment about the y-axis.

Answers

1. (a) $\dfrac{288}{5}$ (b) 6 (c) $\dfrac{189}{4}$ (d) $\dfrac{625}{2}\pi$
2. (a) $(2.46, 3.13)$ (b) $(1.6, 2.49)$ (c) $(2.14, 1.71)$ (d) $(0, 0)$
3. (a) $I_x = \dfrac{5832}{7}, I_y = \dfrac{12636}{35}$ (b) $I_x = 40, I_y = 16$

   (c) $I_x = \dfrac{1377}{8}, I_y = \dfrac{9477}{40}$ (d) $I_x = \dfrac{15625}{6}\pi, I_y = \dfrac{15625}{6}\pi$
4. $75\rho_0\pi$
5. $\dfrac{9}{4}\rho_0$

2 Ordinary Differential Equations

2.1 INTRODUCTION

In this chapter, we introduce the topic of ordinary differential equations. A differential equation is an equation that involves one or more derivatives, or differentials. They are classified as ordinary and partial differential equations, but partial differential equations are not considered in this chapter.

Many physical problems lead to differential equations when formulated in mathematical terms. Differential equations are used in electrical engineering and mechanical systems, among others. We start this topic with a simple mechanical system and show how a simple system leads to solving a differential equation that describes a harmonic motion. We then classify differential equations and introduce simple methods to solve a class of differential equations. This includes separating variables or describing the use of integrating factors to solve first-order linear equations and so on.

The methods that we consider lead us to distinguish between linear and non-linear equations. Non-linear equations are, in general, more complex and their solutions can behave in more varied and strange ways, so we generally do not cover nonlinear differential equations.

The outline of this chapter is as follows:

- Definition of ordinary differential equations (ODEs)
- First-order non-linear equations
- First-order linear equations
- Second-order linear equations
- A numerical method
- Engineering applications

2.2 VIBRATING SPRING

We start this topic by considering a simple mechanical system. Our goal is to show that the behaviour of a simple mechanical system can be described by a differential equation, and its solution provides insight into how the system evolves over time, leading to harmonic motion.

Suppose a spring of length h is attached to point A shown in Figure 2.1, and the weight W of mass m hangs from the other end. The weight W pulls the spring down by a length ℓ so that when the spring is in the state of equilibrium (rest), the length of the stretched spring is $h + \ell$.

DOI: 10.1201/9781032630694-2

Hooke's law describes the tension (the force exerted by the spring) is $k\ell$, where k is the spring constant. At equilibrium, the gravitational force $W = mg$ must cancel out the spring tension; that is

$$mg = k\ell.$$

We now pull the weight farther down, say, by x_0 and release it. In the absence of friction, the spring starts oscillating. Assume the position of the weight W at any instant of time is denoted by $x = x(t)$; thus at $t = 0$, the initial displacement is $x(0) = x_0$, and the initial speed of oscillation is $\frac{dx}{dt}(0) = 0$.

A summary of the preceding can be described as follows:

$$m = \text{mass}$$

$$k = \text{spring stiffness}$$

$$x(t) = \text{position of the mass at time } t$$

$$x(0) = x_0 \text{ (position of the mass at } t = 0\text{)}$$

$$\frac{dx}{dt} = \text{speed of the mass}$$

$$\frac{dx}{dt}(0) = 0 \text{(speed of the mass at } t = 0\text{)}$$

$$\frac{d^2x}{dt^2} = \text{acceleration of the mass}$$

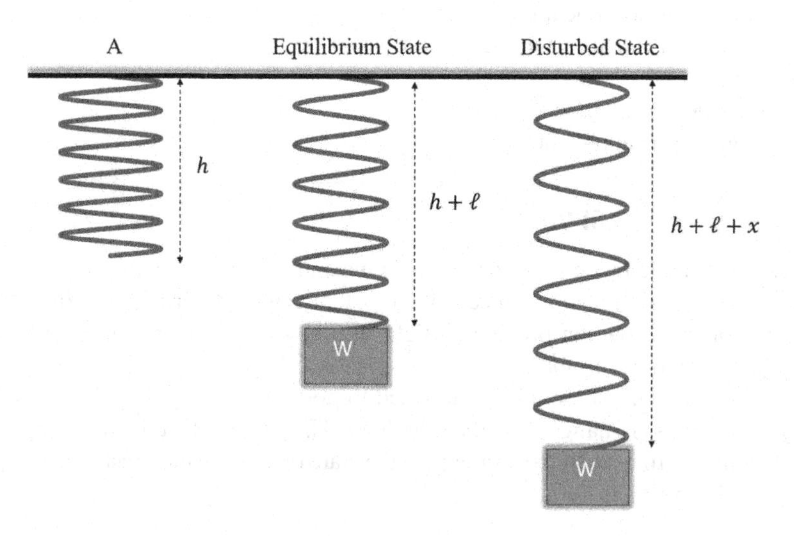

FIGURE 2.1 A spring-mass system.

The forces acting on the mass are

$$-k(\ell + x) = \text{the force on the mass caused by tension of the spring and}$$
$$mg = \text{the gravitation force.}$$

Newton's second law of motion states that Force = Mass × Acceleration, so

$$F = ma \Rightarrow m\frac{d^2x}{dt^2} = mg - k(\ell + x).$$

Since, $mg - k\ell = 0$,

$$m\frac{d^2x}{dt^2} + kx = 0. \tag{2.1}$$

This equation, unlike algebraic equations, is a time dependent "differential equation" and its unknown variable is the function $x(t)$. Equation 2.1 belongs to the class of second-order ordinary differential equations that we consider later in Section 2.7.

Considering the initial conditions,

$$x(0) = x_0 \text{ and } \frac{dx}{dt}(0) = 0.$$

The solution to Equation 2.1 is a simple harmonic motion denoted by the following cosine function:

$$x(t) = x_0 \cos(\omega t),$$

where $\omega = \sqrt{\frac{k}{m}}$ is the angular frequency of the oscillation. The method of solving this type of equation is considered in Section 2.7.

Let us now consider a case where the motion of the weight is slowed down by friction, $c\frac{dx}{dt}$, proportional to the oscillation speed, where c is a positive constant. In this case, the total force applied in the system is as follows:

$$F = mg - k(\ell + x) - c\frac{dx}{dt},$$

and hence the differential equation will take the following form:

$$m\frac{d^2x}{dt^2} + c\frac{dx}{dt} + kx = 0. \tag{2.2}$$

The solution to this equation shows a damped vibrating motion, which we consider later in Section 2.7 by assigning numerical values to the parameters m, k and c. Full details of this equation along with vibration analysis are given in Section 2.11.3.

2.3 ODES – THE BASICS

In Section 2.2, we described a mechanical system that led to a time-dependent equation. We now turn to the formal definition of differential equations and introduce methods for solving them.

DEFINITION 2.1: DIFFERENTIAL EQUATION

A differential equation, in general, is a mathematical equation that involves derivatives of one or more unknown functions with respect to one or more independent variables. In this section, we only consider equations that contain a function $y=y(x)$, its derivatives and the independent variable x.

For example, the following equations are differential equations:

$$\frac{dy}{dx}=3x^2-2x+3$$

$$\frac{d^2y}{dx}-y=0$$

$$a\frac{d^2x}{dt^2}+b\frac{dx}{dt}+cx=f(t).$$

In a technical sense, they are called ordinary differential equations due to their inclusion of ordinary derivatives. Differential equations can be classified into several types, including ODEs and partial differential equations (PDEs). ODEs involve a single independent variable, while PDEs involve multiple independent variables. For example, the following equation

$$\frac{\partial u}{\partial t}=a^2\frac{\partial^2 u}{\partial x^2}$$

is the heat equation which depends on the two state variables t (time) and x (space). This equation can be converted to ordinary differential equations using the method of separation of variables. This book does not cover equations of this type, as our primary focus is solely on ODEs.

Differential equations are used to model a wide range of phenomena in various scientific and engineering fields, including physics, chemistry, biology, economics and engineering. They provide a mathematical framework for understanding and predicting how quantities change over time or with respect to other variables.

Unlike algebraic equations, the solutions of differential equations are functions that are satisfied in the equations. For example, the solution to the equation $\frac{dy}{dx}=y$ is the function $y = Ce^x$, where C is any arbitrary constant. To check this, we differentiate $y = Ce^x \Rightarrow \frac{dy}{dx} = Ce^x$ and then substitute it into the equation $\frac{dy}{dx} = y$, which gives

$$Ce^x = Ce^x.$$

Typically, we lack prior knowledge of the equation's solution and aim to determine the unknown function $y(x)$. Consider a simple ODE:

$$\frac{dy}{dx} = 3x^2 - 2x. \tag{2.3}$$

The solution for this equation can be found by direct integration of both sides of the equation:

$$y = \int \left(3x^2 - 2x\right) dx,$$

and so

$$y(x) = x^3 - x^2 + C. \tag{2.4}$$

Note that C is the arbitrary constant of integration and hence the function shown in Equation 2.4 is not a unique solution of Equation 2.3. The function $y(x)$ shows an infinite number of solutions of the Equation 2.3, depending on the arbitrary constant C, and it is called the "general solution" of Equation 2.3. Solutions with a known constant C are called "particular solutions".

DEFINITION 2.2: ORDER OF ODES

Differential equations can be further categorised based on their *order*, which corresponds to the highest order of derivative appearing in the equation.

The differential equation given in Equation 2.2 is a second-order ODE, and Equation 2.3 is a first-order equation. See also the following equations:

$$\frac{dy}{dx} - y = x^2 - 1 \quad \text{(first order)}$$

$$\left(a\frac{d^2x}{dt^2}\right)^3 + b\frac{dx}{dt} + cx = 0 \quad \text{(second order)}$$

$$a\frac{d^2x}{dt^2} + b\frac{dx}{dt} + cx = 0 \quad \text{(second order).}$$

The number of constants that appear in the general solution depends on the order of ODEs:

- For the first order, only one constant.
- For the second order, two constants and so on.

In general, once the general solution is obtained, for any arbitrary constant of the integration, we need an additional constraint to determine the particular solution, which means that the required number of constraints must be equal to the order of the differential equation. For example, if a constraint $y(0)=1$ is introduced for Equation 2.3; that is

$$\frac{dy}{dx}=3x^2-2x, \qquad y(0)=1.$$

The particular solution can be found as follows:

$$y(x)=x^3-x^2+C \quad (\text{general solution})$$

$$y(0)=0^3-0^2+C=1 \Rightarrow C=1$$

$$y(x)=x^3-x^2+1 \quad (\text{particular solution}).$$

This solution satisfies the equation as well as the condition.

Equation 2.3 is a simple equation that we were able to solve by direct integration. However, solving differential equations can be quite challenging and may require advanced mathematical techniques, depending on the complexity of the equation. Analytical solutions are not always possible, especially for higher order or non-linear equations, in which case numerical methods or approximation techniques are often used to obtain approximate solutions. In the following sections, we consider well-known methods for solving some special classes of differential equations.

2.4 SEPARABLE ODES

The separation of variables is a technique used to solve certain types of first-order ODEs. The idea behind the separation of variables is to rewrite the ODE in a way that allows you to isolate the independent variable on one side of the equation and the dependent variable on the other side, making it easier to integrate both sides separately. The outline of this method is as follows:

- You should have a separable ODE in the form of

$$\frac{dy}{dx}=g(x)f(y),$$

where y is the dependent and x is the independent variable.

- Rearrange the equation such that the variable y is on one side and the variable x is on the other side of the equal sign:

$$\frac{dy}{f(y)} = g(x)dx.$$

- Integrate both sides with respect to their respective variables:

$$\int \frac{dy}{f(y)} = \int g(x)dx.$$

- If we successfully perform both integrations, we can establish a functional correlation between y and x as follows:

$$F(y) = G(x) + C,$$

  where C is the constant of integration. Notice that we only need one constant for the first-order equations.
- If possible, solve for y explicitly. In some cases, this may not be possible, and you'll have an implicit solution.
- Don't forget to include the constant of integration C in your solution. You may need an additional initial condition to determine the arbitrary constant C in a given problem.

Example 2.1 Solve

$$\frac{dy}{dx} = \frac{x^2}{y}.$$

Solution

separate	$ydy = x^2\,dx,$
form integrals	$\int ydy = \int x^2 dx,$
integrate	$\frac{1}{2}y^2 = \frac{1}{3}x^3 + \tilde{C},$
tidy up	$y^2 = \frac{2}{3}x^2 + C, \quad (C = 2\tilde{C}),$
solve for y	$y = \pm\sqrt{\frac{2}{3}x^2 + C}.$

Example 2.2 Solve

$$\frac{dy}{dx} = y(1 - 2x).$$

Solution

$$\text{separate} \qquad \frac{dy}{y} = (1-2x)\,dx,$$

$$\text{form integrals} \qquad \int \frac{dy}{y} = \int (1-2x)\,dx,$$

$$\text{integrate} \qquad \ln|y| = x - x^2 + C,$$

$$\text{solve for y} \qquad y = e^{x-x^2+C}$$

$$= Ae^{x-x^3}, \qquad \left(A = e^C\right).$$

Example 2.3 Solve

$$\frac{dy}{dx} - y = 0.$$

Solution

$$\text{separate} \qquad \frac{dy}{y} = dx,$$

$$\text{form integrals} \qquad \int \frac{dy}{y} = \int 1 \times dx,$$

$$\text{integrate} \qquad \ln|y| = x + C,$$

$$\text{solve for y} \qquad y = e^{x+C}$$

$$= Ae^{x}, \left(A = e^C\right).$$

Example 2.4 Solve

$$\frac{dx}{dt} = te^{-x}.$$

Solution

$$\text{separate} \qquad e^x\,dx = t\,dt,$$

$$\text{form integrals} \qquad \int e^x\,dx = \int t\,dt,$$

$$\text{integrate} \qquad e^x = \frac{1}{2}t^2 + C,$$

$$\text{solve for} \qquad x(t) = \ln\left(\frac{1}{2}t^2 + C\right).$$

Example 2.5 Solve

$$\frac{dy}{dx} = 4x^2 + y^2.$$

Solution

This equation is not separable and cannot be solved using the method of separation of variables.

Example 2.6 Determine the particular solution of the following equation:

$$\frac{dy}{dx} = y(1-2x), \qquad y(0) = 1.$$

Solution

$$\begin{array}{ll} \text{from example 2.2} & \ln|y| = x - x^2 + C, \\ \text{apply condition} & y = 1 \text{ when } x = 0, \\ & \ln|1| = 0 - 0^3 + C, \\ & C = 0, \\ \text{substitute value} & \ln|y| = x - x^3 + 0, \\ \text{solve for } y & y = e^{x-x^3}. \end{array}$$

Example 2.7 Determine the particular solution of the following equation:

$$\frac{dy}{dx} = \frac{2x + e^x}{1 + e^y}, \; y(2) = 0.$$

Solution

$$\begin{array}{ll} \text{separate} & (1+e^y)dy = (2x + e^x)dx, \\ \text{form integrals} & \int (1+e^y)dy = \int (2x + e^x)dx, \\ \text{integrate} & y + e^y = x^2 + e^x + C, \\ \text{apply condition} & y = 0 \text{ when } x = 2, \\ & 0 + e^0 = 2^2 + e^2 + C, \\ & C = -(3 + e^2), \\ \text{substitute value} & y + e^y = x^2 + e^x - (3 + e^2). \end{array}$$

[The preceding equation cannot be solved explicitly for the y variable. So, the solution is left in implicit form.]

2.4.1 Exercises

1. Find the general solution of the following differential equations.

(a) $\dfrac{dy}{dx}-2y=0$ (b) $\dfrac{dy}{dx}+4y^2=0$

(c) $\dfrac{dy}{dx}-2y=4$ (d) $x\dfrac{dy}{dx}+y=0$

(e) $(x+2)\dfrac{dy}{dx}-xy=0$ (f) $\dfrac{dy}{dx}=2x^{-1}\sqrt{y-1}$

(g) $\dfrac{dy}{dx}+e^{2x}y^2=0$ (h) $\dfrac{dy}{dx}+\text{cosec}(y)=0$

(i) $\dfrac{dx}{dt}-(1+t)(1+x^2)=0$ (j) $x\dfrac{dx}{dt}=\sin^2(\omega t)$

(k) $\dfrac{dx}{dt}+3x\sin(\omega t)=0$ (l) $\dfrac{dx}{dt}\sin(2t)=x\cos(2t)$

(m) $\dfrac{dx}{dt}-\sec(x)=0$ (n) $\dfrac{dx}{dt}t\ln(t)=x$

(o) $(1-\cos t)\dfrac{dx}{dt}-x\sin t=0$ (p) $x\dfrac{dx}{dt}=te^{-x^2}$

2. Find the particular solution of the following differential equations.

(a) $(x+2)\dfrac{dy}{dx}=2y,\quad y(0)=1$

(b) $\dfrac{dy}{dx}=y\tan(2x),\quad y(0)=2$

(c) $\dfrac{dx}{dt}t\ln(t)=x,\quad x(2)=\ln(4)$

(d) $2t\dfrac{dx}{dt}=3x,\quad x(1)=4$

Answers

(a) $y=Ae^{2x}$ (b) $y=\dfrac{1}{(4x-C)}$

(c) $y=Ae^{2x}-2$ (d) $y=\dfrac{A}{x}$

(e) $y = \frac{Ae^x}{(x+2)^2}$ (f) $y = [\ln|x| + A]^2 + 1$

(g) $y = \frac{2}{e^{2x} + A}$ (h) $y = \cos^{-1}(x - C)$

(i) $x = \tan\left(\frac{1}{2}t^2 + t + C\right)$ (j) $x = \sqrt{t - \frac{1}{2\omega}\sin(2\omega t) + C}$

(k) $\ln|x| = \frac{3}{\omega}\cos(\omega t) + C$ (l) $x = A\sqrt{\sin(2t)}$

(m) $x = \sin^{-1}(t + C)$ (n) $x = A\ln|t|$

(o) $x = A(1 - \cos t)$ (p) $x = \pm\sqrt{\ln|t^2 + A|}$

2. (a) $y = (x+1)^2$ (b) $y = \frac{2}{\sqrt{\cos(2x)}}$

(c) $x = 2\ln|t|$ (d) $x^2 = 16t^3$

2.5 LINEAR EQUATIONS

A linear differential equation is a type of ODE that can be written in a specific linear form. In the context of ordinary differential equations, a differential equation is linear if the unknown functions and its derivatives appear to the power 1 (products are not allowed); it is non-linear otherwise:

$$\frac{d^2y}{dx^2} - y = e^x \qquad \text{linear 2}^{\text{nd}} \text{ order,}$$

$$\frac{dy}{dx} + y = e^x \qquad \text{linear 1}^{\text{st}} \text{ order,}$$

$$\frac{d^2y}{dx^2} - y = e^x \qquad \text{linear 2}^{\text{nd}} \text{ order,}$$

$$\frac{d^2y}{dx^2} - y\frac{dy}{dx} = \sin x \qquad \text{non – linear 2}^{\text{nd}} \text{ order,}$$

$$\frac{dy}{dx} + y = e^y \qquad \text{non – linear 1}^{\text{st}} \text{ order.}$$

Linear differential equations are important in various fields of science and engineering because they often model physical phenomena, from simple harmonic motion (see Equation 2.2) to electrical circuits and heat conduction, with linear relationships. In this chapter, we introduce first-order and second-order equations that have well-known solutions.

2.6 FIRST-ORDER LINEAR EQUATIONS

First-order linear differential equations are a type of equation that can be written in the following standard form:

$$\frac{dy}{dx}+p(x)y=q(x), \tag{2.5}$$

where p and q are non-zero continuous functions. In general, this equation is not separable. The most common method for solving this equation is the method of "integrating factor". This method involves finding an "integrating factor" that makes the equation simple, allowing solving it using integration techniques. Suppose we have found the integrating factor and it is as follows:

$$r(x)=e^{\int p(x)dx}.$$

We multiply both sides of the Equation 2.5 by $r(x)$:

$$r(x)\frac{dy}{dx}+r(x)p(x)y=r(x)q(x).$$

The equation can be reduced to an integrable equation as follows (see Box 2.1 for an explanation):

$$\frac{d}{dx}\left[r(x)y\right]=r(x)q(x).$$

Integrating both sides implies that

$$r(x)y=\int r(x)q(x)dx$$

$$y=\frac{1}{r(x)}\left[\int r(x)q(x)dx\right].$$

Once the final integration is successfully carried out, we essentially obtain an expression that represents the solution for the unknown function $y=y(x)$.

BOX 2.1: DERIVATION OF THE INTEGRATING FACTOR (THIS CAN BE OMITTED WITHOUT LOSS OF GENERALITY).

To show how we determine the integrating factor $r(x)=e^{\int p(x)dx}$, multiply each side of Equation 2.5 by $r(x)$:

$$r(x)\frac{dy}{dx}+r(x)p(x)y=r(x)q(x)$$

Differentiate $r(x)y$ with respect to x:

$$\frac{d}{dx}\left[r(x)y\right]=r(x)\frac{dy}{dx}+\frac{dr}{dx}y.$$

Comparing it with the left-hand side of $r(x)\frac{dy}{dx}+r(x)p(x)y=r(x)q(x)$ implies that

$$\frac{dr}{dx}=r(x)p(x).$$

This is a separable ODE in terms of r and x and can be solved as described earlier:

$$\frac{dr}{r}=p(x)dx$$

$$\int\frac{dr}{r}=\int p(x)dx$$

$$\ln|r|=\int p(x)dx$$

$$r(x)=e^{\int p(x)dx}.$$

When solving the ODE in Equation 2.5 using the method of integrating factors, it involves two separate integration steps: the first determines the integrating factor, and the second eliminates the derivatives. Initially, it might seem like we will have two constants of integration, which can be problematic when we desire only one. However, it can be demonstrated that any constant arising from the computation of the integrating factor will ultimately cancel out in the final solution. In practice, we assign a zero value to the integration constant arising from calculating the integrating factor. This will result in the retention of only a single arbitrary constant after the final integration.

SUMMARY

We solve the linear first-order differential equations using the following three steps:

Step 1: Write in standard form and compare with the standard form given in (2.5) and identify $p(x)$ and $q(x)$.

Step 2: Determine the integrating factor $r(x)=e^{\int p(x)dx}$.

[Note that we set the constant of integration to zero.]

Step 3: The general solution is $y(x)=\frac{1}{r(x)}\left[\int r(x)q(x)dx\right]$.

Example 2.8 Solve the following ODE:

$$\frac{dy}{dx}+\frac{1}{x}y=x.$$

Solution

Step 1: Compare with the standard form given in Equation 2.5 and identify $p(x)$ and $q(x)$:

$$\frac{dy}{dx}+p(x)y=q(x)\rightarrow p(x)=\frac{1}{x} \quad \text{and} \quad q(x)=x.$$

Step 2: Determine the integrating factor:

$$\begin{aligned} r(x) &= e^{\int p(x)dx} \\ &= e^{\int \frac{1}{x}\times dx} \\ &= e^{\ln x} = x. \end{aligned}$$

[Note that we set the integration constant to zero.]

Step 3: The general solution is as follows:

$$y(x)=\frac{1}{r(x)}\int r(x)q(x)dx.$$

Hence, by substituting the integrating factor and the function $q(x)=x$ in the preceding formula, we obtain

$$\begin{aligned} y(x) &= \frac{1}{x}\left[\int x\times x\,dx\right]=\frac{1}{x}\left[\int x^2\,dx\right] \\ &= \frac{1}{x}\left(\frac{1}{3}x^3+C\right)=\frac{x^2}{3}+\frac{C}{x}. \end{aligned}$$

Example 2.9 Solve the following ODE:

$$3x\frac{dy}{dx}+y=x^2.$$

Solution

Step 1: Write the equation in the standard form (Equation 2.5) and identify $p(x)$ and $q(x)$:

$$3x\frac{dy}{dx}+y=x^2$$

$$\frac{dy}{dx}+\frac{1}{3x}y=\frac{x}{3}.$$

Hence,

$$p(x)=\frac{1}{3x} \text{ and } q(x)=\frac{x}{3}.$$

Step 2: Determine the integrating factor:

$$r=e^{\int p(x)dx}$$

$$=e^{\int \frac{1}{3x}\times dx}$$

$$=e^{\frac{1}{3}\ln x}=e^{\ln x^{\frac{1}{3}}}=x^{\frac{1}{3}}.$$

[Note that we set the integration constant to zero.]

Step 3: The general solution is

$$y(x)=\frac{1}{r(x)}\left[\int r(x)q(x)dx\right].$$

Hence, by substituting the integrating factor and the function $q(x)=\frac{x}{3}$ in the preceding equation, we obtain

$$y(x)=\frac{1}{x^{\frac{1}{3}}}\left[\int x^{\frac{1}{3}}\times\frac{x}{3}dx\right]=\frac{1}{3}x^{-\frac{1}{3}}\int x^{\frac{4}{3}}dx$$

$$=\frac{1}{3}x^{-\frac{1}{3}}\left(\frac{3}{7}x^{\frac{7}{3}}+C\right)=\frac{x^2}{7}+\frac{\frac{1}{3}C}{x^{\frac{1}{3}}}$$

$$=\frac{x^2}{7}+\frac{A}{x^{\frac{1}{3}}}, \quad \left(A=\frac{1}{3}C\right).$$

Example 2.10 Solve the following ODE:

$$\frac{dy}{dx}+y=x.$$

Solution

Step 1: Compare with the standard form given in Equation 2.5 and identify $p(x)$ and $q(x)$:

$$\frac{dy}{dx}+p(x)y=q(x)\rightarrow p(x)=1 \text{ and } q(x)=x.$$

Step 2: Determine the integrating factor:

$$\begin{aligned} r &= e^{\int p(x)dx} \\ &= e^{\int 1\times dx} \\ &= e^{x}. \end{aligned}$$

[Note that we set the integration constant to zero.]

Step 3: The general solution is in the form

$$y=\frac{1}{r(x)}\left[\int r(x)q(x)dx\right].$$

Hence, by substituting the integrating factor and the function $q(x)=x$ in the previous formula, we obtain

$$y=\frac{1}{e^{x}}\left[\int x\times e^{x}dx\right].$$

Integration of the right-hand side (RHS) is obtained using the technique of integration by parts:

Recall the formula for the technique of integration by parts is

$$\int f(x)g'(x)dx=f(x)g(x)-\int f'(x)g(x)dx.$$

Apply integration by parts to the RHS:

$$\begin{aligned} f(x)&=x \quad & g'(x)&=e^{x} \\ f'(x)&=1 \quad & g(x)&=e^{x}. \end{aligned}$$

Thus,

$$y = \frac{1}{e^x}\left(xe^x - \int e^x dx \right) = e^{-x}\left(xe^x - e^x + C\right)$$

$$y = x - 1 + Ce^{-x}.$$

2.6.1 Exercises

1. Determine the general solution of the following first-order linear equations.

(a) $\frac{dy}{dx} - y = 3$ (b) $\frac{dy}{dx} + 2y = 6e^x$

(c) $\frac{dy}{dx} + 2xy = xe^{-x^2}$ (d) $\frac{dy}{dx} + 2xy = 2x$

(e) $x\frac{dy}{dx} + y = x + x^3$ (f) $x\frac{dy}{dx} = y + (x+1)^2$

(g) $\frac{dy}{dx} - \frac{2}{x}y = x^2e^x$ (h) $\frac{dy}{dx} + y = x$

2. Determine the particular solution of the following initial value problems.

(a) $\frac{dy}{dx} + y = (x+1)^2, \quad y(0) = 0$ (b) $\frac{dy}{dx} - 2y = e^{2x} + e^{-2x}, \quad y(0) = -2$

(c) $\frac{dy}{dx} + y = x, \quad y(0) = 0$ (d) $x\frac{dy}{dx} - 3y = x^6, \quad y(1) = 2$

Answers

1. (a) $r(x) = e^{-x}$, $y = Ae^x - 3$ (b) $r(x) = e^{2x}$, $y = 2e^x + Ae^{-2x}$

(c) $r(x) = e^{x^2}$, $y = \left(\frac{1}{2}x^2 + A\right)e^{-x^2}$ (d) $r(x) = e^{x^2}$, $y = 1 + Ae^{-x^2}$

(e) $r(x) = x, \quad y = \frac{1}{2}x + \frac{1}{4}x^3 + \frac{A}{x}$ (f) $r(x) = \frac{1}{x}, \quad y = x^2 + 2x\ \ln|x| + Ax - 1$

(g) $r(x) = \frac{1}{x^2}, \quad y = x^2\left(e^x + A\right)$ (h) $r(x) = e^x, \quad y = (x-1) + Ae^{-x}$

2. (a) $r(x) = e^x, \quad y = x^2 + 1 - e^{-x}$

(b) $r(x) = e^{-2x}, \; y = xe^{2x} - \frac{1}{4}e^{-2x} - \frac{7}{4}e^{2x}$

(c) $r(x) = e^x, \quad y = (x-1) + e^{-x}$

(d) $r(x) = \frac{1}{x^3}, \quad y = \frac{1}{3}x^6 + \frac{5}{3}x^3$

2.7 SECOND-ORDER LINEAR ODES WITH CONSTANT COEFFICIENTS

In the previous section, we covered the first-order linear differential equations. Now we will proceed to describe the second-order linear differential equations. Second-order linear differential equations are a fundamental topic in engineering.

We look at a few specific cases of second-order differential equations that we can solve. However, we need to be more limited about the types of differential equations we will look at. This is needed so that we can really solve them. In this section, we describe special types of second-order ODEs:

$$a\frac{d^2y}{dx^2}+b\frac{dy}{dx}+cy=0 \qquad \text{(Homogeneous)} \tag{2.6}$$

$$a\frac{d^2y}{dx^2}+b\frac{dy}{dx}+cy=f(x) \qquad \text{(Non Homogeneous)} \tag{2.7}$$

where a, b and c are constants. These types of equations have wide applications in physics, engineering, economics and various other disciplines. One example of this kind of equation was considered in Section 2.1, where we described a simple vibrating spring, which led us to the differential equation (Equation 2.2) similar to Equation 2.6. In electronic circuits, equations such as Equations 2.6 and 2.7 can be formulated to describe current or voltage characteristics. Equations of this kind find relevance in numerous areas where oscillations or vibrations occur. Their solutions provide insights into how systems evolve over time and are crucial in solving real-world problems across numerous disciplines.

Before presenting the methods of solving this type of equation, it is necessary to provide the following definitions.

DEFINITION 2.3: LINEAR COMBINATION

If $y_1(x)$ and $y_2(x)$ are solutions of the linear equation (Equation 2.6), then $y(x)=C_1y_1(x)+C_2y_2(x)$ is said to be a "linear combination" of these functions for any arbitrary constants C_1 and C_2, and it is the general solution of ODEs. This is called the "superposition principle".

For example, if $y_1(x)=x$ and $y_2(x)=x^2$, then

$$y(x)=x-x^2 \quad \text{or} \quad y(x)=3x+5x^2$$

are linear combinations of $y_1(x)$ and $y_2(x)$.

DEFINITION 2.4: LINEARLY INDEPENDENCE

Two functions, $y_1(x)$ and $y_2(x)$ are said to be "linearly independent" if neither one of them can be expressed as a scalar multiple of the other function; that is

$$y_1(x) \neq ky_2(x) \text{ or } y_2(x) \neq ky_1(x).$$

For example, $y_1(x) = x$ and $y_2(x) = e^x$ are linearly independent, but $y_1(x) = 3x$ and $y_2(x) = 6x$ are not.

2.7.1 Homogeneous Equations with Constant Coefficients

We now describe a method of solving the following homogeneous equation:

$$a\frac{d^2y}{dx^2} + b\frac{dy}{dx} + cy = 0. \tag{2.8}$$

Instead of using gradual integration approach employed for first-order equations, we will generate solutions by leveraging the intrinsic characteristics of homogeneous equations and specific mathematical functions. Looking at the first-order linear equation $\frac{dy}{dx} + ay = 0$ that has a solution $y = Ae^{-ax}$, we suggest a solution of the form

$$y = e^{\lambda x},$$

where λ is yet to be found. The solution must satisfy Equation 2.8. Therefore, we differentiate it twice:

$$\frac{dy}{dx} = \lambda e^{\lambda x}, \quad \frac{d^2y}{dx^2} = \lambda^2 e^{\lambda x}$$

and substitute them into Equation 2.8 to obtain

$$\left(a\lambda^2 + b\lambda + c\right)e^{\lambda x} = 0.$$

The exponential term is always positive; hence,

$$a\lambda^2 + b\lambda + c = 0. \tag{2.9}$$

Solving this equation for λ means that the function $y = e^{\lambda x}$ is a valid solution of the differential equation (Equation 2.8).

Equation 2.9 is called the "**characteristic equation**" of the differential equation (Equation 2.8). The solutions of the quadratic equation (Equation 2.9) lead to the particular solution and thus to the general solution of the differential equation (Equation 2.8).

Recall that

$$\lambda = \frac{-b \pm \sqrt{b^2 - 4ac}}{2a}.$$

There are three cases to consider:

1. $\Delta = b^2 - 4ac > 0.$

   For $\Delta > 0$, there are two real distinct roots for the quadratic equation: $\lambda = \lambda_1$ and $\lambda = \lambda_2$. Therefore, there will be two particular solutions for the differential equation (Equation 2.8):

   $$y_1 = e^{\lambda_1 x} \text{ and } y_2 = e^{\lambda_2 x}.$$

   These solutions are linearly independent. Using the principle of superposition any linear combination,

   $$y(x) = Ae^{\lambda_1 x} + Be^{\lambda_2 x}$$

   is also a solution, where A and B are arbitrary constants. This can be checked easily, due to the homogeneity and the linearity of Equation 2.8. This is called "general solution" of Equation 2.8 when the characteristic equation has two distinct real roots. Note that the general solution is obtained without using integration.

2. $\Delta = b^2 - 4ac < 0.$

   In this case, the characteristic equation (Equation 2.9) has two complex conjugate roots:

   $$\lambda_1 = \alpha \pm j\beta.$$

   Hence, a solution of Equation 2.8 is

   $$y(x) = e^{(\alpha \pm j\beta)x}.$$

   This is a complex-valued function. However, we are looking for a real-valued function. The general solution in this case is as follows and the full explanation for this is given in Box 2.2:

   $$y(x) = e^{\alpha x}\left(A\cos(\beta x) + B\sin(\beta x)\right),$$

   where A and B are arbitrary constants.

BOX 2.2: WITHOUT LOSS OF GENERALITY, THIS EXPLANATION CAN BE OMITTED.

Using the Euler formulae for the complex numbers, we can write

$$\begin{aligned} y(x) = e^{(\alpha \pm j\beta)x} &= e^{\alpha x}\left[\cos(\beta x) \pm j\sin(\beta x)\right] \quad \text{where} \quad j = \sqrt{-1} \\ &= e^{\alpha x}\cos(\beta x) \pm je^{\alpha x}\sin(\beta x) \\ &= y_1(x) \pm jy_2(x), \end{aligned}$$

where $y_1 := e^{\alpha x}\cos(\beta x)$ and $y_2 := e^{\alpha x}\sin(\beta x)$. The solution $y(x)$ obtained earlier is a complex-valued function. Substituting $y(x) = y_1(x) \pm jy_2(x)$ in Equation 2.8 implies

$$\left(a\frac{d^2y_1}{dx^2} + b\frac{dy_1}{dx} + cy_1\right) \pm j\left(a\frac{d^2y_2}{dx^2} + b\frac{dy_2}{dx} + cy_2\right) = 0.$$

Hence, the real and imaginary parts are zero:

$$a\frac{d^2y_1}{dx^2} + b\frac{dy_1}{dx} + cy_1 = 0 \quad \text{and} \quad a\frac{d^2y_2}{dx^2} + b\frac{dy_2}{dx} + cy_2 = 0.$$

This means that $y_1(x) = e^{\alpha x}\cos(\beta x)$ and $y_2(x) = e^{\alpha x}\sin(\beta x)$ are the particular solutions of Equation 2.8. These solutions are real and linearly independent functions. Using the principle of superposition, any linear combination

$$y(x) = Ay_1(x) + By_2(x) = e^{\alpha x}\left[A\cos(\beta x) + B\sin(\beta x)\right]$$

is also a solution, where A and B are arbitrary constants. This is the "general solution" of Equation 2.8 when the roots are complex conjugate numbers.

3. $\Delta = b^2 - 4ac = 0.$

In this case, the characteristic equation has repeated real roots:

$$\lambda_1 = \lambda_2 = \lambda = -\frac{b}{2a}.$$

The only solution in this case is

$$y_1 = e^{\lambda x} = e^{-\frac{b}{2a}x}. \tag{2.10}$$

It appears insufficient to provide a general solution involving two arbitrary constants similar to the other two cases explained earlier. To construct the general solution, we need another particular solution that is linearly independent from the solution in Equation 2.10. The second solution can be found using the reduction of order explained in Box 2.3. The general solution in this case is

$$y(x) = (A + Bx)e^{\lambda x}. \tag{2.11}$$

BOX 2.3: WITHOUT LOSS OF GENERALITY, THIS PROOF CAN BE OMITTED.

Assume

$$y_2 = v(x)e^{\lambda x}$$

is the solution of the homogeneous equation, so it must satisfy the equation. The derivatives of y_2 are

$$\begin{aligned} y_2' &= v'e^{\lambda x} + \lambda v e^{\lambda x} \\ y_2'' &= v''e^{\lambda x} + \lambda v'e^{\lambda x} + \lambda v'e^{\lambda x} + \lambda^2 v e^{\lambda x} \\ &= v''e^{\lambda x} + 2\lambda v'e^{\lambda x} + \lambda^2 v e^{\lambda x}. \end{aligned}$$

We substitute these into Equation 2.8:

$$av''e^{\lambda x} + 2a\lambda v'e^{\lambda x} + a\lambda^2 v e^{\lambda x} + bv'e^{\lambda x} + b\lambda v e^{\lambda x} + cve^{\lambda x} = 0,$$

which implies

$$av''e^{\lambda x} + 2a\lambda v'e^{\lambda x} + bv'e^{\lambda x} + ve^{\lambda x}\left(a\lambda^2 + b\lambda + c\right) = 0. \tag{2.12}$$

Since $\lambda = -\frac{b}{2a}$, so $2a\lambda v'e^{\lambda x} + bv'e^{\lambda x} = 0$, and since λ is the root of the characteristic equation, so $a\lambda^2 + b\lambda + c = 0$. Substituting these into Equation 2.12 implies that

$$\begin{aligned} &av''e^{\lambda x} = 0 \Rightarrow \\ &v''(x) = 0, \text{ since } ae^{\lambda x} \neq 0. \end{aligned}$$

Integrating the preceding equation twice, we obtain

$$v(x) = w_1 x + w_2,$$

where w_1 and w_2 are the arbitrary constants of integration. Therefore, the proposed solution is

$$y_2 = v(x)e^{\lambda x} = (w_1 x + w_2)e^{\lambda x}$$
$$= w_1 x e^{\lambda x} + w_2 e^{\lambda x}.$$

The second term $w_2 e^{\lambda x}$ in the preceding equation is a scalar multiple of y_1 in Equation 2.10 and is integrated into y_1 whenever we express the final solution. Hence, assuming $w_1 = 1$ and $w_2 = 0$,

$$y_2 = xe^{\lambda x}.$$

In contrast to first-order differential equations, the solution of the second-order equation does not require any integration. Its solution relies on formulating the characteristic equation and applying one of the prescribed cases presented in the following summary.

SUMMARY

(a) Formulate the characteristic equation $a\lambda^2 + b\lambda + c = 0$.
(b) Determine the roots of the characteristic equation.
(c) Considering the roots of the characteristic equation, write the general solution as follows:

1. Two distinct real roots λ_1 and λ_2:

$$y(x) = Ae^{\lambda_1 x} + Be^{\lambda_2 x}.$$

2. Two complex roots $\lambda = \alpha \pm \beta j,\ j = \sqrt{-1}$:

$$y(x) = e^{\alpha x}\left[A\cos(\beta x) + B\sin(\beta x)\right].$$

3. Two repeated real roots λ:

$$y = (A + Bx)e^{\lambda x}.$$

Example 2.11 Solve the following differential equation:

$$\frac{d^2 y}{dx^2} - 2\frac{dy}{dx} - 3y = 0.$$

Solution

The characteristic equation:

$$\lambda^2 - 2\lambda - 3 = 0$$

$$(\lambda - 3)(\lambda + 1) = 0$$

$$\lambda = 3, \ \lambda = -1$$

Two distinct real roots, therefore, the general solution is

$$y = Ae^{3x} + Be^{-x}.$$

Example 2.12 Solve the following differential equation:

$$\frac{d^2y}{dx^2} + 2\frac{dy}{dx} + 5y = 0.$$

Solution

The characteristic equation:

$$\lambda^2 + 2\lambda + 5 = 0$$

$$\lambda = \frac{-2 \pm \sqrt{-16}}{2}$$

$$= -1 \pm 2j := \alpha \pm \beta j \Rightarrow \alpha = -1, \quad \beta = 2.$$

The general solution is

$$y(x) = e^{-x}\left[A\cos(2x) + B\sin(2x)\right].$$

Example 2.13 Solve the following differential equation:

$$\frac{d^2y}{dx^2} - 2\frac{dy}{dx} + y = 0.$$

Solution

The characteristic equation:

$$\lambda^2 - 2\lambda + 1 = 0$$

$$(\lambda - 1)^2 = 0$$

$$\lambda = 1.$$

One repeated root only; therefore,

$$y = (A + Bx)e^x.$$

2.7.2 Exercises

Find the general solution for the following second-order linear differential equations:

(a) $\frac{d^2y}{dx^2} - \frac{dy}{dx} - 6y = 0$

(b) $2\frac{d^2y}{dx^2} - 5\frac{dy}{dx} + 2y = 0$

(c) $\frac{d^2y}{dx^2} - 6\frac{dy}{dx} + 9y = 0$

(d) $4\frac{d^2y}{dx^2} - 12\frac{dy}{dx} + 9y = 0$

(e) $\frac{d^2y}{dx^2} - 2\frac{dy}{dx} + 4y = 0$

(f) $\frac{d^2y}{dx^2} + 2\frac{dy}{dx} + 2y = 0$

(g) $\frac{d^2y}{dx^2} - 9y = 0$

(h) $\frac{d^2y}{dx^2} + 9y = 0$

Answers

(a) $y = Ae^{3x} + Be^{-2x}$

(b) $y = Ae^{0.5x} + Be^{2x}$

(c) $y = (A + Bx)e^{3x}$

(d) $y = (A + Bx)e^{1.5x}$

(e) $y = e^x\left[A\cos\left(\sqrt{3}x\right) + B\sin\left(\sqrt{3}x\right)\right]$

(f) $y = e^{-x}\left[A\cos(x) + B\sin(x)\right]$

(g) $y = Ae^{3x} + Be^{-3x}$

(h) $y = A\cos(3x) + B\sin(3x)$

2.8 HOMOGENEOUS EQUATIONS, THE INITIAL VALUE PROBLEMS

To solve second-order homogeneous equations and determine the arbitrary constants, namely A and B, it is necessary to have two initial conditions. The initial conditions that are usually considered are $y(0) = c_0$ and $\frac{dy}{dx}(0) = c_1$. A solution with known arbitrary constants is called the particular solution. In the following examples, we show two differential equations with their initial conditions to describe particular solution of homogeneous equations.

Example 2.14 Solve

$$\frac{d^2y}{dx^2} + 16y = 0, \quad y(0) = 1, \quad \frac{dy}{dx}(0) = 16.$$

Solution

The characteristic equation:

$$\lambda^2 + 16 = 0$$
$$\lambda^2 = -16$$
$$\lambda = 0 \pm 4j, \quad j = \sqrt{-1}$$
$$\alpha = 0, \ \beta = 4.$$

General solution:

$$y(x) = e^{\alpha x}\left[A\cos(\beta x) + B\sin(\beta x)\right]$$
$$= e^{0x}\left[A\cos(4x) + B\sin(4x)\right]$$
$$= A\cos(4x) + B\sin(4x).$$

Apply $y(0) = 1$:

$$y(0) = A\cos(4\times 0) + B\sin(4\times 9) = 1$$
$$A = 1.$$

Differentiate the general solution and apply the second condition:

$$y = A\cos(4x) + B\sin(4x)$$
$$\frac{dy}{dx} = -4A\sin(4x) + 4B\cos(4x).$$

Now apply $\frac{dy}{dx}(0) = 16$:

$$16 = -4A\sin(4\times 0) + 4B\cos(4\times 0)$$
$$16 = 4B \Rightarrow B = 4.$$

The particular solution is obtained by substituting the values $A = 1$ and $B = 4$ into the general solution:

$$y = \cos(2x) + 4\sin(2x).$$

Example 2.15 Consider the equation of motion given in Equation 2.2, which is described earlier in Section 2.2. In this equation, set $m = 5$, $c = 1$ and $k = 22.1$. Let the initial displacement be $x(0)=10$ units, and subsequently release of the mass from rest. Determine the position of the mass at any instant of time t.

Solution

The ODE is

$$5\frac{d^2x}{dt^2}+\frac{dx}{dt}+22.1x=0.$$

Initial displacement $x(0)=10$ units, Initial speed $\frac{dx}{dt}(0)=0$

Solution of this equation is

$$x(t)=e^{-0.1t}\left[A\cos(2.1t)+B\sin(2.1t)\right].$$

Apply the first condition, $x(0)=10$:

$$10=e^{-0}\left[A\cos(2.1\times 0)+B\sin(2.1\times 0)\right]$$

$$A=10.$$

Substitute

$$x(t)=e^{-0.1t}\left[10\cos(2.1t)+B\sin(2.1t)\right].$$

Now, we differentiate the preceding function (using the product rule) and use the second condition to obtain the arbitrary constant B:

$$\frac{dx}{dt}=-0.1e^{-t}\left[10\cos(2.1t)+B\sin(2.1t)\right]+e^{-0.1t}\left[-21\sin(2.1t)+2.1B\cos(2.1t)\right].$$

Now apply the second condition, $\frac{dx}{dt}(0)=0$:

$$0=-0.1e^{-0}\left[10\cos(0)+B\sin(0)\right]+e^{-0}\left[-21\sin(0)+2.1B\cos(0)\right]$$

$$0=-1+2.1B\Rightarrow B=0.47619.$$

Substitute $A=10$ and $B=0.47619$ in the general solution gives the instantaneous position x:

$$x(t)=e^{-0.1t}\left[10\cos(2.1t)+0.47619\sin(2.1t)\right].$$

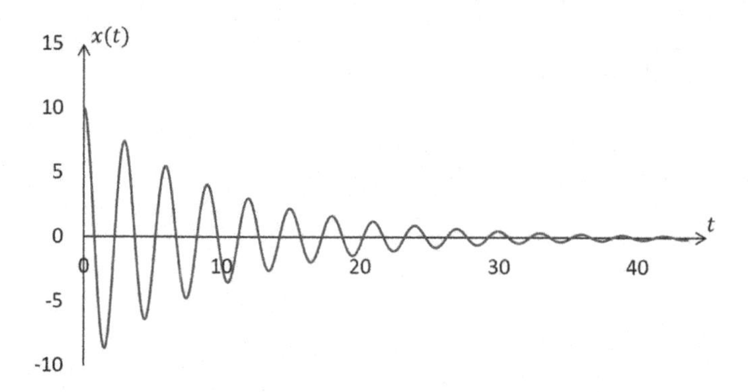

FIGURE 2.2 Damped oscillation.

The function $x(t)$ represents a damped oscillator with the amplitude $e^{-0.1t}$. The oscillation disappears as time t increases, as shown in Figure 2.2.

2.8.1 Exercise

Find the particular solution of the following ordinary differential equations.

(a) $\dfrac{d^2y}{dx^2} - 2\dfrac{dy}{dx} - 8y = 0, \quad y(0) = 0, \quad y'(0) = 6$

(b) $\dfrac{d^2y}{dx^2} + 6\dfrac{dy}{dx} + 13y = 0, \quad y(0) = 2, \quad y'(0) = 0$

(c) $\dfrac{d^2y}{dx^2} + 4\dfrac{dy}{dx} + 4y = 0, \quad y(0) = 1, \quad y'(0) = 0$

(d) $\dfrac{d^2y}{dx^2} + 4y = 0, \quad y(0) = 1, \quad y\left(\dfrac{\pi}{4}\right) = 3$

(e) $\dfrac{d^2y}{dx^2} + 3\dfrac{dy}{dx} + 2y = 0, \quad y(0) = 0, \quad y(1) = 2$

Answers

(a) $y = e^{4x} - e^{-2x}$

(b) $y = e^{-3x}\left[2\cos(2x) + 3\sin(2x)\right]$

(c) $y = (1 + 2x)e^{-2x}$

(d) $y = \cos 2x + 3\sin 2x$

(e) $y = \dfrac{2\left(e^{-x} - e^{-2x}\right)}{e^{-1} - e^{-2}}$

2.9 NON-HOMOGENEOUS EQUATIONS

In Sections 2.7 and 2.8, we considered homogeneous equations that were second-order ODEs with zero RHS. Now we explain how to solve the non-homogeneous differential equation (Equation 2.7).

An equation of the type (Equation 2.7) that we consider in this section can be obtained by modelling a vibrating spring as described in Section 2.2 but with a slight difference in modelling. Suppose, the upper end of the spring in Figure 2.1 (Section 2.2) is not attached to a fixed point *A* but to something that is itself vibrating as a function of time, say $f(t)$. In this case, if the positive direction of $f(t)$ is downward, the equation of motion for the vibrating spring can be written as follows:

$$m\frac{d^2x}{dt^2}+c\frac{dx}{dt}+kx=f(t) \qquad OR$$

$$a\frac{d^2y}{dx^2}+c\frac{dy}{dx}+by=f(x). \tag{2.13}$$

In the next section, we explain the steps to solve this equation.

2.9.1 The Method of Undetermined Coefficients

To solve the non-homogeneous linear second-order ODEs, we follow the following general steps:

Step 1: Write down the equation in the standard form (Equation 2.13).

Step 2: Solve the associated homogeneous equation by setting $f(x)$ to zero and find its general solution. This solution is called "**complimentary function**" and is denoted by $y_c(x)$.

Step 3: Find a particular solution $y_p(x)$ using what is called "method of undetermined coefficients". We can assume that the particular solution $y_p(x)$ has the general form of $f(x)$, with undetermined coefficients and solve for these coefficients using substitution.

Step 4: The general solution to the non-homogeneous equation is the sum of the complementary function $y_c(x)$, and the particular solution $y_p(x)$:

$$y(x)=y_c(x)+y_p(x).$$

Depending on the initial conditions of the specific problem, we may need to solve for the constants of integration using the given conditions to obtain a unique solution. We present Table 2.1 to identify the general form of particular solutions $y_p(x)$.

Here are a few examples to clarify the process.

Example 2.16 Solve the following equation and find its general solution:

$$\frac{d^2y}{dx^2}-\frac{dy}{dx}-2y=8x+2.$$

Solution

Step 1: Find the complementary function $y_c(x)$. For that, we have to make the RHS zero and solve the homogeneous equation. The characteristic equation has two distinct real roots:

$$\lambda^2 - \lambda - 2 = 0$$

$$(\lambda - 2)(\lambda + 1) = 0$$

$$\lambda = 2, \ \lambda = -1.$$

Hence,

$$y_c(x) = Ae^{2x} + Be^{-x}.$$

Step 2: Find the particular solution of $y_p(x)$ using the method of undetermined coefficients. We assume that $y_p(x)$ has the general form of the RHS of the given differential equation, which is a linear term. So, using Table 2.1, the general form for the complimentary function is a linear function; that is

$$y_p(x) = ax + b,$$

where a and b are unknown coefficients. Our aim is to determine these unknowns by substituting $y_p(x)$ and its derivatives into the given differential equation:

$$\frac{dy_p}{dx} = a, \quad \frac{d^2 y_p}{dx^2} = 0$$

$$\Rightarrow 0 - a - 2(ax + b) = 8x + 2$$

$$-2ax - a - 2b = 8x + 2.$$

Equating coefficients of the like terms,

$$-2a = 8$$

$$-a - 2b = 2.$$

Solving these equations for a and b, imply $a = -4$ and $b = 1$, and hence,

$$y_p(x) = -4x + 1.$$

Step 3: Write the general solution of the differential equation as follows:

$$y(x) = y_c(x) + y_p(x) = Ae^{2x} + Be^{-x} - 4x + 1.$$

TABLE 2.1
To Identify the General Form of Particular Solutions $y_p(x)$; This Is a General Form of The Particular Solution for ODEs of the Form $\frac{d^2y}{dx^2}+k^2y=f(x)$ with $f(x)$ Containing $\cos kx$ or $\sin kx$ or Both

$f(x)$	$y_p(x)$
Constant function (e.g., $f(x)=3$ or -5)	a
Linear function ($f(x)=4x$ or $2x-6$)	$ax+b$
Quadratic function ($f(x)=2x^2+3x-7$ or x^2).	ax^2+bx+c
Exponential function ($f(x)=Ae^{kx}$, where k is not a root of the Characteristic Equation (CE): e.g., $f(x)=3e^{5x}$).	ae^{kx}
Exponential function ($f(x)=Ae^{kx}$, where k coincides with a root of the CE).	axe^{kx}
Exponential function ($f(x)=Ae^{kx}$, where k is a repeated root of the CE).	ax^2e^{kx}
$f(x)=\cos(kx)$ or $f(x)=\sin(kx)$ or $f(x)=\sin(kx)+3\cos(kx)$.	$a\cos kx+b\sin kx$ or $x[a\cos kx+b\sin kx]^*$

Example 2.17 Solve the ODE:

$$\frac{d^2y}{dx^2}+3\frac{dy}{dx}+2y=x^2+4.$$

Solution

Step 1: Find the complementary function $y_c(x)$. The characteristic equation has two distinct real roots:

$$\lambda^2+3\lambda+2=0$$

$$(\lambda+1)(\lambda+2)=0$$

$$\lambda=-2,\quad \lambda=-1.$$

Complementary function:

$$y_c(x)=Ae^{-x}+Be^{-2x}.$$

Step 2: Find the particular solution $y_p(x)$ using the method of undetermined coefficients. Considering that the RHS is a quadratic term, so using Table 2.1, we suggest a solution as follows:

$$y_p(x)=ax^2+bx+c,$$

where a, b and c are the undetermined coefficients. We find these unknown parameters by substituting $y_p(x)$ and its derivatives into the given differential equation. The first and second derivatives of $y_p(x)$ are

$$\frac{dy_p}{dx}=2ax+b,\quad \frac{d^2y_p}{dx^2}=2a.$$

Now, we substitute $y_p(x)$ and its derivatives into the given differential equation:

$$(2a)+3(2ax+b)+2\left(ax^2+bx+c\right)=x^2+4.$$

Collecting the like terms on left-hand side of the equation, we get

$$(2a+3b+2c)+(6a+2b)x+2ax^2=x^2+0x+4.$$

Equate the coefficients of the like terms from both sides:

Coefficients of $x^2: 2a=1$
Coefficients of x: $6a+2b=0$
Constant term: $2a+3b+2c=4$

Solving these 3×3 equations imply

$$a=\frac{1}{2} \quad b=-\frac{3}{2} \quad c=\frac{15}{4}.$$

Hence, the particular solution is

$$y_p(x)=\frac{1}{2}x^2-\frac{3}{2}x+\frac{15}{4}.$$

Step 3: The general solution of the differential equation is

$$y(x)=y_c(x)+y_p(x)=Ae^{-x}+Be^{-2x}+\frac{1}{2}x^2-\frac{3}{2}x+\frac{15}{4}.$$

Example 2.18 Solve the following ODE and find its general solution:

$$\frac{d^2y}{dx^2}+4\frac{dy}{dx}+4y=2e^{-2x}.$$

Solution

Step 1: Find the complementary function $y_c(x)$. The characteristic equation has a repeated root:

$$\lambda^2+4\lambda+4=0$$

$$(\lambda+2)^2=0$$

$$\lambda=-2 \qquad \text{(repeated).}$$

Complementary function

$$y_c=(A+Bx)e^{-2x}.$$

Step 2: Find the particular solution of $y_p(x)$ using the method of undetermined coefficients. The RHS of the ODE is an exponential function $f(x) = e^{-2x}$. This is a solution of the homogeneous equation, hence using Table 2.1 (for the case $k = \lambda = -2$), we suggest a particular solution of the form

$$y_p = ax^2e^{-2x}.$$

Differentiate using the product rule:

$$y_p' = a\left(2xe^{-2x} - 2x^2e^{-2x}\right)$$

$$y_p'' = a(2e^{-x} - 4xe^{-2x} - 2\left(2xe^{-2x} - 2x^2e^{-2x}\right)$$

$$= a\left(2e^{-2x} - 8xe^{-2x} + 4x^2\mathrm{e}^{-2x}\right).$$

Substitute into the ODE:

$$a\left(2e^{-2x} - 8xe^{-2x} + 4x^2\mathrm{e}^{-2x}\right) + 4a\left(2xe^{-2x} - 2x^2e^{-2x}\right) + 4ax^2e^{-2x} = e^{-2x}.$$

Simplifying implies

$$2ae^{-2x} = 2e^{-2x}$$

$$2a = 2$$

$$a = 1.$$

Hence,

$$y_p = x^2e^{-2x}.$$

Step 3: The general solution can be written by combining y_c and y_p:

$$y = (A + Bx)e^{-2x} + x^2e^{-2x}.$$

Example 2.19 Solve the following differential equation and find its general solution.

$$\frac{d^2y}{dx^2} + 4y = \sin x.$$

Solution

Step 1: Find the complementary function $y_c(x)$. The characteristic equation has imaginary roots:

$$\lambda^2 + 4 = 0$$

$$\lambda^2 = -4$$

$$\lambda = \pm 2j = 0 \pm 2j, \ j = \sqrt{-1}.$$

Complementary function is

$$y_c(\mathrm{x}) = A\cos 2x + B\sin 2x.$$

Step 2: Find the particular solution of $y_p(x)$ using the method of undetermined coefficients. The RHS of the equation is $\sin(x)$ with the angular frequency 1. Since the angular frequency of $\sin(x)$ and the angular frequency of y_c are different, so using Table 2.1, we suggest a particular solution of the form

$$y_p = a\cos x + b\sin x.$$

Differentiate:

$$y'_p = -a\sin x + b\cos x$$

$$y''_p = -a\cos x - b\sin x.$$

Substitute into the ODE:

$$[-a\cos x - b\sin x] + 4[a\cos x + b\sin x] = \sin x.$$

Simplify:

$$3a\cos x + 3b\sin x = \sin x.$$

Equate coefficients of the like terms:

Cosine terms: $3a = 0 \rightarrow a = 0$
Sine terms: $3b = 1 \rightarrow b = \frac{1}{3}$

Hence,

$$y_p = \frac{1}{3}\sin x.$$

Step 3: The general solution is obtained by combining y_c and y_p:

$$y(x) = y_c(x) + y_p(x) = A\cos 2x + B\sin 2x + \frac{1}{3}\sin x.$$

Example 2.20 Solve the following ODE and find its general solution:

$$\frac{d^2y}{dx^2} + y = \sin x.$$

Solution

Step 1: Find the complementary function $y_c(x)$. The characteristic equation has two imaginary roots:

$$\lambda^2 + 1 = 0$$
$$\lambda^2 = -1$$
$$\lambda = \pm j = 0 \pm j, \quad j = \sqrt{-1}.$$

The complementary function is

$$y_c(x) = A\cos x + B\sin x.$$

Step 2: Find the particular solution of $y_p(x)$ using the method of undetermined coefficients. The frequency of the function $f(x) = \sin x$ on the RHS of the ODE is the same as the frequency of the complementary function, so using Table 2.1, we suggest a particular solution of the form

$$y_p = x[\mathrm{a}\sin\ x + b\cos x].$$

Differentiate

$$y_p' = a\sin x + b\cos x + x[a\cos x - b\sin x]$$
$$y_p'' = 2\mathrm{a}\cos\ x - 2b\sin x + x[-a\sin x - b\cos x].$$

Substituting into the differential equation and equating the coefficients of the like terms leads to the following solution:

$$y_p = -\frac{1}{2}x\cos x.$$

Step 3: The general solution is as follows:

$$y = A\ \cos x + B\ \sin x - \frac{1}{2}x\ \cos x.$$

2.9.2 Exercises

1. Solve the following ODEs and find their general solutions.

(a) $\dfrac{d^2y}{dx^2} + \dfrac{dy}{dx} - 2y = 1 + x^2$ (b) $\dfrac{d^2y}{dx^2} + 3\dfrac{dy}{dx} + 2y = 2e^{2x}$

(c) $\frac{d^2y}{dx^2}+3\frac{dy}{dx}+2y=2e^{-2x}$

(d) $\frac{d^2y}{dx^2}+4\frac{dy}{dx}+4y=e^{-2x}$

(e) $2\frac{d^2y}{dx^2}-\frac{dy}{dx}-y=85\sin 2x$

(f) $2\frac{d^2y}{dx^2}-\frac{dy}{dx}-y=85\cos 2x$

(g) $\frac{d^2y}{dx^2}+4y=\sin 3x$

(h) $\frac{d^2y}{dx^2}+4y=\sin 2x$

2. Find the particular solution of the following equations.

(a) $\frac{d^2y}{dx^2}-y=2,\quad y(0)=0,\quad y'(0)=0$

(b) $\frac{d^2y}{dx^2}+4y=\cos x,\quad y(0)=1,\quad y\left(\frac{\pi}{4}\right)=\frac{1}{3\sqrt{2}}$

(c) $\frac{d^2y}{dx^2}-\frac{dy}{dx}-2y=e^{-2x},\quad y(0)=1,$ y finite as $x\to\infty$

Answers

1. (a) $y=Ae^{x}+Be^{-2x}-\frac{1}{4}\left(2x^2+2x+5\right)$

(b) $y=Ae^{-x}+Be^{-2x}+\frac{1}{6}e^{2x}$

(c) $y=Ae^{-x}+Be^{-2x}-2xe^{-2x}$

(d) $y=\left(A+Bx\right)e^{-2x}+\frac{1}{2}x^2e^{-2x}$

(e) $y=Ae^{-0.5x}+Be^{x}+2\cos 2x-9\sin 2x$

(f) $y=Ae^{-0.5x}+Be^{x}-9\cos 2x-2\sin 2x$

(g) $y=A\cos 2x+B\sin 3x-\frac{1}{5}\sin 3x$

(h) $y=A\cos 2x+B\sin 2x-\frac{1}{4}x\cos 2x$

2. (a) $y=e^{x}+e^{-x}-2$

(b) $y=\frac{2}{3}\cos 2x+\frac{1}{3}\cos x$

(c) $y=\frac{3}{4}e^{-x}+\frac{1}{4}e^{-2x}$

2.10 NUMERICAL SOLUTIONS OF DIFFERENTIAL EQUATIONS

The methods we have covered so far apply to special classes of differential equations. The solutions we obtained described the exact relationship between the unknown variable y and the dependent variable x. These solutions are said to be "analytical" or "exact" solutions. However, not all equations have analytical (exact) solutions. When exact solutions are not available, we use numerical methods to obtain an approximate solution. Numerical methods offer control over the desired level of accuracy. Numerical methods discretise differential equations into a finite set of equations, which can be solved using computational techniques. This process allows for the application of computer algorithms and numerical software to find approximate solutions efficiently.

In this section, we introduce a numerical method called the "Euler method". This method works well on first-order initial value problems and produces approximate solutions at specific points. Implementing this method is a straightforward process, and it offers exceptionally accurate solutions. Numerous alternative techniques for solving ODEs are available within textbooks dedicated to the field of numerical analysis.

2.10.1 Euler Method

Consider the following first-order ODE:

$$\frac{dy}{dx} = f(x, y)$$

subject to an initial condition $y(x_0) = y_0$. The exact solution will appear as a continuous graph like Figure 2.3a. For our specific intent, an approximate solution will consist of a set of discrete points positioned in close proximity to the true curve shown in Figure 2.3b.

To describe the Euler method, we start from the initial point (x_0, y_0) and draw a tangent line to the curve of $y = y(x)$ at this point (even though the exact solution is unknown). The gradient of the tangent line, m, to the exact curve $y = y(x)$ at the point (x_0, y_0) is defined by the RHS of the differential equation as follows:

$$m = \left(\frac{dy}{dx}\right)_{(x_0, y_0)} = f(x_0, y_0).$$

We now select $x_1 = x_0 + h$, where h is a small step size, and find its associated value y_1 (obtained at the point x_1) using the tangent line shown in Figure 2.3a. This is the first approximation value to the exact value $y(x_1) \approx y_1$. The accuracy of the approximation depends on the value of the step size h. Note that in order to explain the method, h in Figure 2.3a is very large, but in practice, we choose a very small step size.

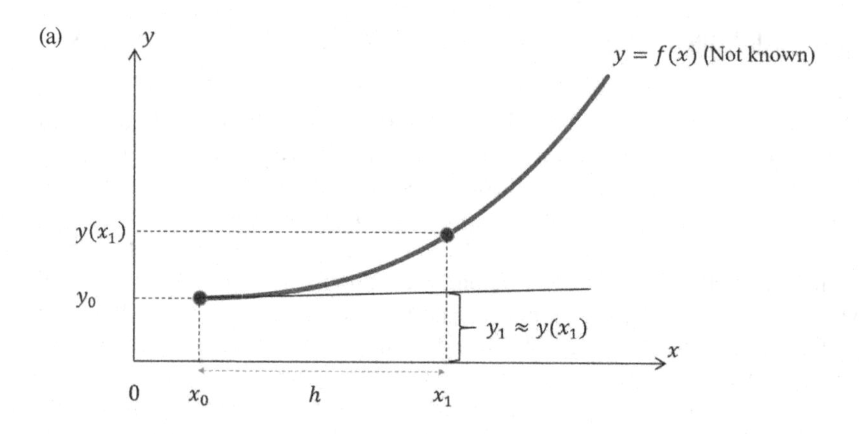

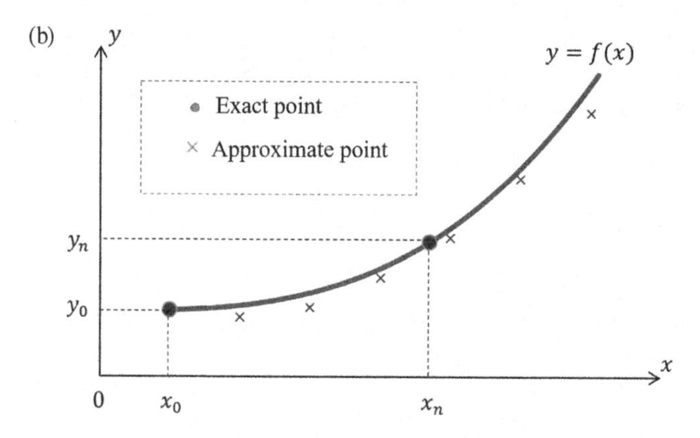

FIGURE 2.3 (a) The graph of the exact solution $y = y(x)$. The slope of the tangent line to the curve at the point (x_0, y_0) is denoted by m. (b) A set of discrete approximate points lying near the true solution.

The value of y_1 can be obtained using the slope, m of the tangent line at the point (x_0, y_0) as follows:

$$m = \left(\frac{dy}{dx}\right)_{x=x_0} = f(x_0, y_0) = \frac{y_1 - y_0}{x_1 - x_0} = \frac{y_1 - y_0}{h}.$$

Rearranging the formula, we get

$$y(x_1) \approx y_1 = y_0 + hf(x_0, y_0).$$

Repeating the same process but starting with the approximate point (x_1, y_1) instead of (x_0, y_0), results in the following approximation:

$$y(x_2) \approx y_2 = y_1 + hf(x_1, y_1).$$

Note for repeating the process we have assumed the approximate solution y_1 to be exact, as we do not have any information about the exact solution $y(x_1)$. Subsequent approximations at $x_n, n = 3,4,5,\cdots$ are produced in a similar way starting with approximate solutions. This leads to the following general recursive relation:

$$y(x_{n+1}) \approx y_{n+1} = y_n + h\,f(x_n, y_n), \quad n = 0,1,2,\ldots$$

The accuracy of the Euler method for solving differential equations depends on several factors, including the step size (h) used in the numerical approximation. The Euler method is a first-order numerical method. This means that the error in the solution typically decreases linearly with decreasing step size. However, this method can accumulate errors over time, especially for complex differential equations. In such cases, more sophisticated higher order methods like the Runge–Kutta method may be more suitable. The Euler method can provide reasonable approximations for simple differential equations and in situations in which high accuracy is not critical.

We continue this section by presenting some examples to clarify the method, which entirely is numerical.

Example 2.21 Use the Euler method to find the approximate solutions of the following differential equation:

$$\frac{dy}{dx} = x^2 + y, \quad y(0) = 3$$

at the points $x = 0.02,\ 0.04,\ 0.06,\ 0.08$ and 0.10.

Solution

This equation is a first-order linear equation, which has the exact solution

$$y(x) = -\left(2 + 2x + x^2\right) + 5e^x.$$

Now we apply the Euler method and compare the numerical solutions with exact solutions at the given points:

$$n = 0,\ x_0 = 0 \quad y_0 = 3 \quad h = 0.02 \quad f(x,y) = x^2 + y$$

$$\begin{aligned} y_1 &= y_0 + h\,f(x_0, y_0) \\ &= y_0 + h\left(x_0^2 + y_0\right) \\ &= 3 + 0.02\left(0^2 + 3\right) \\ &= 3.06 \end{aligned}$$

$$n=1,\quad x_1=0.02\quad y_1=3.06\quad h=0.02\quad f(x,y)=x^2+y$$

$$\begin{aligned} y_2 &= y_1+hf(x_1,y_1)\\ &= y_1+h\left(x_1^2+y_1\right)\\ &= 3.06+0.02\left(0.02^2+3.06\right)\\ &= 3.12121 \end{aligned}$$

$$n=2,\quad x_2=0.04\quad y_2=3.12121\quad h=0.02\quad f(x,y)=x^2+y$$

$$\begin{aligned} y_3 &= y_2+hf(x_2,y_2)\\ &= y_2+h\left(x_2^2+y_2\right)\\ &= 3.12121+0.04\left(0.04^2+3.12121\right)\\ &= 3.18366 \end{aligned}$$

Calculating the remaining approximations is straightforward and can be obtained in the same way by incrementing all subscripts by one with each application of Euler's formula. A full approximate value and comparison with the exact solutions are given in Table 2.2. Note that the numbers are rounded correctly to 5 decimal places.

TABLE 2.2
Numerical Solutions y_n Versus Exact Values $y(x_n)$

x_n	y_n	$y(x_n)$
0	3	3
0.02	3.06000	3.06061
0.04	3.12121	3.12245
0.06	3.18366	3.18558
0.08	3.24741	3.25004
0.1	3.31249	3.31585

2.10.2 Implementation of the Euler Method Using Excel

There are several computer packages such as MATLAB®, Mathematica and Maple with differential equation solvers; however, the numerical methods they apply are hidden from users. Nevertheless, due to the iterative process of the Euler method, it can easily be implemented by using Microsoft Excel. In Table 2.3, we present a step-by-step implementation of Euler's method using Excel to solve the equation in Example 2.22. We complete this section with the second example of solving differential equations, in which we use Excel along with Euler's method to find its approximate solution.

TABLE 2.3
Implementation of Excel to Solve the Equation in Example 2.21; Start from Cell D4 (Where the Value x_0 Is Entered as 0)

x_n	y_n	$y(x_n)$
0	3	=5*EXP(D4)-(D4^2+2*D4+2)
=D4+0.02	=E4+0.02*(D4^2+E4)	=5*EXP(D5)-(D5^2+2*D5+2)
=D5+0.02	=E5+0.02*(D5^2+E5)	=5*EXP(D6)-(D6^2+2*D6+2)
=D6+0.02	=E6+0.02*(D6^2+E6)	=5*EXP(D7)-(D7^2+2*D7+2)
=D7+0.02	=E7+0.02*(D7^2+E7)	=5*EXP(D8)-(D8^2+2*D8+2)
=D8+0.02	=E8+0.02*(D8^2+E8)	=5*EXP(D9)-(D9^2+2*D9+2)

Example 2.22 Given

$$\frac{dy}{dx} = (1-2x)y$$

subject to the initial condition $y(0) = 1$. Use Excel to determine the approximate solutions of the differential equation at the points $x = 0.4,\ 0.8,\ 1.2,\ 1.6,\ 2.0,\ 2.4$ and 2.8. Find its analytical solution and compare your answer with exact values.

Solution

The analytical solution can be found using the method of separation of variables:

$$y(x) = e^{(x-x^2)}.$$

Now, we use the same cell reference as in Example 2.22 and present Table 2.4 for the Excel implementation.

TABLE 2.4
Excel Implementation of the Euler Method

x_n	y_n	$y(x_n)$
0	1	=EXP(D4-D4^2)
=D4+0.4	=E4+0.4*(1–2*D4)*E4	=EXP(D5-D5^2)
=D5+0.4	=E5+0.4*(1–2*D5)*E5	=EXP(D6-D6^2)
=D6+0.4	=E6+0.4*(1–2*D6)*E6	=EXP(D7-D7^2)
=D7+0.4	=E7+0.4*(1–2*D7)*E7	=EXP(D8-D8^2)
=D8+0.4	=E8+0.4*(1–2*D8)*E8	=EXP(D9-D9^2)
=D9+0.4	=E9+0.4*(1–2*D9)*E9	=EXP(D10-D10^2)
=D10+0.4	=E10+0.4*(1–2*D10)*E10	=EXP(D11-D11^2)

The numerical values shown in Table 2.5 represent reasonable approximate values of the exact solutions.

TABLE 2.5
Numerical Approximation Versus the Exact Solutions

x_n	y_n	$y(x_n)$
0	1.000000	1
0.4	1.400000	1.27125
0.8	1.512000	1.17351
1.2	1.149120	0.78663
1.6	0.505613	0.38289
2	0.060674	0.13534
2.4	−0.012135	0.03474
2.8	0.006310	0.00647

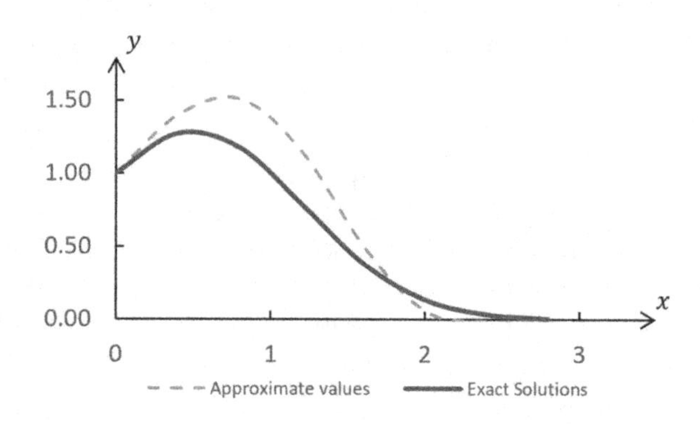

FIGURE 2.4 Graphical presentation of Table 2.5.

TABLE 2.6
Approximate Values for h = 0.2

x_n	y_n	$y(x_n)$
0	1	1
0.2	1.2	1.173511
0.4	1.344	1.271249
0.6	1.39776	1.271249
0.8	1.34185	1.173511
1	1.180828	1
1.2	0.944662	0.786628
1.4	0.680157	0.571209
1.6	0.4353	0.382893
1.8	0.243768	0.236928
2	0.117009	0.135335
2.2	0.046803	0.071361
2.4	0.014977	0.034735
2.6	0.003595	0.015608
2.8	0.000575	0.006474

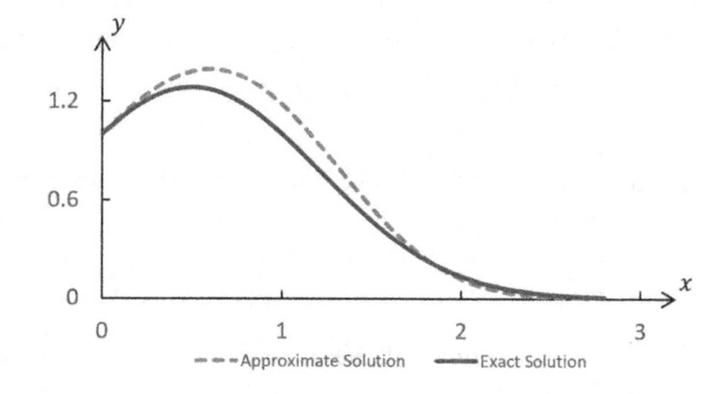

FIGURE 2.5 Graphical presentation of Table 2.6.

As we observe from Table 2.5 and Figure 2.4, the approximated values are not quite accurate. As mentioned earlier, the accuracy of the method depends on several factors, including the step size h. Now we repeat this process with $h = 0.2$ (the half of the step size in the previous case) and compare the approximate values presented in Table 2.6

and the associated Figure 2.5 with the values obtained for $h = 0.4$ in Table 2.5 with the associated Figure 2.4. We see significant improvement for the smaller step size h.

2.10.3 Exercises

1. Solve the following first-order ODE using Euler's method:

$$(x+1)\frac{dy}{dx} = 2y, \qquad y(0) = 1.$$

Use $h = 0.01$ and estimate $y(x)$ for $x = 0$ to $x = 0.04$ manually. Use $h = 0.01$ and estimate $y(0.1)$ using Excel. Solve for the analytical solution and compare the two sets of solutions.

2. Use Excel to find an approximate solution of the following first-order ODE:

$$\left(x^2+1\right)\frac{dy}{dx} = 2xy, \qquad y(0) = 1.$$

Use $h = 0.1$ and estimate $y(0.1)$ up to $y(1)$,and compare the numerical solutions with the exact values.

TABLE 2.7
Numerical Solutions y_n Versus Exact Values $y(x_n)$

x_n	y_n	$y(x_n)$
0	1	1
0.01	1.02	1.0201
0.02	1.040198	1.0404
0.03	1.060594	1.0609
0.04	1.081188	1.0816

TABLE 2.8
Numerical Solutions y_n Versus Exact Values $y(x_n)$

x_n	y_n	$y(x_n)$
0	1	1
0.1	1	1.004988
0.2	1.009901	1.019804
0.3	1.029322	1.044031
0.4	1.057652	1.077033
0.5	1.094123	1.118034
0.6	1.137888	1.16619
0.7	1.188089	1.220656
0.8	1.243905	1.280625
0.9	1.304583	1.345362
1	1.369452	1.414214

Answers

1. See Table 2.7.
2. See Table 2.8.

2.11 ENGINEERING APPLICATIONS

In this section, we consider the application of differential equations in various engineering topics. This includes falling objects, the mixture of solutions, fluid mechanics, heat transfer, vibration analysis and analysis of electrical circuits.

2.11.1 Falling Objects

Modelling a falling object involves considering principles of classical mechanics, particularly Newton's laws of motion. The primary force acting on the falling object is gravity, which imparts a constant acceleration (g) towards the centre of the Earth. Depending on the context, you might choose to include or neglect air resistance. In the absence of air resistance, the only force acting on the object is gravity.

Consider a falling object under the influence of gravity and air resistance proportional to the velocity v of the object. The forces applied on the object are

$$\text{Gravitational force } W = mg$$

$$\text{Air resistance} = -kv(t),$$

where k is the proportionality constant. Using the Newton's second law of motion, we can write

$$F = ma$$

$$m\frac{dv}{dt} = mg - kv$$

$$\Rightarrow$$

$$m\frac{dv}{dt} + kv = mg$$

$$\frac{dv}{dt} + \frac{k}{m}v = g.$$

This is a first-order linear differential equation that we can solve to find the velocity of the object at any instant of time, as well as when the object hits the ground, given the initial conditions.

Example 2.23 A 65-kg man clings to a parachute and is suspended in space. He falls in space under the pull of gravity $g = 9.8\,m/s^2$. If air resistance creates a decelerating force $5v$ proportional to the man's speed v, assuming the initial speed is zero, find the distance he falls when $t = 10\,s$.

Solution

The forces applied are

$$\text{Gravitational force} = mg = 65 \times 9.8 = 637\,N$$

$$\text{Air resistance} = -5v(t)$$

$$v(0) = 0.$$

Using Newton's second law ($F = ma$), we can write the following differential equation:

$$65\frac{dv}{dt} = 637 - 5v(t) \Rightarrow$$

$$65\frac{dv}{dt} + 5v(t) = 637.$$

This is a linear equation that can be solved in three steps:

Step 1: Write the equation in the standard form (Equation 2.5) and identify $p(t)$ and $q(t)$:

$$\frac{dv}{dt} + \frac{1}{13}v(t) = 9.8.$$

Hence,

$$p(t) = \frac{1}{13} \quad \text{and} \quad q(t) = 9.8.$$

Step 2: Determine the integrating factor:

$$\begin{aligned} r &= e^{\int p(t)dt} \\ &= e^{\int \frac{1}{13} \times dt} \\ &= e^{\frac{1}{13}t}. \end{aligned}$$

[Note that we set the integration constant to zero.]

Step 3: The general solution is

$$v(t) = \frac{1}{r(t)}\left[\int r(t)q(t)dt\right].$$

Hence, by substituting the integrating factor and the function $q(t) = 9.8$ in the previous equation, we obtain

$$v(t) = \frac{1}{e^{\frac{1}{13}t}}\left[\int e^{\frac{1}{13}t} \times 9.8\,dt\right] = 9.8e^{-\frac{1}{13}t}\int e^{\frac{1}{13}t}\,dt$$

$$= 9.8e^{-\frac{1}{13}t}\left(13e^{\frac{1}{13}t} + C\right) = 127.4 + 9.8Ce^{-\frac{1}{13}t}.$$

This is the general solution. We need to determine C using the initial condition $v(0) = 0$:

$$v(0) = 0 \Rightarrow 0 = 127.4 + 9.8Ce^{-\frac{1}{13}\times 0}$$

$$C = -\frac{127.4}{9.8} = -13.$$

Hence,

$$v(t) = 127.4 - 127.4e^{-\frac{1}{13}t}.$$

The position $s(t)$ of the man is obtained by integrating $v(t)$:

$$s(t) = 127.4t + 1656.2e^{-\frac{1}{13}t} + A.$$

Apply $s(0) = 0$ to determine A:

$$0 = 0 + 1656.2 + A$$

$$A = -1656.2.$$

Hence,

$$s(t) = 127.4t + 1656.2e^{-\frac{1}{13}t} - 1656.2 \Rightarrow$$

$$s(10) = 1274 + 1656.2e^{-\frac{10}{13}} - 1656.2 = 385.232\,m.$$

2.11.2 Mixture of Solutions

Modelling a liquid mixture in a tank involves considering principles of fluid dynamics, mass balance and the behaviour of the components within the mixture. Here is a brief overview of how one might approach this model:

(i) Mass Balance Equation:
The fundamental principle is the mass balance equation, which states that the rate of change of mass m in the tank equals the difference between the mass flow rate in and the mass flow rate out. Mathematically, this is expressed as

$$\frac{dm}{dt} = \text{inflow} - \text{outflow}.$$

(ii) Tank Volume:
The tank's volume and shape play a crucial role. The volume V can be constant or variable, depending on the tank's design. If the tank has a constant cross-sectional area, the volume is simply the area times the height.

(iii) Concentration Profiles:
For a mixture, you would introduce the concept of concentrations. If there are multiple components in the mixture, you will have concentration profiles for each component. The concentration C can be expressed as mass per unit volume.

(iv) Flow Rates:
Inflow and outflow rates depend on the system. For instance, if there is an inlet pipe, the flow rate Q_{in} might be a function of time. Outflow Q_{out}, can be controlled by valves or other factors.

(v) Differential Equations:
Develop a set of differential equations based on the principles mentioned. For a single-component system, it might look like

$$\frac{d(CV)}{dt} = C_1 \times Q_{in} - C_2 \times Q_{out},$$

where C_1 and C_2 are the concentration of inflow and outflow, respectively.

Example 2.24 In a very large tank, there is an initial mixture comprising 1000 gallons of brine containing 200 pounds of salt. A solution of salt water is continuously added to the tank at a rate of 5 gallons per minute, which contains 3 pounds of salt per gallon. Simultaneously, the well-mixed solution is pumped out at the same rate of 5 gallons per minute. What is the quantity of salt in the tank after a duration of 1 hour? Additionally, as time approaches infinity what is the salt content in the tank?

Solution

Assume $S(t)$ is the amount of salt in the tank at any instant of time. Hence,

$$\frac{dS}{dt} = \text{inflow} - \text{outflow}$$

$$\text{inflow} = C_1 \times Q_{in} = \left(3\frac{\text{lbs}}{\text{gal}}\right) \times \left(5\frac{\text{gal}}{\text{min}}\right) = 15\frac{\text{lbs}}{\text{min}}$$

$$\text{outflow} = C_2 \times Q_{out} = \left(\frac{S}{1000}\frac{\text{lbs}}{\text{gal}}\right) \times \left(5\frac{\text{gal}}{\text{min}}\right) = \frac{S}{200}\frac{\text{lbs}}{\text{min}}$$

$$\Rightarrow$$

$$\frac{dS}{dt} = 15 - \frac{S}{200} \qquad \text{OR}$$

$$\frac{dS}{dt} + \frac{S}{200} = 15.$$

The integrating factor is

$$r = e^{\int p(t)dt} = e^{\int \frac{1}{200} \times dt} = e^{\frac{1}{200}t}.$$

Hence, the solution is

$$S(t) = \frac{1}{e^{\frac{1}{200}t}}\left[\int e^{\frac{1}{200}t} \times 15dt\right] = 15e^{-\frac{1}{200}t}\int e^{\frac{1}{200}t}dt$$

$$= 15e^{-\frac{1}{200}t}\left(200e^{\frac{1}{200}t} + C\right) = 3000 + 15Ce^{-\frac{1}{200}t}.$$

Using the initial condition $S(0) = 200$:

$$S(t) = 3000 + 15C = 200 \Rightarrow C = -\frac{2800}{15}$$

$$\Rightarrow S(t) = 3000 - 2800e^{-\frac{1}{200}t}$$

$$\Rightarrow S(60) = 3000 - 2800e^{-\frac{60}{200}} \approx 925.71\text{lbs}.$$

When $t \to \infty$

$$S = 3000\,\text{lbs}.$$

2.11.3 Vibration of a Spring-Mass System

A number of physical systems subjected to vibration, such as a washing machine in operation, vehicle suspension system, vibrating structures/machinery and so on, can be modelled as spring-mass-damper systems. As discussed in Sections 2.2 and 2.9, a spring-mass-damper system can be described by second-order differential equations. We explain this topic in more detail in the following sections.

(a) Damping Behaviour of the System

If the system (in the absence of an external force) vibrates by itself, due to the physics of the spring, the second-order differential equation governing the mechanical system is the following homogeneous equation:

$$m\frac{d^2x}{dt^2} + c\frac{dx}{dt} + kx = 0,$$

where $m > 0$, $c > 0$ and $k > 0$. Dividing the equation by m gives

$$\frac{d^2x}{dt^2} + \beta\frac{dx}{dt} + \omega^2 x = 0, \quad \beta = \frac{c}{m}, \quad \omega = \sqrt{\frac{k}{m}}.$$

Note that ω is the frequency of harmonic motion when the system is not experiencing any damping. The roots of the characteristic equation $\lambda^2 + \beta\lambda + \omega^2 = 0$ are

$$\lambda = \frac{-\beta \pm \sqrt{\beta^2 - 4\omega^2}}{2}.$$

There are three cases to consider.

Case 1: $\beta^2 - 4\omega^2 = 0$ or $\beta = 2\omega$ in this case the solution is $x(t) = (A + Bt)e^{-\lambda t}$ which is not oscillatory and presents **critical damping** behaviour. For example, for the equation $\frac{d^2x}{dt^2} + 6\frac{dx}{dt} + 9x = 0$ with the initial conditions $x(0) = 1$, $x'(0) = 0$, the parameters are $m = 1$, $c = 6$ and $k = 9$. The solution $x(t) = (1 + 3t)e^{-3t}$ is shown in Figure 2.6a, which exhibits critical damping behaviour and rapidly approaches zero at $t \approx 2.5$.

Case 2: $\beta^2 - 4\omega^2 > 0$ or $(\beta > 2\omega)$, in this case the solution is $x(t) = Ae^{\lambda_1 t} + Be^{\lambda_2 t}$, which is not oscillatory and presents **overcritical damping** behaviour

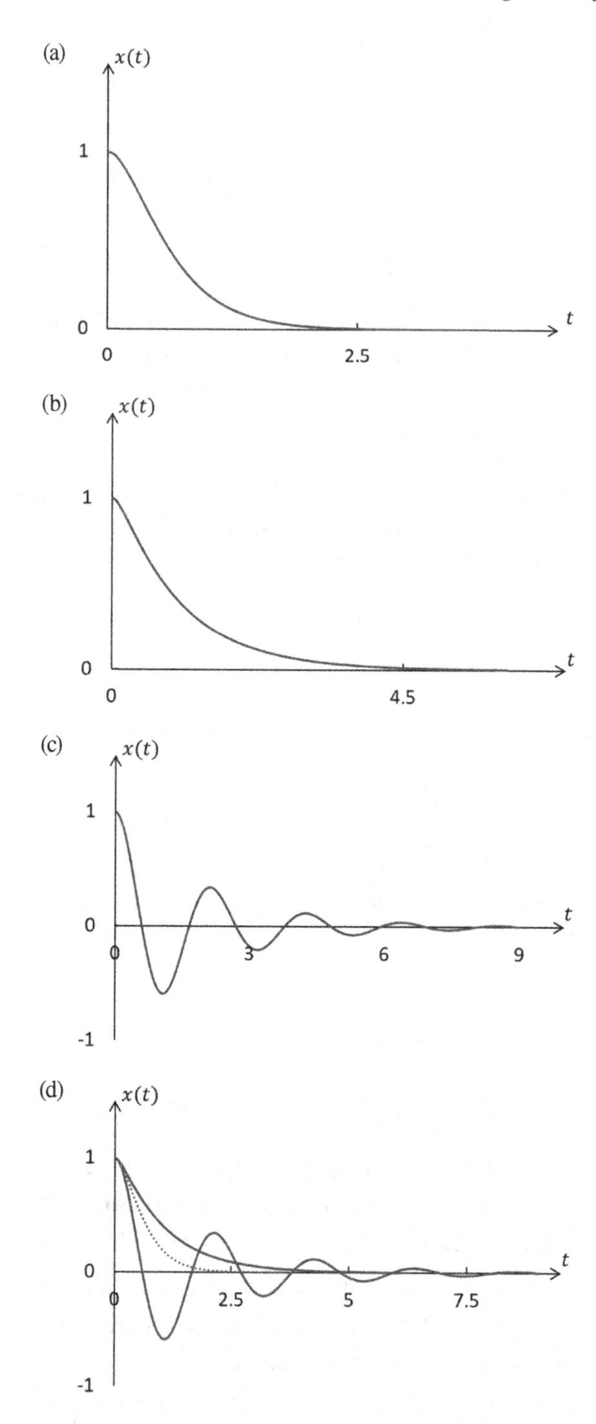

FIGURE 2.6 (a) Critical damping behaviour. (b) Overcritical damping behaviour. (c) Under-critical damping behaviour. (d) The dashed curve exhibits critical damping behaviour, and the two solid curves present over- and under-damping behaviour.

given that λ_1 and λ_2 are both negative. Note that since β and ω are positive constants $\beta^2 > \beta^2 - 4\omega^2$, which implies $\sqrt{\beta^2 - 4\omega^2} < \beta$ and hence $\lambda = \frac{-\beta \pm \sqrt{\beta^2 - 4\omega^2}}{2} < 0$. For example, for the following equation:

$$\frac{d^2x}{dt^2} + 10\frac{dx}{dt} + 9x = 0, \quad x(0) = 1, \quad x'(0) = 0.$$

The parameters are $m = 1$, $c = 10$ and $k = 9$. The solution $x(t) = \frac{1}{8}\left(9e^{-t} - e^{-9t}\right)$ is shown in Figure 2.6b, which exhibits overcritical damping behaviour and approaches zero in a longer time $t \approx 4.5$.

Case 3: $\beta^2 - 4\omega^2 < 0$, or $(\beta < 2\omega)$ in this case the solution is oscillatory and exhibits **under-critical damping** behaviour. For example, for the following equation,

$$\frac{d^2x}{dt^2} + \frac{dx}{dt} + 9x = 0, \quad x(0) = 1, \quad x'(0) = 0,$$

the parameters are $m = 1$, $c = 1$ and $k = 9$. The solution

$$x(t) = \frac{1}{35}e^{-0.5t}\left[35\cos\left(\frac{\sqrt{35}}{2}t\right) + \sqrt{35}\sin\left(\frac{\sqrt{35}}{2}t\right)\right]$$

is shown in Figure 2.6c, which is oscillatory vibrations presenting under-critical damping behaviour and approaches zero (after several oscillations) at $t \approx 9$.

SUMMARY

The equation $\frac{d^2x}{dt^2} + \beta\frac{dx}{dt} + 9x = 0, \omega = 3$ is subject to the initial condition, $x(0) = 1$, $x'(0) = 0$ is considered, its solution depending on the parameter $\beta = \frac{c}{m}$ exhibits critical damping, overdamping and under-damping behaviour. For the values $\beta = 6$ or $\beta = 2\omega$ (critically damped), $\beta = 10$ or $\beta = 1$ (over damped) and $\beta = 1$ or $\beta < 2\omega$ (underdamped) are obtained. The solutions are plotted in Figure 2.6d, and it is clear that the solution with critical damping (dashed curve) decays faster than the other two solutions.

(b) System with External Force without Damping

As mentioned in Section 2.9, when the upper end of spring (given in Figure 2.1) is subjected to an external force $f(t)$, the differential equation governing the mechanical system can be written as follows:

$$m\frac{d^2x}{dt^2} + c\frac{dx}{dt} + kx = f(t).$$

Assume $f(t) = F\cos(\omega_0 t)$ or $f(t) = F\sin(\omega_0 t)$, where F is the amplitude of the external force and ω is the frequency. In the absence of damping $(c = 0)$, the system oscillates naturally due to the physics of the spring, but in addition to the natural vibrations, the mass experiences an external periodic force that causes further oscillations. Dividing both sides of the equation by m gives

$$\frac{d^2x}{dt^2} + \omega^2 x = F_0\cos(\omega_0 t),\ F_0 = \frac{F}{m},\ \omega = \sqrt{\frac{k}{m}}. \tag{2.14}$$

Using the method of undetermined coefficients explained in Section 2.9, the solution of Equation 2.14 can be written as

$$x(t) = A\cos(\omega t) + B\sin(\omega t) + \frac{F_0}{\left(\omega^2 - \omega_0^2\right)}\cos(\omega_0 t).$$

The system's behaviour depends on the relationship between the natural frequency of the system $\left(\frac{\omega}{2\pi}\right)$ and the frequency of the applied force $\left(\frac{\omega_0}{2\pi}\right)$.

For example, given $m = 1, k = 4$ and $f(t) = 5\sin(3t)$ with the initial conditions $x(0) = x'(0) = 0$, the solution is $x(t) = \frac{3}{2}\sin(2t) - \sin(3t)$. The natural frequency and time period of the system are $f_1 = \frac{2}{2\pi} = \frac{1}{\pi}, T_1 = \pi$, the frequency and time period due to the forced oscillation are $f_2 = \frac{3}{2\pi}, T_2 = \frac{2\pi}{3}$. In this example, $x(t)$ is composed of two sinusoidal functions with different frequencies, and this topic is further explained in detail in Chapter 5; however, the frequency of the combined vibrations is $f_0 = \frac{\gcd(2,3)}{2\pi} = \frac{1}{2\pi}$, as shown in Figure 2.7a for two cycles.

If ω is close to ω_0 ($\omega_0 \neq \omega$) and initial conditions are $x(0) = x'(0) = 0$, then the solution of Equation 2.14 can be written as

$$x(t) = \frac{F_0}{\left(\omega^2 - \omega_0^2\right)}\left[\cos(\omega_0 t) - \cos(\omega t)\right],$$

which leads to a very interesting result that is called the **phenomenon of beat**. The beat phenomenon is considered in detail in Chapter 5, which consists of a rapid oscillation with a slowly varying amplitude.

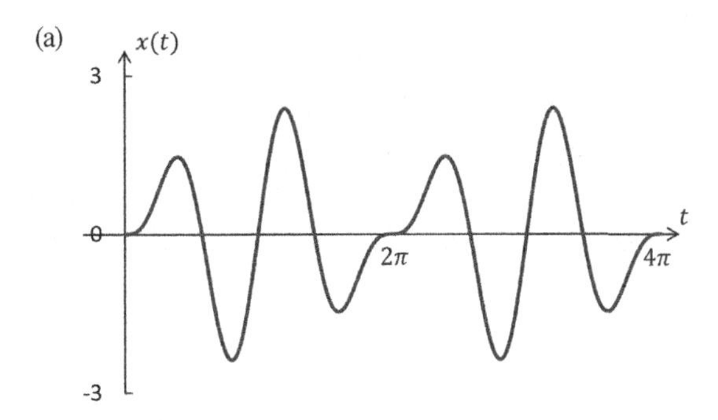

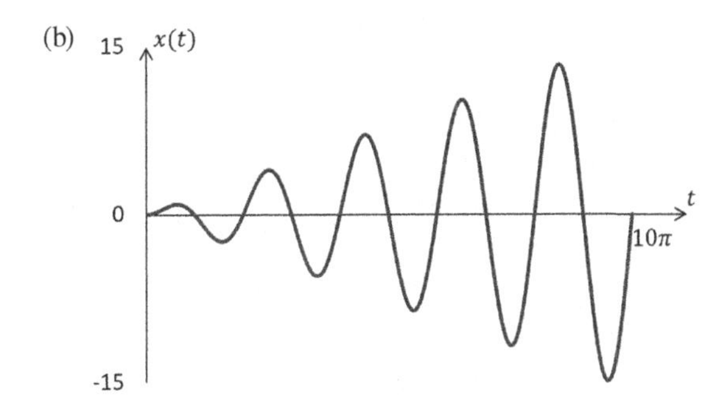

FIGURE 2.7 (a) The system's behaviour as a result of applying an external force. (b) The phenomenon of resonance.

Another interesting result is that if $\omega_0 = \omega$ and the initial conditions are $x(0) = x'(0) = 0$, in that case, the solution of Equation 2.14 can be written as follows:

$$x(t) = \frac{F_0}{2\omega} t \sin(\omega t).$$

In this case, the amplitude of the sinusoidal function increases with time without any boundary, and this result is called the **resonance phenomenon**, in which the external vibrations reinforce the natural frequency of the system with the same frequency.

An example of this phenomenon is $\frac{d^2x}{dt^2} + x = \cos t$ with the initial condition $x(0) = x'(0) = 0$, which has solution $x(t) = \frac{1}{2} t \sin t$. As shown in Figure 2.7b, the amplitude of the oscillation grows without any bounds as time goes on.

(c) Systems with External Force with Damping

As mentioned in Section 2.11.3.2 when the upper end of spring (given in Figure 2.1) is subjected to an external force $f(t)$, the damped systems with an external force can be described by

$$m\frac{d^2x}{dt^2} + c\frac{dx}{dt} + kx = f(t),$$

which includes both a damping term $\left(c\frac{dx}{dt}\right)$ and an external force $f(t)$. One common form is when $f(t)=F\cos(\omega_0 t)$ or $f(t)=F\sin(\omega_0 t)$, where F is the amplitude of the external force and ω is the frequency. This equation combines the effects of damping $\left(c\frac{dx}{dt}\right)$ with an external sinusoidal force. The solution involves both the complementary function $x_c(t)$ and the particular solution $x_p(t)$ (response to the external force). The complementary function $x_c(t)$ takes one of the following forms:

- $x_c(t)=(A+Bt)e^{-\lambda t}$ (Critical damping)
- $x_c(t)=Ae^{\lambda_1 t}+Be^{\lambda_2 t}$ (Overdamping)
- $x_c(t)=e^{\alpha t}\left[A\cos(\gamma t)+B\sin(\gamma t)\right]$ (Under-damping).

In all three cases, when $t\to\infty$, the damped solutions $x_c(t)\to 0$, which means that the vibration of the system decays by the damper, and thus, the complementary function becomes a transient solution while the response to the external force $x_p(t)$ becomes a periodic steady-state solution.

For example, if $m=1, c=2, k=5$, and $f(t)=2\cos t$ the differential equation with the initial conditions $x(0)=0, x'(0)=0$ has the following solution:

$$
\begin{aligned}
x(t) &= x_c(t)+x_p(t)\\
&= -e^{-t}\left[\frac{2}{5}\cos(2t)+\frac{3}{10}\sin(2t)\right]+\frac{1}{5}(2\cos t+\sin t)\\
&= -e^{-t}\left[\frac{2}{5}\cos(2t)+\frac{3}{10}\sin(2t)\right]+\frac{\sqrt{5}}{5}\cos\left(t-0.46^c\right).
\end{aligned}
$$

When $t\to\infty$, the solutions $x_c(t)=-e^{-t}\left[\frac{2}{5}\cos(2t)+\frac{3}{10}\sin(2t)\right]\to 0$ (decays) and the response to the external force; that is $x_p(t)$ becomes the periodic solution $x_p(t)=\frac{\sqrt{5}}{5}\cos\left(t-0.46^c\right)$, with the amplitude $F_0=\frac{\sqrt{5}}{5}$ and the frequency $f_0=\frac{1}{2\pi}$.

SUMMARY

When an external force is applied to the spring-mass system, the behaviour of the system depends on the relationship between the damping coefficient, the natural frequency of the system and the frequency of the applied force. It can exhibit different responses, including resonance and beat phenomenon depending on these factors.

2.11.4 Fluid Flow Streamlines

In fluid mechanics, the path of the fluid elements in steady flow is described using a family of curves called streamlines. For a fluid flow in two dimensions, streamlines can be described mathematically using differential equations as follows:

$$\frac{dy}{dx} = f(x, y).$$

To visualise the streamlines, we need to solve the differential equation.

Example 2.25 Suppose that the streamlines of a fluid flow are given by

$$\frac{dy}{dx} = -\frac{x}{y}.$$

Plot the streamlines of the fluid flow.

Solution

In order to plot the streamlines, we will need to solve the given differential equation using the method of separation of variables:

$$y\,dy = -x\,dx$$

$$\int y dy = \int -x dx$$

$$\frac{y^2}{2} = -\frac{x^2}{2} + C$$

$$y^2 = -x^2 + 2C$$

$$x^2 + y^2 = 2C,$$

where C is the arbitrary constant of integration. Assuming $2C = r^2$, the preceding equation represents circles of radius r:

$$x^2 + y^2 = r^2.$$

Depending on the value of r, there are an infinite number of streamlines in the form of circles of radius r and centre at the origin.

2.11.5 Heat Transfer – Fourier's Law of Heat Conduction

Fourier's law of heat conduction is a fundamental principle that describes how heat is conducted through a material. Fourier's law is a basic equation governing one-dimensional heat conduction.

According to Fourier's law of heat conduction, heat flux (rate of heat transfer per unit area) through a material is typically expressed as

$$q = -k\frac{dT}{dx}.$$

The negative sign indicates that heat flows from a higher temperature to a lower temperature, where k is the thermal conductivity of the material and $\frac{dT}{dx}$ is the temperature gradient in the direction of heat transfer. The temperature gradient represents the rate at which temperature changes with respect to distance. A steeper temperature gradient indicates a higher rate of heat transfer.

Example 2.26 Consider a flat plate of thickness 0.1 m. The temperature of the hot side of the plate is 500 K, and the heat flux through the plate is 500 W/m^2. Given that the thermal conductivity of the material is 20 W/(mK), determine the temperature of the cold side of the plate.

Solution

Using Fourier's law of heat conduction:

$$q = -k\frac{dT}{dx}.$$

We can substitute the values for q and k. Therefore,

$$500 = -20\frac{dT}{dx}.$$

Separating the variables, we obtain

$$500\,dx = -20\,dT.$$

Because the thickness of the plate is 0.1m, x varies between 0 and 0.1. The temperature at $x = 0$ is 500K (hot side of the plate), assuming the temperature at $x = 0.1$ (cold side of the plate) is T_c (not yet found), we can construct the following integral using the preceding differential equation and obtain a value for T_c:

$$\int_{x=0}^{x=0.1} 500\,dx = \int_{T=500}^{T=T_c} -20\,dT$$

$$500[x]_0^{0.1} = -20[T]_{500}^{T_c}$$

$$500[0.1-0] = -20[T_c - 500]$$

$$50 = -20T_c + 10000$$

$$T_c = \frac{10000 - 50}{20} = 497.5\,\text{K}.$$

2.11.6 Electrical Circuit

Differential equations can be used in the modelling of Resistor Inductor Capacitor (RLC) circuits which are composed of resistors, inductors and capacitors. Figure 2.7 shows an electrical circuit with a resistor of resistance R, an inductor with inductance L and a capacitor with conductance C arranged in series in an electrical circuit powered by an input voltage V which could be a constant or variable voltage.

Voltage drop across a resistor is given by

$$V_R = Ri.$$

Voltage drop across an inductor is given by

$$V_L = L\frac{di}{dt}.$$

Voltage drop across a capacitor is given by

$$V_C = \frac{q}{C},$$

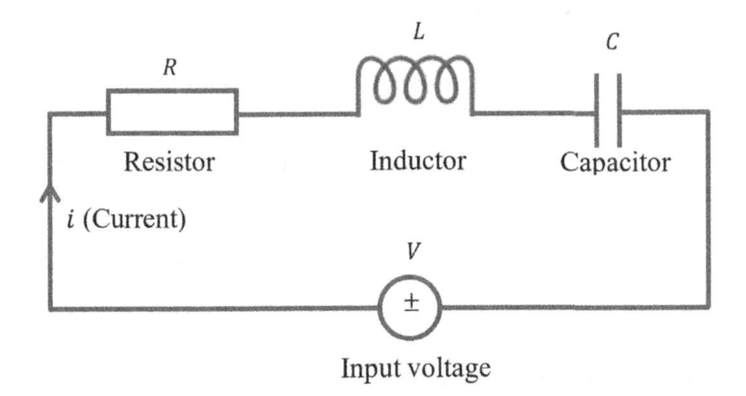

FIGURE 2.8 RLC circuit.

where q is the charge on the capacitor, which is a function of time, so current through the capacitor:

$$i = \frac{dq}{dt} \Rightarrow \frac{di}{dt} = \frac{d^2q}{dt^2}.$$

Using Kirchhoff's voltage law, which states that the net voltage drops around a closed circuit is zero, we will get a first-order differential equation as follows:

$$V_R + V_L + V_C = V$$

$$L\frac{di}{dt} + Ri + \frac{q}{C} = V. \tag{2.15}$$

We can write Equation 2.15 in terms of q as

$$L\frac{d^2q}{dt^2} + R\frac{dq}{dt} + \frac{q}{C} = V.$$

Equation 2.15 can be written in terms of i by differentiating it with respect to t:

$$L\frac{d^2i}{dt^2} + R\frac{di}{dt} + \frac{1}{C}i = \frac{dV}{dt}. \tag{2.16}$$

We now consider circuits with different components and source voltages in the following examples.

Example 2.27 Suppose a simple electrical circuit contains a resistor, an inductor and a constant input voltage V in series. Using Equation 2.15 (dropping the term for the capacitor), we obtain the following differential equation:

$$L\frac{di}{dt} + Ri = V.$$

Assuming $i(0) = i_0$, determine the current i at any instant of time t. What value does i approach when $t \to \infty$?

Solution

By writing the equation in the standard form $\frac{dI}{dt} + \frac{R}{L}I = \frac{V}{L}$ and using the integrating factor $r = e^{\frac{R}{L}t}$ the general solution can be found as follows:

$$i(t) = \frac{1}{e^{\frac{R}{L}t}}\left[\int e^{\frac{R}{L}t} \times \frac{V}{L}dt\right] = \frac{V}{R} + Ce^{-\frac{R}{L}t}.$$

Applying the given initial condition $i(0) = i_0$, we obtain

$$i(t) = \frac{V}{R} + \left(i_0 - \frac{V}{R}\right)e^{-\frac{R}{L}t}.$$

When $t \to \infty$, $i = \frac{V}{R}$, which means the final value of current is constant and the final voltage across the inductor is zero, hence the voltage drop occurs only in the resistor.

Example 2.28 For a series RLC circuit subjected to a constant voltage, suppose we obtain the following differential equation using Equation 2.16:

$$\frac{d^2i}{dt^2} + 7\frac{di}{dt} + 12i = 0.$$

Determine the current at any instant of time t, assuming the initial conditions are $i(0) = 0$ and $\frac{di}{dt}(0) = 5$.

Solution

This is a second-order homogeneous differential equation. The roots of the characteristic are the real values $\lambda = -3$ and $\lambda = -4$, so the general solution is $i(t) = Ae^{-3t} + Be^{-4t}$. Using the initial conditions, we obtain

$$i(t) = 5e^{-3t} - 5e^{-4t}.$$

Example 2.29 Suppose a series RLC circuit is subjected to a variable voltage $V(t) = e^{-2t}$, in this case using Equation 2.16, we can write the following differential equation:

$$\frac{d^2i}{dt^2} + 7\frac{di}{dt} + 12i = -2e^{-2t}.$$

Given the initial conditions $i(0) = 0$ and $\frac{di}{dt}(0) = 5$, determine the current at time t.

Solution

The complementary function and the particular solution are $i_c(t) = Ae^{-3t} + Be^{-4t}$ and $i_p(\mathrm{t}) = -e^{-2t}$, respectively. Hence, the general solution can be written as follows:

$$i(t) = i_c(t) + i_p(\mathrm{t}) = Ae^{-3t} + Be^{-4t} - e^{-2t}.$$

Using the initial conditions,

$$i(t) = e^{-3t} - e^{-2t}.$$

2.11.7 Exercises

1. A 100-kg man jumps out of an airplane at an altitude of 5000 metres above the ground. Air resistance is $10v(t)$ proportional to the man's speed $v(t)$. How long does it take for the man to reach the ground? What is his speed at this time? What is his final speed?
2. A 2-kg object falls from a tower at a height 2000 m with an initial speed of zero. If the air resistance creates a decelerating force $4v(t)$ proportional to the object's speed $v(t)$. What is the final speed of the object? How long does it take for the object to reach the ground?
3. A 2-kg object is thrown straight up from a bridge 15 metres above the ground with an initial speed $10\,\text{m/s}$. Assuming there is no air resistance, calculate the time required for the object to return to the ground.
4. A 10-kg object is shot straight up from a 60-metre-high cliff with an initial speed $10\,\text{m/s}$. If the air resistance is $2v(t)N$, determine the speed of the object when it hits the ground.
5. In a very large barrel, there is an initial 1000 litres of wine containing 12% alcohol. The wine is continuously added to the barrel at a rate of 4 litres per minute, which contains 5% alcohol. Simultaneously, the well-mixed wine leaves the barrel at the same rate of 4 litres per minute. What is the amount of alcohol in the barrel after 10 minutes? What is the percentage of alcohol in the barrel after 10 minutes? What is the final percentage of alcohol in the barrel?
6. Water flows into a pool at $10\,\text{m}^3$/min and out in the same rate. The pool's volume is $1{,}000\ \text{m}^3$. Initially, the pool has 100 g of pollutants, and there are additional pollutants flowing into the pool at a rate 2 g/m^3. When the new pollutant enters the pool, they mix quickly so that the concentration in the pool is uniform. What is the amount of pollutant after 20 minutes? What is the percentage of pollutant in the pool after 20 minutes?
7. A low-concentration wine containing 5% alcohol is poured into a large wine barrel at a rate of 4 litres per minute. The barrel contains 1,000 litres of wine with 12% alcohol in it. When the low-concentration wine enters the barrel, it is quickly mixed with the wine in the barrel to even out the concentration in the barrel. Simultaneously, the well-mixed wine exits the barrel at a reduced rate of 2 litres per minute. What is the amount of alcohol in the barrel after 10 minutes? What is the percentage of alcohol in the barrel after 10 minutes? What is the final percentage of alcohol in the barrel? How long does it take for the amount alcohol in the tank to double? Moreover, what is the percentage of alcohol in the barrel at this time?
8. A large tank contains 500 gallons of water with 20 pounds of fertiliser dissolved in it. A mixture of water and fertiliser is continuously added to the tank at a rate of 3 gal/hour, which contains 10 pounds/gal fertiliser. Simultaneously, the well-mixed solution is pumped out of the tank at the same rate of 3 gal/

hour. When the amount of fertiliser in the tank reaches 200 pounds, the entry of the mixture of water and fertiliser is stopped, and instead, fresh water enters the tank at a rate of 2 gal/hour while the mixture leaves the tank at a rate of 4 gal/hour. Determine the amount of fertiliser in the tank at any instant of time.

9. The streamlines of a fluid flow are given by

$$\frac{dy}{dx} = \frac{y-2}{x-1}.$$

Show that $y = C(x-1)+2$, where C is a constant.

10. The streamlines of a fluid flow are given by

$$\frac{dy}{dx} = \frac{y}{x}.$$

Solve the differential equation for the streamlines.

11. Consider a flat plate of thickness 0.05 m. The temperature of the cold side of the plate is 280 K, and the heat flux through the plate is 1000 W/m^2. Given that the thermal conductivity of the material is 5 W/(m.K), determine the temperature of the hot side of the plate.
12. Suppose the wall of a furnace is made of such a material whose thermal conductivity is a function of temperature given by $k = 1 + 3\times10^{-6}T^2$, where T is the temperature in Kelvin. The inner side of the wall is at a temperature of 1000 K and the outer side is at a temperature of 300 K. Determine the heat flux through the wall if its thickness is 1 m.
13. A spring-mass-damper system which has a mass, $m = 1$ kg, spring constant, $k = 25$ N/m and a damping constant, $c = 10$ Ns/m is vibrating under an oscillating force, $f(t) = 5\cos(5t)$ N. Given that $x(0) = 0$ and $\dot{x}(0) = 0$, write down the differential equation and determine the position of the mass at any instant of time. What is the steady-state solution?
14. Suppose a spring-mass-damper system, consists of a mass $m = 500$ kg, spring constant $k = 10000$ N/m and a damping constant $c = 4500$ Ns/m, is vibrating under a force $f(t) = 2000\cos(2t)$ N. Write down the differential equation governing this system and determine the position of the mass at any instant of time t
15. A simple electrical circuit contains a resistor, a capacitor and a constant input voltage V in series. The differential equation governing this circuit is as follows

$$RC\frac{dV_C}{dt} + V_C = V,$$

where V_C is the voltage drop across the capacitor. If $V_C(0) = 0$, determine $V_C(t)$.

16. In a parallel RLC circuit where a DC current i is applied, we can obtain the following differential equation:

$$LC\frac{d^2i_L}{dt^2} + \frac{L}{R}\frac{di_L}{dt} + i_L = i,$$

where i_L is the current flowing through the inductor. Assuming the capacitance is 10×10^{-6} F, the inductance is 0.1H and the resistance is 50, determine $i_L(t)$. What is the final value of i_L?

Answers

1. $t\approx 60.998$s, $v\approx 98$m/s, and the final speed is 98m/s.
2. $t\approx 408.66$s, $v\approx 4.9$m/s
3. $t\approx 3.1s$
4. $v\approx 27.5$m/s
5. 11.7%,5%
6. $s(20)\approx 444$g, 44.4%
7. 11.9%, $t = 29$ hours, 3480, 7%
8. $S(t)=\begin{cases} 20e^{-\frac{3t}{500}}\left(-249+250e^{\frac{3t}{500}}\right), & 0<t\le 6 \\ \frac{2}{625}(-256+t)^2, & 6<t\le 256 \end{cases}$

10. $y = Cx$
11. 290 K
12. 1673 W/m²
13. $x=-\frac{1}{2}te^{-5t}+\frac{1}{10}\sin 5t$; $x=\frac{1}{10}\sin 5t$ as $t\to\infty$
14. $x=-\frac{4}{5}e^{-4t}+\frac{20}{29}e^{-5t}+\frac{16}{145}\cos 2t+\frac{18}{145}\sin 2t$;

    $x=\frac{16}{145}\cos 2t+\frac{18}{145}\sin 2t$ as $t\to\infty$
15. $V_C = V\left(1-e^{-\frac{1}{RC}t}\right)$
16. $i_L(t)=-(10+10000t)e^{-1000t}+10$; $i_L=10$ A as $t\to\infty$

3 Laplace Transform

3.1 INTRODUCTION

Mathematicians are always looking for new approaches to transform a difficult problem into an easy one and the Laplace transform is one of the approaches to presenting problems in the time domain in the algebraic domain that can be solved faster and require only algebraic manipulation.

The Laplace transform is similar to using logarithms to solve or simplify certain types of mathematical equations. For example, applying logarithms to some exponential equations turns them into simple algebraic equations by getting rid of exponential expressions. By comparison, the components of AC electrical circuits can be converted from the time domain to the simpler phasor domain, where mathematical calculations require only the addition, subtraction, multiplication and division of phasors. The Laplace transform is often used in engineering and physics. In engineering, many systems can be described by time-dependent differential equations and similar approaches can be adopted to overcome complexities in the time domain.

The Laplace transform converts equations in the time domain to the algebraic domain by expressing them in terms of the variable "s" or "ω", where the original problem can be quickly manipulated and solved using only algebraic manipulations.

The Laplace transform can be considered as an integral transformation in solving physical problems. The Laplace transform is particularly useful in solving linear ordinary differential equations.

One important application occurs in linear time-invariant systems. In such systems, the output of the system is determined by the convolution integral of the input signal and the impulse response of the system in the time domain, which can lead to a complex integration. Nevertheless, by using the Laplace transform, the calculation of convolution integral turns to a multiplication.

The outline of this chapter is as follows:

- Definition of the Laplace transform with examples of using the table of Laplace transform
- First and second shifting theorems
- Solving differential equations using Laplace transform
- The unit step and Dirac functions
- Convolution integral and its visual explanation
- Application in control engineering systems

DOI: 10.1201/9781032630694-3

3.2 DEFINITION OF THE LAPLACE TRANSFORM

DEFINITION: LAPLACE TRANSFORM

The Laplace transform of a function $f(t)$ for $0 \le t < \infty$ is defined by the following improper integral:

$$\mathcal{L}\{f(t)\} = \int_0^{\infty} e^{-st} f(t)\,dt = F(s), \qquad s > 0.$$

In this transformation, the starting point is a time-dependent function $f(t)$, but after applying the Laplace transformation, we finally have a function with an independent variable s. The variable s is generally a complex variable, but it is not so important to us because the function $f(t)$ is usually a function of time.

Example 3.1 Find the Laplace transform of the following functions:

(a) $f(t) = 1$.

(b) $f(t) = e^{at}$.

Solution

(a) $\mathcal{L}\{1\} = \int_0^{\infty} e^{-st} \times 1 \times dt = \int_0^{\infty} e^{-st} dt$

$$= \left[-\frac{e^{-st}}{s} \right]_{t=0}^{t=\infty} = \left[-\frac{e^{-s\times\infty}}{s} + \frac{e^{0}}{s} \right] = \left[-\frac{0}{s} + \frac{1}{s} \right] = \frac{1}{s}, \qquad s > 0.$$

Strictly speaking, the variable t cannot take the value of infinity, and we cannot directly substitute $t = \infty$ into the earlier exponential term; however, it is customary to take the limit as $t \to \infty$. We assume $s > 0$ to ensure that the exponent is negative, and thus, the exponential term is zero when the variable t approaches infinity.

(b) $\mathcal{L}\{e^{at}\} = \int_0^{\infty} e^{-st} \times e^{at} dt = \int_0^{\infty} e^{-(s-a)t} dt$

$$= -\left[\frac{e^{-(s-a)t}}{s-a} \right]_{t=0}^{t=\infty} = -\left[\frac{e^{-\infty}}{s-a} - \frac{e^{0}}{s-a} \right] = \frac{1}{s-a}, \qquad s > a.$$

3.3 LINEAR PROPERTY OF THE LAPLACE TRANSFORM

The linear property of the Laplace transform can be written as follows:

$$\mathcal{L}\{c_1 f(t) + c_2 g(t)\} = c_1 \mathcal{L}\{f(t)\} + c_2 \mathcal{L}\{g(t)\}, \text{ where } c_1 \text{ and } c_2 \text{ are constants.}$$

Example 3.2 In this example, we use the Laplace transform of the exponential function and the Euler formula to obtain the Laplace transform of the trigonometrical functions *sine* and *cosine.*

(a) Find the Laplace transform of the function

$$f(t) = e^{jat}; \qquad j^2 = -1.$$

(b) Use part (a) and the linear property and find the Laplace transform of the following functions:

$$f(t) = \cos(at) \qquad \text{and} \qquad g(t) = \sin(at).$$

Solution

(a) Using Example 3.1, we obtain the following Laplace transform:

$$\mathcal{L}\left\{e^{jat}\right\} = \frac{1}{s - ja} = \frac{1}{s - ja} \times \frac{s + ja}{s + ja} = \frac{s}{s^2 + a^2} + j\frac{a}{s^2 + a^2}.$$

(b) Using the Euler formula, $e^{jat} = \cos(at) + j\sin(at)$, and the linear property of the Laplace transform, we obtain

$$\begin{aligned}\mathcal{L}\left\{e^{jat}\right\} &= \mathcal{L}\{\cos(at) + j\sin(at)\} \\ &= \mathcal{L}\{\cos(at)\} + j\mathcal{L}\{\sin(at)\}.\end{aligned}$$

Comparing this result with the result obtained in part (a) implies the Laplace transform of the trigonometric functions is

$$\mathcal{L}\{\cos(at)\} + j\mathcal{L}\{\sin(at)\} = \frac{s}{s^2 + a^2} + j\frac{a}{s^2 + a^2}$$

$$\Rightarrow$$

$$\mathcal{L}\{\cos(at)\} = \frac{s}{s^2 + a^2}$$

$$\mathcal{L}\{\sin(at)\} = \frac{a}{s^2 + a^2}.$$

Note: The Laplace transform of cosine or sine function can be calculated using direct integration but requires integration by parts twice. In general, finding the Laplace transform of functions using direct integration is not ideal and in practice very quickly becomes cumbersome and difficult. For this reason, we usually use the predefined table of Laplace transforms to find the Laplace transform of some known functions (see the Laplace transforms table in Section 3.13).

Example 3.3 Find the Laplace transform of the following functions using the table of Laplace transforms in Section 3.13:

(a) $f(t) = t^4$.

Using the table: $\mathcal{L}\{t^n\} = \dfrac{n!}{s^{n+1}}$

$$n = 4 \Rightarrow \mathcal{L}\{t^4\} = \frac{4!}{s^5} = \frac{4\times3\times2\times1}{s^5} = \frac{24}{s^5}.$$

(b) $f(t) = e^{-3t}$.

Using the table: $\mathcal{L}\{e^{-\alpha t}\} = \dfrac{1}{s+\alpha}$

$$\alpha = 3 \Rightarrow \mathcal{L}\{e^{-3t}\} = \frac{1}{s+3}.$$

(c) $f(t) = \cos(5t)$.

Using the table: $\mathcal{L}\{\cos(\omega t)\} = \dfrac{s}{s^2+\omega^2}$

$$\omega = 5 \Rightarrow \mathcal{L}\{\cos(5t)\} = \frac{s}{s^2+5^2} = \frac{5}{s^2+25}.$$

(d) $\mathcal{L}\{3\cos(4t)+2\sin(4t)\} = 3\times\dfrac{s}{s^2+4^2} + 2\times\dfrac{4}{s^2+4^2}$

$$= \frac{3s+8}{s^2+16}.$$

3.4 INVERSE LAPLACE TRANSFORM

In the previous sections, we defined the Laplace transform and used the Laplace transforms table to find the transforms of some common functions. We have shown that applying the Laplace transform to time-dependent functions transforms them into functions of "s". In other words, the time variable "t" reaches the algebraic variable "s", which is sometimes called the frequency domain.

In solving some problems, especially in finding solutions to differential equations using the Laplace transform, we need to be able to invert functions of s into functions of t. This process is called inverse transformation and hence to complement the Laplace transform, we define the inverse Laplace transform as shown.

DEFINITON: INVERSE LAPLACE TRANSFORM

Suppose the Laplace transform of the function $f(t)$ is defined as $F(s)$, then the inverse transform is defined as follows:

$$\mathcal{L}^{-1}\{F(s)\} = f(t).$$

The inverse Laplace transform is also a linear operator:

$$\mathcal{L}^{-1}\{c_1F(s)+c_2G(s)\}=c_1\mathcal{L}^{-1}\{F(s)\}+c_2\mathcal{L}^{-1}\{G(s)\},$$

where c_1 and c_2 are constants. This characteristic assists in calculating the inverse transform.

Example 3.4 Determine the following inverse transformations using the Laplace transform table:

(a) $\mathcal{L}^{-1}\left\{\frac{1}{s-1}\right\}$ (b) $\mathcal{L}^{-1}\left\{\frac{1}{s^2}\right\}$

(c) $\mathcal{L}^{-1}\left\{\frac{3}{s^7}\right\}$ (d) $\mathcal{L}^{-1}\left\{\frac{1}{s+4}+\frac{s}{s^2+9}\right\}$

(e) $\mathcal{L}^{-1}\left\{\frac{s-3}{s^2-6s+8}\right\}$.

Solution

The inverse transformations in parts (a)–(c) can easily be obtained from the table of Laplace Transforms:

(a) $\mathcal{L}^{-1}\left\{\frac{1}{s-1}\right\}=e^t$

(b) $\mathcal{L}^{-1}\left\{\frac{1}{s^2}\right\}=t$

(c) $\mathcal{L}^{-1}\left\{\frac{3}{s^7}\right\}=\frac{3}{6!}L^{-1}\left\{\frac{6!}{s^7}\right\}=\frac{1}{240}t^6$

(d) First use the linear property followed by the Laplace transforms table:

$$\mathcal{L}^{-1}\left\{\frac{1}{s+4}+\frac{s}{s^2+9}\right\}=\mathcal{L}^{-1}\left\{\frac{1}{s+4}\right\}+\mathcal{L}^{-1}\left\{\frac{s}{s^2+9}\right\}$$
$$=e^{-4t}+\cos(3t).$$

(e) The inverse Laplace transform of the function in part (e) is not in the table of Laplace transforms. This question can be solved using the partial fraction decomposition, which involves factorising the denominator, but the details of this process are omitted here.

$$\frac{s-3}{s^2-6s+8}=\frac{s-3}{(s-2)(s-4)}=\frac{1}{2}\left(\frac{1}{s-4}+\frac{1}{s-2}\right).$$

Hence,

$$\mathcal{L}^{-1}\left\{\frac{s-3}{s^2-6s+8}\right\}=\frac{1}{2}\mathcal{L}^{-1}\left\{\frac{1}{s-4}+\frac{1}{s-2}\right\}$$

$$=\frac{1}{2}\mathcal{L}^{-1}\left\{\frac{1}{s-4}\right\}+\frac{1}{2}\mathcal{L}^{-1}\left\{\frac{1}{s-2}\right\}$$

$$=\frac{1}{2}\left(e^{4t}+e^{2t}\right).$$

3.5 THE FIRST SHIFTING THEOREM

Suppose $g(t)=e^{-\alpha t}f(t)$, where $f(t)$ is a given function with a known (possibly from the table of Laplace transforms) Laplace transform $F(s)=\mathcal{L}\{f(t)\}$, and α is a given constant. From the definition of the Laplace transform, the following transformation can be easily obtained:

$$\mathcal{L}\left\{e^{-\alpha t}f(t)\right\}=F(s+\alpha)$$

$$\mathcal{L}^{-1}\left\{F(s+\alpha)\right\}=e^{-\alpha t}f(t).$$

This transformation is called the first shifting theorem, and it means that when the function $f(t)$ is multiplied by $e^{-\alpha t}$, its Laplace transform is shifted by α units.

Such transformation can be obtained directly using the definition of the Laplace transform integral, but this theorem is an easy way to find the Laplace transform of functions in the form of $e^{-\alpha t}f(t)$.

Example 3.5 Using the first shifting theorem determine the following transformations:

(a) $\mathcal{L}\left\{e^{-2t}t^3\right\}$

(b) $\mathcal{L}\left\{e^{t}(t+1)^2\right\}$

(c) $\mathcal{L}^{-1}\left\{\frac{1}{(s+1)^2}\right\}$.

Solution

We solve all parts in three steps.

(a)

Step 1: $e^{-2t}t^3=e^{-\alpha t}f(t)\Rightarrow \alpha=2$ and $f(t)=t^3$.

Step 2: $F(s)=\mathcal{L}\left\{t^3\right\}=\frac{3!}{s^4}=\frac{6}{s^4}$ (table).

Step 3: Shift $F(s)$ by $\alpha = 2$: $\mathcal{L}\{e^{-2t}t^3\} = F(s+2) = \dfrac{6}{(s+2)^4}$.

(b)

Step 1: $e^t(t+1)^2 = e^{-\alpha t} f(t) \Rightarrow \alpha = -1$ and $f(t) = (t+1)^2$.

Step 2: $F(s) = \mathcal{L}\{f(t)\} = \mathcal{L}\{(t+1)^2\} = \mathcal{L}\{t^2 + 2t + 1\}$

$$= \mathcal{L}\{t^2\} + 2\mathcal{L}\{t\} + \mathcal{L}\{1\} = \frac{2!}{s^3} + \frac{2}{s^2} + \frac{1}{s}.$$

Step 3: Shift by $\alpha = -1$:

$$\mathcal{L}\{(t+1)^2 e^t\} = F(s-1) = \frac{2}{(s-1)^3} + \frac{2}{(s-1)^2} + \frac{1}{(s-1)}.$$

(c)

Step 1: $F(s+\alpha) = \dfrac{1}{(s+1)^2} \Rightarrow \alpha = 1$ and $F(s) = \dfrac{1}{s^2}$.

Step 2: $f(t) = \mathcal{L}^{-1}\{F(s)\} = \mathcal{L}^{-1}\left\{\dfrac{1}{s^2}\right\} = t$.

Step 3: $\mathcal{L}^{-1}\left\{\dfrac{1}{(s+1)^2}\right\} = e^{-t}t$.

3.5.1 Exercises

Using the table of Laplace transforms and the first shifting theorem, find the Laplace transform of the following functions:

(a) $f(t) = t + 3$

(b) $f(t) = at^2 + bt + c$

(c) $f(t) = t^2 + 3t + 1$

(d) $f(t) = (a - bt)^2$

(e) $f(t) = 4te^{-t}$

(f) $f(t) = t^2 e^{2t}$

(g) $f(t) = e^{-2t+1}$

(h) $f(t) = e^t \sin(2t)$

(i) $f(t) = \cos\left(3t + \frac{\pi}{2}\right)$

(j) $f(t) = \cos(\omega t + \theta)$

(k) $f(t) = 3\sin^2(t)$

(l) $f(t) = 2\cos^2(t)$

(m) $f(t) = e^{-2t}\sin(3t)$

(n) $f(t) = e^t \sin\left(3t - \frac{\pi}{2}\right)$

Answers

(a) $F(s)=\frac{1}{s^2}+\frac{3}{s}$ (b) $F(s)=\frac{2a}{s^3}+\frac{b}{s^2}+\frac{c}{s}$

(c) $F(s)=\frac{2}{s^3}+\frac{3}{s^2}+\frac{1}{s}$ (d) $F(s)=\frac{2b^2}{s^3}-\frac{2ab}{s^2}+\frac{a^2}{s}$

(e) $F(s)=\frac{4}{(s+1)^2}$ (f) $F(s)=\frac{2}{(s-2)^3}$

(g) $F(s)=\frac{e}{(s+2)}$ (h) $F(s)=\frac{2}{(s-1)^2+4}$

(i) $F(s)=-\frac{3}{s^2+9}$ (j) $F(s)=\frac{s\cos(\theta)-\omega\sin(\theta)}{s^2+\omega^2}$

(k) $F(s)=\frac{6}{s(s^2+4)}$ (l) $F(s)=\frac{2s^2+4}{s(s^2+4)}$

(m) $F(s)=\frac{3}{(s+2)^2+9}$ (n) $F(s)=-\frac{s-1}{(s-1)^2+9}$

3.6 INVERSE LAPLACE TRANSFORM USING COMPLETING THE SQUARE

Referring to Example 3.4e, in some cases, the denominator cannot be factorised, so the partial fraction decomposition method cannot be used to find inverse Laplace transform. In these cases, we use the method of completing the square. We give some examples for further explanation.

Example 3.6 Using the technique of completing the square calculate the following inverse transforms:

(a) $\mathcal{L}^{-1}\left\{\frac{s-2}{s^2-4s+5}\right\}$ (b) $\mathcal{L}^{-1}\left\{\frac{1}{s^2+2s+5}\right\}$

(c) $\mathcal{L}^{-1}\left\{\frac{s-3}{s^2-2s-8}\right\}$.

Solution

The denominator of parts (a)–(b) cannot be factorised, and hence, they require the technique of "completing the square" in the denominator followed by using the first shifting theorem:

(a)

$$\frac{s-2}{s^2-4s+5}=\frac{s-2}{s^2-4s+4+1}$$

$$=\frac{s-2}{(s-2)^2+1}\Rightarrow\mathcal{L}^{-1}\left\{\frac{s-2}{(s-2)^2+1}\right\}=e^{2t}\cos t.$$

This result is listed in the Laplace transform table with $\alpha=2$ and $\omega=1$.

(b)

$$\frac{1}{s^2+2s+5}=\frac{1}{(s+1)^2+4}=\frac{1}{2}\frac{2}{(s+1)^2+4}$$

Hence, using the table of Laplace transform, we get

$$\mathcal{L}^{-1}\left\{\frac{1}{s^2+2s+5}\right\}=\frac{1}{2}e^{-t}\sin 2t.$$

(c) This part can be done using partial fraction decomposition; however, completing the square leads to

$$\frac{s-3}{s^2-2s-8}=\frac{s-3}{(s-1)^2-9}$$

$$=\frac{s-1}{(s-1)^2-3^2}-\frac{2}{(s-1)^2-3^2}$$

$$=\frac{s-1}{(s-1)^2-3^2}-\frac{2}{3}\frac{3}{(s-1)^2-3^2}.$$

Hence, using the table of Laplace transform, we get

$$\mathcal{L}^{-1}\left\{\frac{s-3}{s^2-2s-8}\right\}=e^t\left[\cosh(3t)-\frac{2}{3}\sinh(3t)\right]=\frac{1}{6}e^{4t}+\frac{5}{6}e^{-2t}.$$

3.6.1 Exercises

Using the table of Laplace transforms, find the inverse Laplace transform of the following functions:

(a) $F(s)=\frac{3}{s-3}$

(b) $F(s)=\frac{5}{s^2+4}$

(c) $F(s)=\frac{s-1}{s^2+9}$

(d) $F(s)=\frac{s-2}{s^2+4}$

(e) $F(s)=\frac{s+4}{s^2+9}$

(f) $F(s)=\frac{1}{(s-2)^2}$

(g) $F(s)=\frac{s-3}{(s-3)^2+4}$

(h) $F(s)=\frac{s+5}{(s+2)^2+9}$

(i) $F(s)=\frac{1}{s^4}$

(j) $F(s)=\frac{4}{s^3}$

(k) $F(s)=\frac{s}{(s-1)^2-9}$

(l) $F(s)=\frac{s+4}{(s^2+2s+2)(s-1)}$

(m) $F(s)=\frac{s+1}{(s+2)(s-4)}$

(n) $F(s)=\frac{32}{s(s^2+16)}$

Answers

(a) $f(t) = 3e^{3t}$

(b) $f(t) = \frac{5}{2}\sin(2t)$

(c) $f(t) = \cos(3t) - \frac{1}{3}\sin(3t)$

(d) $f(t) = \cos(2t) - \sin(2t)$

(e) $f(t) = \cos(3t) + \frac{4}{3}\sin(3t)$

(f) $f(t) = te^{2t}$

(g) $f(t) = e^{3t}\cos(2t)$

(h) $f(t) = e^{-2t}\left[\cos(3t) + \sin(3t)\right]$

(i) $f(t) = \frac{1}{6}t^3$

(j) $f(t) = 2t^2$

(k) $f(t) = e^{t}\left[\cosh(3t) + \frac{1}{3}\sinh(3t)\right]$

(l) $f(t) = e^{t} - e^{-t}\left[\cos(t) + \sin(t)\right]$

(m) $f(t) = \frac{1}{6}\left(5e^{4t} + e^{-2t}\right)$

(n) $f(t) = 2\left[1 - \cos(4t)\right]$

3.7 DERIVATIVES AND THE LAPLACE TRANSFORM

As mentioned before, the function $f(t)$ is transformed into the function $F(s)$ by applying the Laplace transform. In this section, the objective is to find the Laplace transform of the derivatives, $f'(t)$ and $f''(t)$, of the function $f(t)$, which will be useful in solving ordinary differential equations. Using the definition of Laplace transform,

$$\mathcal{L}\{f'(t)\} = \int_0^\infty f'(t)e^{-st}\,dt.$$

This can be integrated with the use of integration by parts (in the order of differentiating e^{-st} and integrating $f'(t)$):

$$\mathcal{L}\{f'(t)\} = \left[f(t)\,e^{-st}\right]_0^\infty - \int_0^\infty f(t)(-s)e^{-st}dt$$

$$= 0 - f(0) + s\int_0^\infty f(t)e^{-st}dt$$

$$= -f(0) + s\,F(s).$$

Therefore, by applying the Laplace transform on the derivative of the function $f(t)$, we get the following transformation:

$$\mathcal{L}\{f'(t)\} = sF(s) - f(0).$$

This means that applying the Laplace transform to $f'(t)$ expresses it in terms of the Laplace transform of $f(t)$ and its value at $t = 0$. Moreover, it transforms the derivative operation into a simpler operation that involves multiplication and addition in the s domain. Strictly speaking, it gets rid of the derivative term.

Now we do some pattern matching to find the Laplace transform of the second derivative:

$$\begin{aligned}\mathcal{L}\{f''(t)\} &= \mathcal{L}\{(f'(t))'\} \\ &= s\mathcal{L}\{f'(t)\} - f'(0) \\ &= s\left[sF(s) - f(0)\right] - f'(0) \\ &= s^2F(s) - sf(0) - f'(0).\end{aligned}$$

The results obtained are very useful properties of the Laplace transform because they eliminate derivatives when applied to differential equations, which is what we really need to do when we solve a certain type of ordinary differential equations (ODEs).

3.7.1 Using Laplace Transform to Solve ODEs

In the previous section, we showed that applying the Laplace transform on the derivatives of the function $f(t)$ removes the derivative term. In this section, we use this property to solve the first- and second-order linear differential equations. The method we describe in this section applies to a special class of differential equations involving step or Dirac function, with known initial conditions that cannot be solved by traditional methods.

The notation used to represent the Laplace transform of functions of time is not suitable for solving differential equations like $a\ddot{x} + b\dot{x} + cx = f(t)$. Therefore, we introduce the following simple alternative notation to represent the Laplace transform of time-dependent functions:

$$\begin{aligned}&\mathcal{L}\{x(t)\} = \overline{x} \\ &\mathcal{L}\{\dot{x}(t)\} = s\overline{x} - x(0) \\ &\mathcal{L}\{\ddot{x}(t)\} = s^2\overline{x} - sx(0) - \dot{x}(0).\end{aligned}$$

The process of solving the differential equation using the Laplace transform can be divided into the following three simple steps:

Step 1: Apply the Laplace transform to each individual term of the ODE and substitute the initial values into the transformed equation.

Step 2: Solve the equation in step 1 for $\overline{x}$ and manipulate the resulting expression so that it can be inverted using the table of Laplace transforms.

Step 3: Finally, use the inverse transformation to obtain the solution of the equation.

In order to understand the theory of solving differential equations using the Laplace transform, we will look at some examples. However, knowledge of the results obtained in the previous sections is also required. Before attempting to solve the differential equations, it is necessary to closely examine the inverse process of obtaining the original functions using their given Laplace transform.

Example 3.7 In this example, we provide two linear second-order differential equations with given initial values. These equations can be solved using the method of undetermined coefficients as explained in Chapter 2; however, we transform these time-dependent equations into algebraic equations using the Laplace transform. This method is useful when the right-hand side of equations are a step or Dirac function.

(a) Solve the following differential equation:

$$\frac{d^2x}{dt^2}+9x=9$$

with the initial conditions $x(0)=5$ and $\frac{dx}{dt}(0)=0$.

(b) Solve the following differential equation:

$$\frac{d^2x}{dt^2}+4\frac{dx}{dt}+8x=29e^{3t}$$

with the initial conditions $x(0)=\frac{dx}{dt}(0)=0$.

Solution

We solve these equations in three steps.

(a)

Step 1: Transform and substitute the initial conditions:

$$[s^2\overline{x}-sx(0)-\dot{x}(0)]+9\overline{x}=\frac{9}{s}$$

$$\left[s^2\overline{x}-5s-0\right]+9\overline{x}=\frac{9}{s}.$$

Step 2: Simplify and solve for $\overline{x}$:

$$s^2\overline{x}+9\overline{x}=\frac{9}{s}+5s$$

$$\overline{x}=\frac{9}{s\left(s^2+9\right)}+\frac{5s}{\left(s^2+9\right)}.$$

Step 3: Apply the inverse transformation to get the solution of the equation:

$$x(t) = 1 - \cos(3t) + 5\cos(3t) = 1 + 4\cos(3t).$$

(b)

Step 1: Transform both sides of the equation and substitute the initial conditions:

$$\left[s^2\overline{x} - sx(0) - \dot{x}(0)\right] + 4\left[s\overline{x} - x(0)\right] + 8\overline{x} = \frac{29}{s-3}$$

$$\left[s^2\overline{x} - s\times 0 - 0\right] + 4\left[s\,\overline{x} - 0\right] + 8\,\overline{x} = \frac{29}{s-3}$$

$$\left(s^2 + 4s + 8\right)\overline{x} = \frac{29}{s-3}.$$

Step 2: Solve for $\overline{x}$:

$$\overline{x} = \frac{29}{(s-3)\left(s^2 + 4s + 8\right)}.$$

In order to be able to obtain the solution of the differential equation using the inverse transformation in the last step, we must manipulate the previous equation so that its inverse transformation can be easily found from the table of Laplace transforms. First, we need to apply the partial fraction decomposition and then complete the square of the denominator with a quadratic term and adjust its numerator. We omit the detailed explanation and analysis of the partial fraction decomposition and completing the square, hence

$$\overline{x} = \left[\left(\frac{1}{s-3}\right) - \left(\frac{s+2}{(s+2)^2 + 2^2}\right) - \frac{5}{2}\left(\frac{2}{(s+2)^2 + 2^2}\right)\right].$$

Step 3: Finally invert each fraction in the preceding equation using the table and simplify the final result:

$$x = e^{3t} - e^{-2t}\cos(2t) - \frac{5}{2}e^{-2t}\sin(2t)$$

$$= e^{3t} - e^{-2t}\left[\cos(2t) + \frac{5}{2}\sin(2t)\right].$$

Note: The use of Laplace transform helps solve equations like

$$a\frac{d^2x}{dt^2}+b\frac{dx}{dt}+cx=f(t).$$

When traditional methods fail, this can be due to the complexity of the function $f(t)$. We consider such cases in Sections 3.8 and 3.10.

3.7.2 Exercises

Find the solution for the following initial value problems using the Laplace transform.

(a) $2\dot{x}-x=1,\qquad x(0)=1$

(b) $2\dot{x}-x=t,\qquad x(0)=1$

(c) $\dot{x}-x=\cos(t),\qquad x(0)=1$

(d) $\ddot{x}-x=1,\qquad x(0)=1$ and $\dot{x}(0)=0$

(e) $\ddot{x}+4x=4,\qquad x(0)=1$ and $\dot{x}(0)=2$

(f) $\ddot{x}+\dot{x}+4x=4,\qquad x(0)=1$ and $\dot{x}(0)=0$

(g) $\ddot{x}-2\dot{x}+x=0,\qquad x(0)=1$ and $\dot{x}(0)=-1$

(h) $\ddot{x}+4\dot{x}=0,\qquad x(0)=0$ and $\dot{x}(0)=1$

Answers

(a) $x(t)=-1+2e^{\frac{t}{2}}$

(b) $x(t)=-2+3e^{\frac{t}{2}}-t$

(c) $x(t)=\frac{3}{2}e^{t}-\frac{1}{2}\left[\cos(t)-\sin(t)\right]$

(d) $x(t)=-1+e^{-t}+e^{t}$

(e) $x(t)=1+\sin(2t)$

(f) $x(t)=1$

(g) $x(t)=e^{t}(-2t+1)$

(h) $x(t)=\frac{1}{4}\left(1-e^{-4t}\right)$

3.8 THE UNIT STEP FUNCTION

In engineering or physical applications, we often encounter sudden changes in the behaviour of systems. An example of this scenario appears in electrical systems when a voltage is turned on or off at a certain time.

The mathematical description of a sudden change in systems can be explained by a well-known unit step (Heaviside) function.

DEFINITION: UNIT STEP FUNCTION

The unit step function is defined as

$$u(t-\alpha)=\begin{cases}0, & t<\alpha\\ 1, & t>\alpha\end{cases},$$

where $\alpha \geq 0$ is a constant and t is usually time. The variable α describes the instant of time at which the value of the function changes abruptly and is known as the "critical value" of the step function, as shown in Figure 3.1a for $\alpha = 3$. In this definition, the value of the step function at $t=\alpha$ is not defined; in some literature, $u(t-\alpha)$ is a known value at $t=\alpha$, but it doesn't matter to us.

Note: Assuming $\alpha = 0$, the critical value occurs at $t=0$, and the step function is represented as $u(t)$. Therefore, in the earlier definition, $u(t-\alpha)$ can be considered a shifted function for $u(t)$ and does not represent the multiplication of u by $(t-\alpha)$. Assuming $t \geq 0$ and $\alpha = 0$, then $u(t-0)=u(t)\equiv 1$.

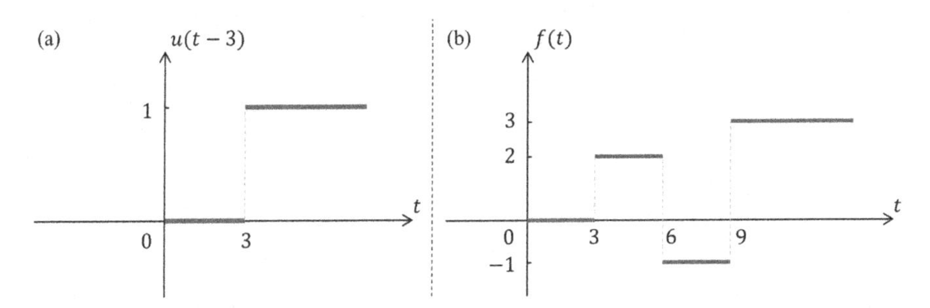

FIGURE 3.1 (a) The step function with a critical value at $\alpha = 3$. (b) Combined step function. This graph can also be explained as follows: For $t < 3$ all step functions are 0, at $t = 3$, the function f jumps up by 2 units and at $t = 6$, the function f jumps down by 3 units and, again, at $t = 9$, jumps up by 4 units.

Step functions can be used to model sudden changes in mechanical systems, electrical circuits, or any other system in the event of a sudden change of state. For example, in a mechanical system, the sudden application of a force or its removal, and in an electrical circuit, the momentary change of a voltage or turning on an electrical circuit. These scenarios can be represented by combining step functions with unequal critical values or multiplying a step function by some special function, as shown in the following examples.

Example 3.8 Graph the following function:

$$f(t)=2u(t-3)-3u(t-6)+4u(t-9).$$

Solution

$$f(t) = 2u(t-3) - 3u(t-6) + 4u(t-9)$$

$$= \begin{cases} 0+0+0, & t<3 \\ 2+0+0, & 3 \le t < 6 \\ 2-3+0, & 6 \le t < 9 \\ 2-3+4, & t \ge 9 \end{cases} = \begin{cases} 0, & t<3 \\ +2, & 3 \le t < 6 \\ -1, & 6 \le t < 9 \\ +3, & t \ge 9 \end{cases}.$$

An alternative explanation of the preceding example is to consider step functions as arithmetic jumps at critical values. All step functions are stationary for $t < 3$ and have zero values, so $f(t) = 2\times 0 - 3\times 0 + 4\times 0 = 0$. For $3 < t < 6$, the first step function jumps up 2 steps, and the others are still stationary, so $f(t) = 2\times 1 - 3\times 0 + 4\times 0 = 2$. For $6 < t < 9$, the second step function jumps down 3 steps from its current location and so $f(t) = 2\times 1 - 3\times 1 + 4\times 0 = -1$. At $t > 9$, the third step function jumps up 4 steps, giving rise to $f(t) = 2\times 1 - 3\times 1 + 4\times 1 = 3$. Note that the jump up or down depends on the sign of the coefficients of the step functions; when the sign is positive, it jumps up; otherwise, it jumps down. So, $f(t)$ can easily be plotted, which is shown in Figure 3.1b.

3.8.1 Products Involving Unit Step Functions

When the step function is combined with other functions defined for $t > 0$, it "truncates" a portion of their graph and makes it zero.

Three types of products with the step function are in our interest:

Type 1: $g(t) = f(t)\times u(t-a)$, where

$$u(t-\alpha) = \begin{cases} 0, & t<\alpha \\ 1, & t>\alpha \end{cases}.$$

Hence,

$$g(t) = f(t)\times u(t-\alpha) = \begin{cases} 0, & t<\alpha \\ f(t), & t>\alpha \end{cases}.$$

To plot $g(t)$, draw the graph of the function $f(t)$ and set it to zero before $t = \alpha$.

Type 2: $g(t) = f(t)\times\left[u(t-\alpha) - u(t-\beta)\right], \quad \beta > \alpha$

$$u(t-\alpha) - u(t-\beta) = \begin{cases} 0, & t<\alpha \\ 1, & \alpha < t < \beta \\ 0, & t>\beta \end{cases}.$$

Hence $g(t) = f(t)\times\left[u(t-\alpha) - u(t-\beta)\right] = \begin{cases} 0, & t<\alpha \\ f(t), & \alpha < t < \beta \\ 0, & t>\beta \end{cases}.$

To plot $g(t)$, first draw the graph of the function $f(t)$ and set it to zero before $t=\alpha$ and after $t=\beta$.

Type 3: $g(t)=f(t-a)\times u(t-\alpha)$

Shift $f(t)$ by $t=\alpha$ and set it to zero before $t=\alpha$.

Example 3.9 Sketch the graph of the following functions:

Type 1: $g(t)=(t+3)u(t-2)$.

Type 2: $g(t)=(t-3)(6-t)\left[u(t-3)-u(t-6)\right]$.

Type 3: $g(t)=8\cos\left(2t-\frac{\pi}{4}\right)u\left(t-\frac{\pi}{8}\right)$.

Solution

In this example, three different functions are multiplied by the step functions:

Type 1: $g(t)=(t+3)u(t-2)\Rightarrow f(t)=t+3$.

We sketch the graph of $f(t)=t+3$ and set it to zero before the critical point $t=2$, as shown in Figure 3.2a.

Type 2: $g(t)=(t-3)(6-t)\left[u(t-3)-u(t-6)\right]\Rightarrow$

$f(t)=-t^2+9t-18$.

Since

$$u(t-3)-u(t-6)=\begin{cases}0, & t<3\\ 1, & 3<t<6,\\ 0, & t>6\end{cases}$$

hence

$$g(t)=\begin{cases}0, & t<3\\ -t^2+9t-18, & 3<t<6.\\ 0, & t>3\end{cases}$$

So, sketch the graph of the parabola and set it to zero before $t=3$ and after $t=6$ (see Figure 3.2b). Note that $f(t)$ is a parabola whose t-intercepts are at $t=3$ and $t=6$ and whose vertex occurs at (4.5, 2.25).

Type 3: $g(t)=8\cos\left(2t-\frac{\pi}{4}\right)u\left(t-\frac{\pi}{8}\right)\Rightarrow f(t)=8\cos(2t)$.

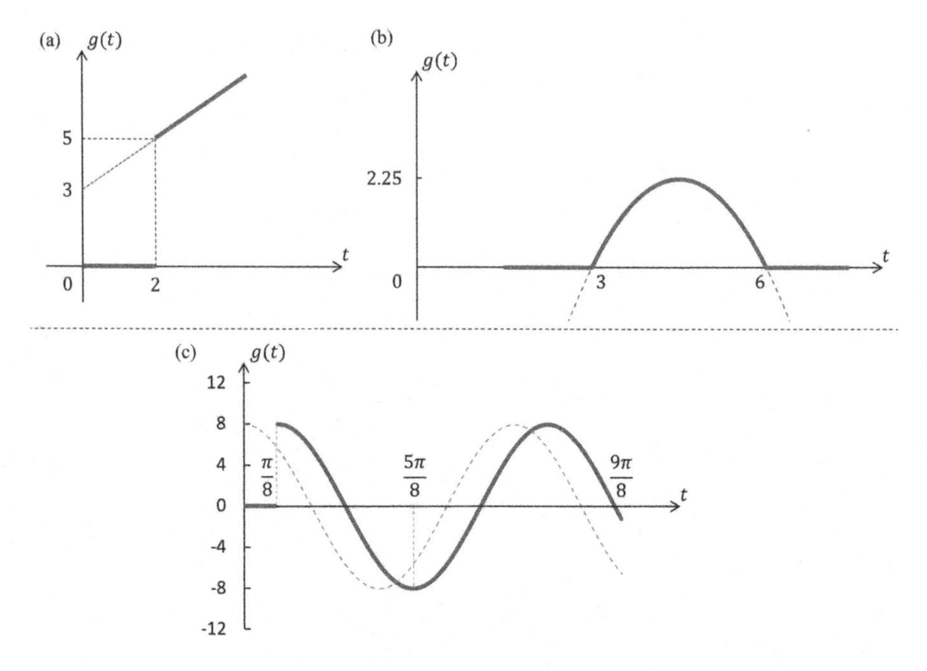

FIGURE 3.2 The graph of three product types with step functions.

To sketch the graph of $g(t)$, shift $f(t)$ by $t = \frac{\pi}{8}$ and set it to zero before the shift. See Figure 3.2c.

Example 3.10

(a) Determine the equation for the function $f(t)$ using Figure 3.3a and write it in terms of combination of step functions.
(b) Determine the equation for $f(t)$ using Figure 3.3b and write it in terms of combination of step functions.

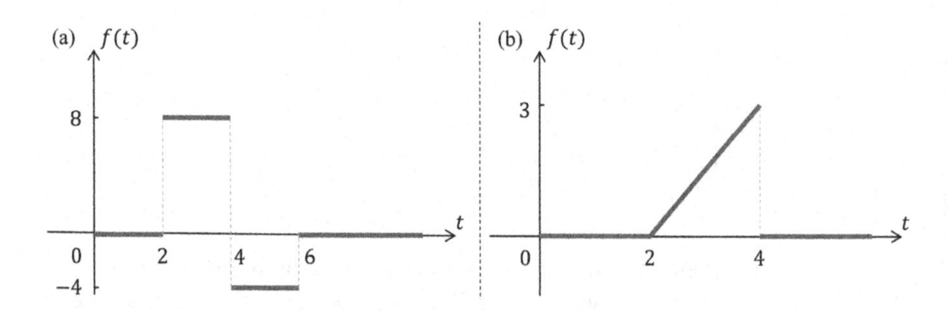

FIGURE 3.3 Combination of the step functions.

Solution

(a) There are three critical values, at which the function $f(t)$ jumps up or down. Hence, it shows a combination of three step functions whose critical values occur at $t=2, t=4$ and $t=6$, respectively. These functions are $u(t-2)$, $u(t-4)$ and $u(t-6)$. Using the graph, we need to get the coefficients of these functions before combining them.

Working from left to right, the function $f(t)$ is zero for $t<2$. At $t=2$, it jumps up 8 units and continues until $t=4$, which corresponds to $+8\times u(t-2)$ for $2<t<4$. At $t=4$, it jumps down 12 units and continues until $t=6$, which corresponds to $-12\times u(t-4)$ for $4<t<6$. At $t=6$, it jumps up again 4 units and continues, which corresponds to $+4\times u(t-6)$ for $t>6$. Hence, the function $f(t)$ is a combination of 3 step functions $+8u(t-2)$, $-12u(t-4)$ and $+4u(t-6)$. Combining these step functions, we obtain

$$f(t)=8u(t-2)-12u(t-4)+4u(t-6).$$

(b) Using two points (2, 0) and (4, 3), the equation of the straight line passing through these points can be easily found:

$$y(t)=\frac{3}{2}t-3.$$

From the Figure 3.3b, this line is truncated before $t=2$ and after $t=4$ and set to zero. So,

$$f(t)=\begin{cases}0, & t<2\\ \dfrac{3}{2}t-3, & 2<t<4\\ 0, & t>4\end{cases}$$

$$=\left(\frac{3}{2}t-3\right)\left[u(t-2)-u(t-4)\right].$$

3.8.2 Laplace Transform of the Unit Step Function

In this section, we introduce the Laplace transform of the unit step function which is easily obtained by substituting $f(t)=u(t-\alpha)$ in the Laplace transform integral and integrating it in a suitable interval as explained next. This is an important result for solving differential equations involving step functions.

$$\mathcal{L}\left[u(t-\alpha)\right]=\int_0^{\infty}u(t-\alpha)e^{-st}\,dt$$

$$= \int_0^{\alpha} 0 \times e^{-st}\,dt + \int_{\alpha}^{\infty} 1 \times e^{-st}\,dt$$

$$= 0 + \left[-\frac{e^{-st}}{s} \right]_{t=\alpha}^{t=\infty}$$

$$= \frac{1}{s} e^{-\alpha s} \quad (s > 0).$$

3.9 THE SECOND SHIFTING THEOREM

For the first shifting theorem in Section 3.5, it was the variable s that was shifted by α, but for the second shifting theorem, instead of shifting the variable s, we shift the time variable t and find the Laplace transform of the product $f(t-\alpha)u(t-\alpha)$. The graph of this function was considered in Section 3.8.1. If $\mathcal{L}\{f(t)\} = F(s)$, then

$$\mathcal{L}\left[f(t-\alpha)u(t-\alpha)\right] = F(s)e^{-\alpha s}$$

$$\mathcal{L}^{-1}\left\{F(s)e^{-\alpha s}\right\} = f(t-\alpha)u(t-\alpha).$$

We do not provide any proof of this theorem, but the results can be easily obtained from the definition of the Laplace transform. The inverse transformation described earlier is useful for solving differential equations.

Now we give some examples to clarify the second shifting theorem and its application in solving differential equations.

Example 3.11 We apply three steps to determine the Laplace Transform of the functions defined in parts (a)–(c):

(a) Determine the Laplace transform of $g(t) = (t-4)u(t-4)$.

Step 1: $(t-4)u(t-4) = f(t-\alpha)u(t-\alpha) \Rightarrow \alpha = 4$ and $f(t) = t$.

Step 2: $F(s) = \mathcal{L}\{t\} = \frac{1}{s^2}$.

Step 3: $\mathcal{L}\{(t-4)u(t-4)\} = e^{-4s} \times \frac{1}{s^2}$.

(b) Determine the Laplace transform of $g(t) = (t-4)^2\, u(t-4)$.

Step 1: $(t-4)^2 u(t-4) = f(t-\alpha)(u(t-\alpha) \Rightarrow \alpha = 4$ and $f(t) = t^2$.

Step 2: $F(s) = \mathcal{L}\{t^2\} = \frac{2}{s^3}$.

Step 3: $\mathcal{L}\{(t-4)^2 u(t-4)\} = e^{-4s} \times \frac{2}{s^3}$.

(c) Determine $\mathcal{L}\left\{t^2 u(t-5)\right\}$.

Step 1: $t^2 u(t-5) = f(t-\alpha)u(t-\alpha) \Rightarrow \alpha = 5$ and $f(t-5) = t^2 \Rightarrow$

$$f(t) = (t+5)^2 = t^2 + 10t + 25.$$

Step 2: $F(s) = \frac{2}{s^3} + \frac{10}{s^2} + \frac{25}{s}$.

Step 3: $\mathcal{L}\left\{t^2 u(t-5)\right\} = \left[\frac{2}{s^3} + \frac{10}{s^2} + \frac{25}{s}\right] e^{-5s}$.

Example 3.12

(a) Determine the inverse Laplace transform of $G(s) = \frac{1-e^{-2s}}{s^2}$.

Split the fraction into two fractions and use the table to invert the first part:

$$\mathcal{L}^{-1}\{G(s)\} = \mathcal{L}^{-1}\left\{\frac{1-e^{-2s}}{s^2}\right\} = \mathcal{L}^{-1}\left\{\frac{1}{s^2}\right\} - \mathcal{L}^{-1}\left\{\frac{e^{-2s}}{s^2}\right\}$$

$$= t - \mathcal{L}^{-1}\left\{\frac{e^{-2s}}{s^2}\right\}.$$

To find $\mathcal{L}^{-1}\left\{\frac{e^{-2s}}{s^2}\right\}$, we use the second shifting theorem:

Step 1: $e^{-\alpha s} F(s) = \frac{e^{-2s}}{s^2} \Rightarrow \alpha = 2, \; F(s) = \frac{1}{s^2}$.

Step 2: $f(t) = \mathcal{L}^{-1}\left\{\frac{1}{s^2}\right\} = t$.

Step 3: Using $\mathcal{L}^{-1}\left\{F(s)e^{-s\alpha}\right\} = f(t-\alpha)u(t-\alpha) \Rightarrow$

$$\mathcal{L}^{-1}\left\{\frac{e^{-2s}}{s^2}\right\} = (t-2)u(t-2).$$

Hence, $\mathcal{L}^{-1}\{G(s)\} = \mathcal{L}^{-1}\left\{\frac{1-e^{-2s}}{s^2}\right\} = \mathcal{L}^{-1}\left\{\frac{1}{s^2}\right\} - \mathcal{L}^{-1}\left\{\frac{e^{-2s}}{s^2}\right\}$

$$= t - (t-2)u(t-2).$$

(b) Determine the inverse Laplace transform of $\frac{1}{s(s-1)} e^{-s}$.

Step 1: $e^{-\alpha s} F(s) = \frac{e^{-s}}{s(s-1)} \Rightarrow \alpha = 1, \; F(s) = \frac{1}{s(s-1)}$.

Step 2: Applying inverse transformation to $F(s) \Rightarrow$

$$f(t) = \left(-1 + e^{t}\right) \quad (\text{table}).$$

Step 3: Using $\mathcal{L}^{-1}\left\{F(s)e^{-s\alpha}\right\} = f(t-\alpha)u(t-\alpha)$ implies that

$$\mathcal{L}^{-1}\left\{\frac{1}{s(s-1)}e^{-s}\right\} = \left\{-1 + e^{(t-1)}\right\}u(t-1).$$

Now we solve a differential equation whose right-hand side has a step function. This differential equation cannot be solved by traditional methods.

Example 3.13 Determine the solutions of the following differential equation:

$$\frac{d^2x}{dt^2} + 4x = u(t-4), \qquad x(0) = \frac{dx}{dt}(0) = 0.$$

Solution

As explained in the earlier section, we apply three steps to solve the differential equations using the Laplace transform:

Step 1: Apply the Laplace transform and substitute the initial conditions:

$$\left[s^2\bar{x} - sx(0) - \dot{x}(0)\right] + 4\bar{x} = \frac{1}{s}e^{-4s}$$

$$\left(s^2 + 4\right)\bar{x} = \frac{1}{s}e^{-4s}.$$

Step 2: Solve for $\bar{x}$:

$$\bar{x} = \frac{1}{s\left(s^2 + 4\right)}e^{-4s}.$$

Step 3: Apply the inverse transform:

$$x(t) = \mathcal{L}^{-1}\left\{\frac{1}{s\left(s^2 + 4\right)}e^{-4s}\right\}.$$

To obtain the inverse transformation of the expression on the right-hand side, we use the second shifting theorem:

- $e^{-\alpha s}F(s) = \frac{e^{-4s}}{s\left(s^2+4\right)} \Rightarrow \alpha = 4,\ F(s) = \frac{1}{s\left(s^2+4\right)} = \frac{1}{4}\left[\frac{2^2}{s\left(s^2+2^2\right)}\right].$

- $\mathcal{L}^{-1}\{F(s)\} = f(t) = \frac{1}{4}[1-\cos(2t)]$.
- $\mathcal{L}^{-1}\{F(s)e^{-s\alpha}\} = f(t-\alpha)u(t-\alpha) \Rightarrow$

$$x(t) = \frac{1}{4}[1-\cos(2t-8)]u(t-4).$$

3.9.1 Exercises

1. The following functions are defined using step functions. Draw their graphs and find their Laplace transform.

(a) $f(t) = 3u(t-4)$ (b) $f(t) = -2u(t-1)$

(c) $f(t) = 2[u(t-3) - u(t-6)]$

(d) $f(t) = u(t-5) - 3u(t-10) + 2u(t-15)$

(e) $f(t) = u(t-2)u(t-3)$ (f) $f(t) = u(t-1)u(t-4)$

(g) $g(t) = 3(t-5)u(t-5)$ (h) $g(t) = (t-3)^2 u(t-3)$

(i) $g(t) = (t+1)u(t-6)$ (j) $f(t) = t^2 u(t-4)$

(k) $g(t) = [u(t-3) - u(t-5)](t-3)^2$

(l) $g(t) = [u(t-3) - u(t-5)](t-3)^2 + 2[u(t-5) - u(t-7)](7-t)$

2. Find the equation of the graphs shown in Figure 3.4 in terms of a combination of step functions and determine their Laplace transform.
3. Using the table of Laplace transforms and the methods described in the previous sections, find the inverse Laplace transform of the following functions.

(a) $F(s) = \frac{1}{s}e^{-5s}$ (b) $F(s) = \frac{2}{s}e^{-4s}$

(c) $F(s) = \frac{3}{s^2}e^{-3s}$ (d) $F(s) = \frac{1}{s^3}e^{-2s}$

(e) $F(s) = \frac{1}{s-2}e^{-4s}$ (f) $F(s) = \frac{s}{s^2+25}e^{-2s}$

(g) $F(s) = \frac{3}{s^2+9}e^{-4s}$ (h) $F(s) = \frac{s}{s^2+6s+13}e^{-s}$

(i) $F(s) = \frac{(s-5)}{s^2-25}e^{-2s}$

4. In an electrical circuit, the behaviour of a time-dependent current denoted by $i(t)$ is described by the following differential equation:

$$\frac{di}{dt} + i = v(t),$$

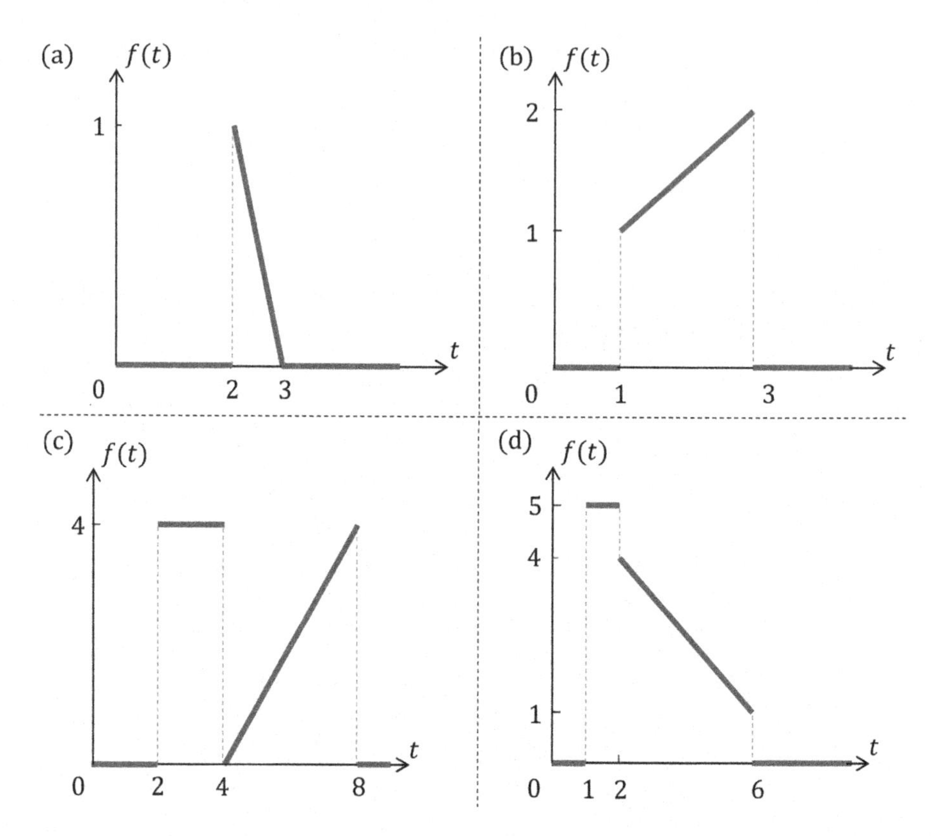

FIGURE 3.4 Graphs of the combined step functions.

where $v(t)$ is the supply voltage. Two voltages are defined as follows:

(i) $v(t)=\begin{cases}1, & t<1\\ 0, & t>1\end{cases},\qquad i(0)=1.$

(ii) $v(t)=\begin{cases}0, & t<1\\ t-1, & 1<t<2,\\ 0, & t>2\end{cases}\qquad i(0)=0.$

Answer the following questions:

(a) For each voltage, sketch its graph.
(b) Represent the voltages in terms of the unit step function.
(c) Solve the differential equation and determine the current $i(t)$.
(d) Sketch the graph of the current $i(t)$.

5. Suppose a vibrating spring with a stiffness $k=1$ N/m is connected to a member that is subject to a force that is a time-dependent function $f(t)$.

The mass $m = 1$ kg is suspended from the spring, and it is pulled from its equilibrium and released. Let the position of the mass be denoted by $x(t)$ at any instant of time and consider that the oscillation is retarded by a frictional force, $2\,\frac{dx}{dt}$, proportional to the speed of the oscillation. Assuming the initial displacement $x(0) = 1$ and the initial speed $\dot{x} = 0$, the equation governing this mechanical system can be written as follows (see Section 2.2 in Chapter 2):

$$\frac{d^2x}{dt^2} + 2\frac{dx}{dt} + x = f(t), \quad x(0) = 1 \text{ and } \frac{dx}{dt}(0) = 0.$$

The force $f(t)$ is given in the following different forms:

(i)

$$f(t) = \begin{cases} 0, & t < 1 \\ 1, & t > 1 \end{cases}.$$

(ii)

$$f(t) = \begin{cases} \text{t}, & t < 1 \\ 0, & t > 1 \end{cases}.$$

Answer the following questions:

(a) For each force, $f(t)$, sketch its graph.
(b) Represent the force $f(t)$ in terms of step functions.
(c) Solve the differential equation and determine the position, $x(t)$, of the mass.

Answers

1.

(a) $F(s) = \frac{3}{s} e^{-4s}$

(b) $F(s) = -\frac{2}{s} e^{-s}$

(c) $F(s) = \frac{2}{s}\left(e^{-3s} - e^{-6s}\right)$

(d) $F(s) = \frac{1}{s}\left(e^{-5s} - 3e^{-10s} + 2e^{-15s}\right)$

(e) $F(s) = \frac{1}{s} e^{-3s}$

(f) $F(s) = \frac{1}{s} e^{-4s}$

(g) $F(s) = \frac{3}{s^2} e^{-5s}$

(h) $F(s) = \frac{2}{s^3} e^{-3s}$

(i) $F(s) = \left(\frac{7}{s} + \frac{1}{s^2}\right) e^{-6s}$

(j) $F(s) = \left(\frac{2}{s^3} + \frac{8}{s^2} + \frac{16}{s}\right) e^{-4s}$

(k) $F(s) = \frac{2}{s^3} e^{-3s} - \left(\frac{2}{s^3} + \frac{4}{s^2} + \frac{4}{s}\right) e^{-5s}$

(l) $F(s) = \frac{2}{s^3} e^{-3s} - \left(\frac{2}{s^3} + \frac{4}{s^2} + \frac{4}{s}\right) e^{-5s} + 2e^{-5s}\left(-\frac{1}{s^2} + \frac{2}{s}\right) + 2e^{-7s}\frac{1}{s^2}$

2.

(a) $f(t)=(-t+3)\left[u(t-2)-u(t-3)\right]$,

$$F(s)=\left(-\frac{1}{s^2}+\frac{1}{s}\right)e^{-2s}+\frac{1}{s^2}e^{-3s}$$

(b) $f(t)=\frac{1}{2}(t+1)\left[u(t-1)-u(t-3)\right]$,

$$F(s)=\frac{1}{2}\left(\frac{1}{s^2}+\frac{1}{s}\right)\left[e^{-s}-e^{-3s}\right]-\frac{1}{s}e^{-3s}a$$

(c) $f(t)=4u(t-2)+(t-8)u(t-4)+(t-4)u(t-8)$,

$$F(s)=4\frac{e^{-2s}}{s}+\left(\frac{1}{s^2}-\frac{4}{s}\right)e^{-4s}+\left(\frac{1}{s^2}+\frac{4}{s}\right)e^{-8s}$$

(d) $f(t)=5u(t-1)+\left(-\frac{3}{4}t+\frac{1}{2}\right)u(t-2)-\left(-\frac{3}{4}t+\frac{11}{2}\right)u(t-6)$,

$$F(s)=\frac{5}{s}e^{-s}+\left(-\frac{3}{4s^2}-\frac{1}{s}\right)e^{-2s}-\left(-\frac{3}{4s^2}+\frac{1}{s}\right)e^{-6s}$$

3.

(a) $f(t)=u(t-5)$

(b) $f(t)=2u(t-4)$

(c) $f(t)=3(t-3)u(t-3)$

(d) $f(t)=\frac{1}{2}(t-2)^2u(t-2)$

(e) $f(t)=e^{2(t-4)}u(t-4)$

(f) $f(t)=\cos\left[5(t-2)\right]u(t-2)$

(g) $f(t)=\sin\left[3(t-4)\right]u(t-4)$

(h) $f(t)=e^{-3(t-1)}\cos\left[2(t-1)\right]-\frac{3}{2}\sin\left[2(t-1)\right]u(t-1)$

(i) $f(t)=e^{-5(t-2)}u(t-2)$

4. For the voltage defined in part (i):

(a) See Figure 3.5a.

(b) $v(t)=u(t)-u(t-1)$

(c) $i(t)=1-\left(1-e^{-t+1}\right)u(t-1)=\begin{cases}1, & t<1\\ e^{-t+1}, & t>1\end{cases}$

(d) Sketch the graph of $e \times e^{-t}$, truncate it for $t<1$ and set it to $v(t)=1$. See Figure 3.5d.

For the voltage defined in part (ii):

(a) See Figure 3.5b.

(b) $v(t)=(t-1)\left[u(t-1)-u(t-2)\right]$

(c) $i(t)=\left(t-2+e^{-t+1}\right)u(t-1)-(t-2)u(t-2)$

(d) For drawing $i(t)$ in part (c), by eliminating the step functions we write it as follows and sketch the graph (see Figure 3.5e):

$$i(t)=\left(t-2+e^{-t+1}\right)u(t-1)-(t-2)u(t-2)$$

$$=\begin{cases} 0, & t<1 \\ t-2+e^{-t+1}, & 1<t<2 \\ e^{-t+1}, & t>2 \end{cases}$$

5. For the force defined in part (i):

(a) It is the step function with the critical point at $t=1$.

(b) $f(t)=u(t-1)$

(c)

$$x(t)=e^{-t}+te^{-t}+\left[1-te^{-t+1}\right]u(t-1)$$

$$=\begin{cases} e^{-t}+te^{-t}, & t<1 \\ e^{-t}(1+t-et)+1, & t>1 \end{cases}$$

For the force defined in part (ii):

(a) See Figure 3.5c.

(b) $f(t)=t\left[u(t)-u(t-1)\right]$

(c)

$$x(t)=e^{-t}(2t+3)+(t-2)\left[1-u(t-1)\right]-e^{-t+1}u(t-1)$$

$$=\begin{cases} e^{-t}(2t+3)+t-2, & t<1 \\ e^{-t}(2t+3)-e^{-t+1}, & t>1 \end{cases}$$

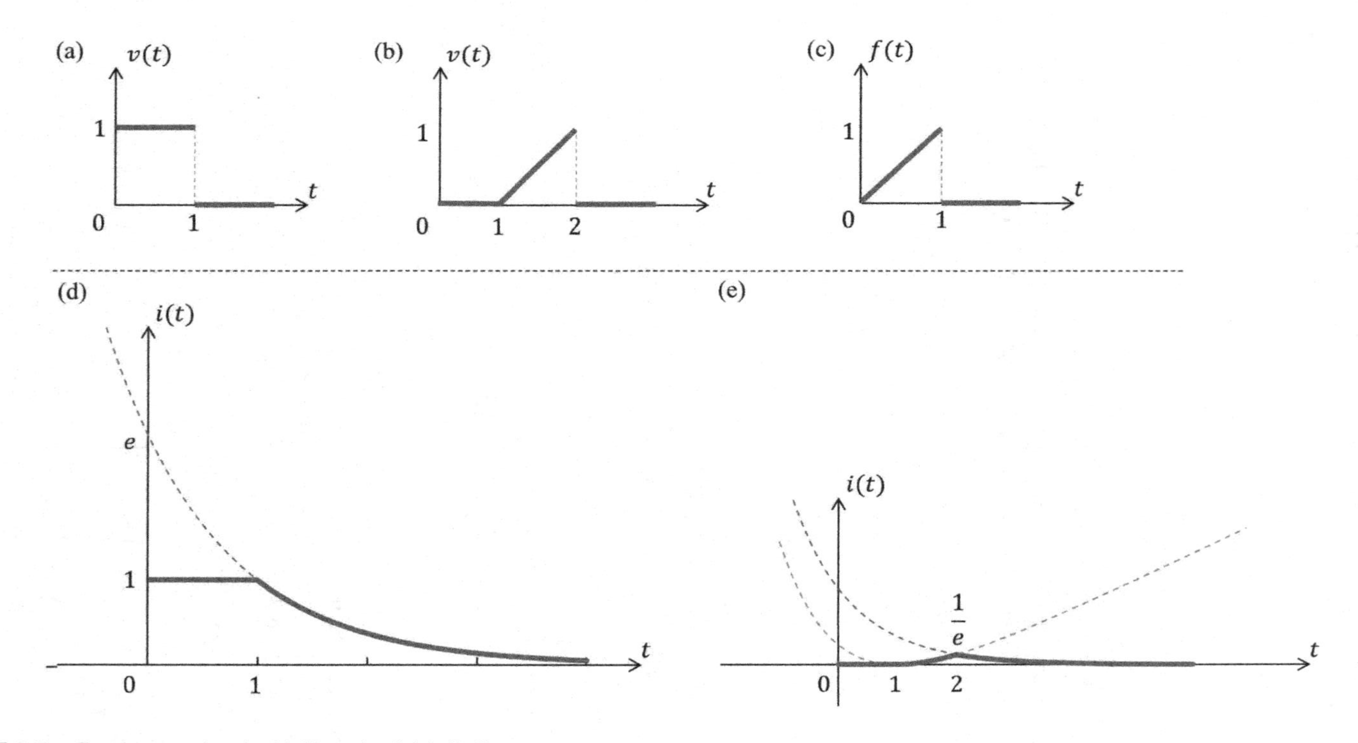

FIGURE 3.5 Graphs associated with Exercise 3.9.1 (4–5).

3.10 DIRAC DELTA FUNCTION

The Dirac function (also called the impulse function) is often used in engineering or physics problems when an event occurs over a small time interval. For example, a strong force is applied and then removed, or a voltage is turned on and then immediately turned off. These types of events are compressed into a single point for mathematical implementation. The mathematical definition of this function follows.

DEFINITION: DIRAC DELTA FUNCTION

Dirac delta function is denoted by the Greek letter δ and is defined as

$$\delta(t-\alpha)=0 \qquad \text{for } t \neq \alpha.$$

It is a function which is zero everywhere except at point $t=\alpha$. The weight of this function is concentrated in one point.

To understand this function let us define a function $D_\epsilon(t-\alpha)$ something like the following:

$$D_\epsilon(t-\alpha)=\begin{cases} 0, & t<\alpha-\epsilon \\ \dfrac{1}{2\epsilon}, & \alpha-\epsilon \leq t \leq \alpha+\epsilon \\ 0, & t>\alpha+\epsilon \end{cases}.$$

The graph of this function is given in Figure 3.6a. The area under the rectangular part is 1 unit^2. Now if we bring the ε closer to zero, then peak will get longer ($\frac{1}{2\epsilon}$ grows) as shown in Figure 3.6b; however, the area remains one unit^2.

If $\epsilon \to 0$, the interval around α (the width of the rectangle) shrinks and the height of the rectangle increases, but the area of the rectangle remains 1 unit^2. In the limit, the height of the rectangle becomes very large like a vertical line whose one end is

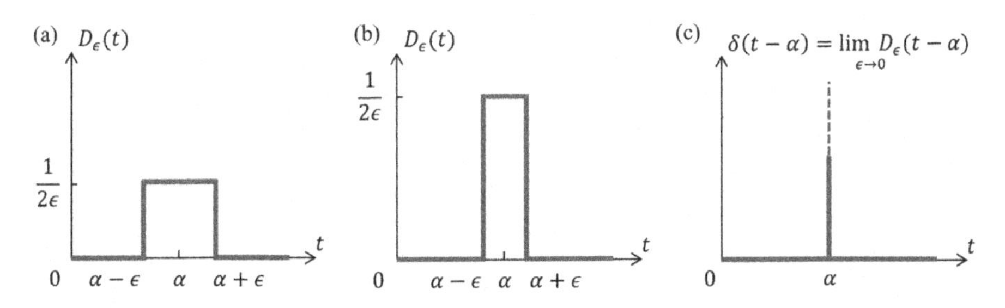

FIGURE 3.6 Graph of $D_\epsilon(t-\alpha)$ is a Dirac delta function when $\epsilon \to 0$.

on the t-axis, and the other end is infinite. This "infinite line" at $t=\alpha$ is called the **Dirac delta function**, which is zero everywhere but $t=\alpha$, and as stated earlier, it is denoted by $\delta(t-\alpha)$; that is $\lim_{\epsilon\to 0} D_\epsilon(t-\alpha)=\delta(t-\alpha)$. The graph of the Dirac function is shown in Figure 3.6c.

The Dirac function satisfies the following conditions:

$$\int_{-\infty}^{\infty}\delta(t-\alpha)dt=1$$

$$\int_{-\infty}^{\infty}f(t)\delta(t-\alpha)dt=f(\alpha) \tag{3.1}$$

$$\frac{d}{dt}u(t-\alpha)=\delta(t-\alpha). \tag{3.2}$$

Using the Condition 3.1 earlier, we can obtain the Laplace transform of the Dirac function:

$$\mathcal{L}\{\delta(t-\alpha)\}=\int_0^{\infty}\delta(t-\alpha)e^{-st}dt=e^{-\alpha s}.$$

Note: In the context of the differential equation

$$a\ddot{x}+b\dot{x}+cx=f(t),$$

the function $f(t)$ is said to be input and $x(t)$ is said to be the output of the system. If we replace $f(t)$ with an impulse function $\delta(t)$, the solution $x(t)$ is called the impulse response of that system. The impulse response can be used to find the system response for any arbitrary input $f(t)$ using the convolution integral described in Section 3.11.

Now we consider two examples in which we use the Dirac function and the Laplace transform.

Example 3.14 In this example, we consider three differential equations including the Dirac function on their right-hand sides. These equations cannot be solved using traditional methods, highlighting the importance of the Laplace transform.

(a) Find the solution of the following differential equation:

$$\frac{dx}{dt}+2x=\delta(t), \qquad x(0)=0.$$

(b) Assume we have a system that consists of a mass of m (kg) attached to one end of a spring with the spring constant k (N/m) shown, as shown in

Figure 3.7b. If the mass m is subjected to the impulse force $\delta(t)$ at $t=0$, the governing equation of the system can be written as follows:

$$m\frac{d^2x}{dt^2}+kx=\delta(t), \qquad x(0)=\frac{dx}{dt}(0)=0,$$

where $x=x(t)$ is the position of mass at each instant of time and the system is initially at rest. Evaluate $x(t)$.

(c) In an electric circuit, the differential equation governing the time-dependent current, $i(t)$, is expressed as

$$\frac{d^2i}{dt^2}+4i=\frac{dv}{dt}, \qquad i(0)=\frac{di}{dt}(0)=0,$$

where $v(t)$ is the supply voltage and $v(t)=120\left[u(t-3)\right]$. Using the Laplace transform find the solution, $i(t)$, of the equation.

Solution

(a)

Step 1: Apply the Laplace transform, use the initial condition and simplify:

$$\left[s\overline{x}-x(0)\right]+2\overline{x}=1$$
$$(s\overline{x}-0)+2\overline{x}=1.$$

Step 2: Solve for $\overline{x}$:

$$\overline{x}=\frac{1}{(s+2)}.$$

Step 3: Use the inverse transformation to find the solution:

$$\frac{1}{(s+2)} \underset{\rightarrow}{\text{Invert}} \quad e^{-2t}.$$

Hence,

$$x(t)=e^{-2t}.$$

The problem arises when we want to check the initial condition $x(0)=0$, we observe $x(0)=e^{-2\times 0}=1$. The physical justification is that the inclusion of Dirac functions in the differential equations that governs the behaviour of systems is associated with the application of a shock on the systems, which caused a jump of discontinuity on the initial condition of the present equation.

The mathematical justification is as follows: The behaviour of the system associated with the earlier differential equation is not really defined by $x(t)=e^{-2t}$, since $x(-1)=e^{2}$ contradicts the resting behaviour of the system prior to $t=0$. Therefore, we modify our solution as follows:

$$x(t)=\begin{cases}0, & t<0\\ e^{-2t}, & t>0\end{cases}$$
$$=e^{-2t}u(t).$$

The solution is plotted in Figure 3.7a, which shows that the system is at rest for $t<0$ and that there is a jump of discontinuity at $t=0$, due to the impulse $\delta(t)$. The derivative of the modified function is undefined at $t=0$, and the problem of our solution still remains; however $x(0^-)=0$ and $x(0^+)=e^{-2\times 0}=1$. Therefore, we also modify the differential equation as follows:

$$\frac{dx}{dt}+2x=\delta(t), \quad x(0^-)=0.$$

(b)

Step 1: Apply the Laplace transform, use the initial condition and simplify:

$$m\left[s^2\bar{x}-sx(0)-\bar{x}(0)\right]+k\bar{x}=1$$
$$m\left(s^2\bar{x}-s\times 0-0\right)+k\bar{x}=1.$$

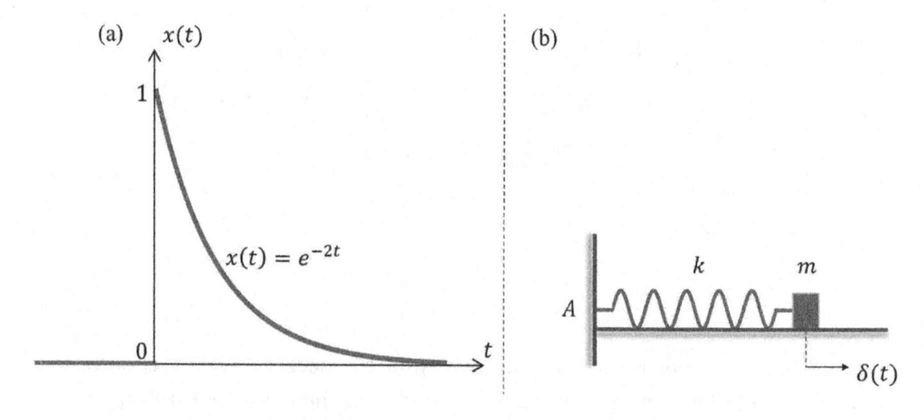

FIGURE 3.7 (a) The answer to the differential equation involving the Dirac delta function. The jump of discontinuity is due the shock at $t=0$. (b) A spring-mass diagram.

Step 2: Solve for $\overline{x}$:

$$\overline{x} = \frac{1}{ms^2+k} = \frac{1}{m}\frac{1}{s^2+\frac{k}{m}} = \frac{1}{m\omega}\frac{\omega}{s^2+\omega^2}, \qquad \left[\omega^2 = \frac{k}{m}\right]$$

$$= \frac{1}{\sqrt{mk}}\frac{\omega}{s^2+\omega^2}.$$

Step 3: Use the inverse transformation to find the solution:

$$\frac{\omega}{s^2+\omega^2} \underset{\rightarrow}{\text{Invert}} \sin(\omega t).$$

Hence,

$$x(t) = \frac{1}{\sqrt{mk}}\sin(\omega t) = \frac{1}{\sqrt{mk}}\sin\left(\sqrt{\frac{k}{m}}t\right).$$

The initial displacement is $x(0) = \frac{1}{\sqrt{mk}}\sin\left(\sqrt{\frac{k}{m}}\times 0\right) = 0$; however, $\dot{x}(0) = \frac{1}{m}$, and this contradicts our assumption $\dot{x}(0) = 0$. The physical justification is that the impulse force causes a discontinuity in the speed of oscillation. It means the speed of the mass is zero before the impulse and is at rest, but after the impulse force is applied, there is a sudden jump of $\frac{1}{m}$ in the speed of the mass.

The mathematical justifications follow the explanation given in part (a). Therefore, we need to modify the equation as

$$m\frac{d^2x}{dt^2} + kx = \delta(t), \qquad x(0^-) = \frac{dx}{dt}(0^-) = 0$$

and the solution as

$$x(t) = \begin{cases} 0, & t < 0 \\ \frac{1}{\sqrt{mk}}\sin\left(\sqrt{\frac{k}{m}}t\right), & t > 0 \end{cases}$$

$$= \frac{1}{\sqrt{mk}}\sin\left(\sqrt{\frac{k}{m}}t\right)u(t).$$

The introduction of the step function in our solution is to emphasise that the behaviour of the system changes after $t = 0$. The system is at rest for $t < 0$, but when an impulse is applied, it is no longer at rest.

The purpose of this section is to solve differential equations involving Dirac functions and to face mathematical problems like the two equations above, where we introduced the step function in our final solution. In the next example, we choose an impulse at a different time scale to avoid a mathematical problem at initial values.

(c)

Using Equation 3.2, the right side of the equation can be written in terms of Dirac functions:

$$\frac{dv}{dt} = 120\left[\delta(t-3)\right].$$

This leads to the following differential equation:

$$\frac{d^2i}{dt^2} + 4i = 120\left[\delta(t-3)\right].$$

Applying the Laplace transform and following the process explained in Section 3.7.1, we get the following result for the Laplace transform of the current:

$$\overline{i} = \frac{120}{s^2+4}\left[e^{-3s}\right].$$

The required current $i(t)$ is obtained by inverting the previous expression using the second shifting theorem:

$$i(t) = 60\sin\left[2(t-3)\right]u(t-3).$$

We now check the initial conditions, and we observe $i(0)=0$ since the step function is zero for $t<3$. To check the second condition $\frac{di}{dt}(0)=0$, we should first differentiate $i(t)$ using the product rule.

$$i'(t) = 120\cos\left[2(t-3)\right]u(t-3) + 60\sin\left[2(t-3)\right]\delta(t-3) \Rightarrow$$
$$i'(0) = 0.$$

Note that the impulse is applied at $t=3$, and hence, there will be no change in the rate of change of current for $t<3$.

3.10.1 Exercises

1. Find the value of the following integrals:

(a) $\int_0^3 \delta(t-1)\,dt$ (b) $\int_0^2 \delta(t-4)\,dt$ (c) $\int_2^4 \delta(t-1)\,dt$

(d) $\int_{1}^{5} t^2\delta(t-2)dt$ (e) $\int_{2}^{5} e^{-t}\delta(t-3)dt$ (f) $\int_{0}^{2} te^{t}\delta(t-3)dt$

2. Write down the Laplace transform of the following Dirac functions:

(a) $4\delta(t-3)$ (b) $3\delta(t-1)$ (c) $\delta(t)$

3. Using the second shifting theorem and the Laplace transform Dirac function, invert the following functions:

(a) $\frac{s^2+1}{s^2}$ (b) $\frac{s^2-1}{s^2}e^{-s}$ (c) $\frac{s+2}{s+1}e^{-s}$ (d) $\frac{s^2+2s}{s^2+4}e^{-2s}$

4. Solve the following differential equations:

(a) $\frac{dx}{dt}+3x=\delta(t-1),\quad x(0)=0$

$\frac{d^2x}{dt^2}+3\frac{dx}{dt}+2x=\delta(t),\quad x(0)=0 \text{ and } \frac{dx}{dt}(0)=0$

Comment on the initial values of the solution.

(b) $\frac{d^2x}{dt}+x=A\delta\left(t-\frac{\pi}{2}\right),\quad x(0)=1 \text{ and } \frac{dx}{dt}(0)=0$

Explain the solution for $A>1, A<1$ and $A=1$.

(c) $\frac{d^2x}{dt^2}+x=3\delta(t-1)+2\delta(t-4),\quad x(0)=0 \text{ and } \frac{dx}{dt}(0)=0$

What is the effect of the second impulse on the solution?

5. In an electrical circuit, the behaviour of current, $i(t)$, is described by the following differential equation:

$$\frac{di}{dt}+i=\frac{dv}{dt},\quad i(0)=0$$

The function $v(t)$ is the supply voltage. Two voltages are defined as follows:

(a) $v(t)=\begin{cases}0, & t<2\\ 40, & t>2\end{cases}$

(b) $v(t)=\begin{cases}0, & t<1\\ 10, & 1<t<2\\ 0, & t>2\end{cases}$

Answer the following questions:

(i) For each voltage, find its derivative.
(ii) Find the current $i(t)$.
(iii) Sketch the graph of the current $i(t)$.

Answers

1.

(a) 1 (b) 0 (c) 0
(d) 4 (e) e^{-3} (f) 0

2. (a) $4e^{-3s}$ (b) $3e^{-s}$ (c) 1

3.

(a) $\delta(t)+t$
(b) $\delta(t-1)-(t-1)u(t-1)$
(c) $\delta(t-1)+e^{-t+1}u(t-1)$
(d) $\delta(t-2)+2[\cos(2t-4)-\sin(2t-4)]u(t-2)$

4.

(a) $x(t)=e^{-3(t-1)}u(t-1)$
(b) $x(t)=e^{-t}\left(1-e^{-t}\right)$

$x(0)=0$ but $\dot{x}(0)=1$, which contradicts the initial speed $\dot{x}(0)=0$. This means that applying the impulse function at time $t=0$ causes a jump of 1 on the speed.

(c) $x(t)=A\sin\left(t-\frac{\pi}{2}\right)u\left(t-\frac{\pi}{2}\right)+\cos(t)$

For $A>1$ (say, $A=3$) the impulse reverses the direction of oscillation from $x(t)=\cos(t)$ to $x(t)=-2\cos(t)$. For $A<1$ the oscillation will continue as $x(t)=\cos(t)$ but with a different amplitude, and for $A=1$, the impulse will not affect the output of the system.

(d) $x(t)=3\sin(t-1)u(t-1)+2\sin(t-4)u(t-4)$

Due to the shift of the "sine" functions at $t=1$ and $t=4$, the functions $x_1=3\sin(t-1)$ and $x_2=2\sin(t-4)$ has opposite signs for $t>4$, which causes the second impulse to slow down the oscillations.

5. (a) (i) $v'(t)=40\delta(t-2)$

(ii) $i(t)=40e^{-(t-2)}u(t-2)$

(iii) See Figure 3.8a.

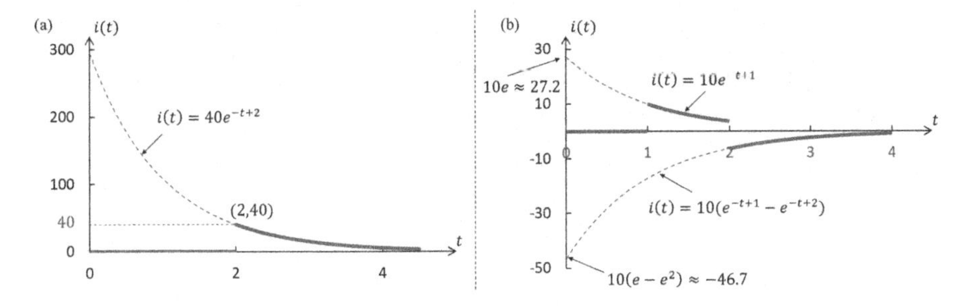

FIGURE 3.8 Current for different voltages.

(b) (i) $v'(t) = 10\left[\delta(t-1) - \delta(t-2)\right]$

(ii) $i(t) = 10\left[e^{-(t-1)}u(t-1) - e^{-(t-2)}u(t-2)\right]$

$$= \begin{cases} 0, & 0 \le t < 1 \\ 10e^{-t+1}, & 1 < t < 2 \\ 10e^{-t}\left(e - e^2\right), & t > 2 \end{cases}$$

(iii) See Figure 3.8b.

3.11 CONVOLUTION

Convolution is used in many engineering fields such as signal processing, digital communication, control systems and so on. For example, in linear time-invariant systems if the input is $f(t)$ and the impulse response is $g(t)$, then the output $h(t)$ of the system can be expressed as the convolution of the input $f(t)$ with the impulse response $g(t)$. Following is the mathematical definition of the convolution of two functions.

DEFINITION: CONVOLUTION OF TWO FUNCTIONS

The convolution of two functions $f(t)$ and $g(t)$ can be described with an integral that expresses the overlap of these two functions. Strictly speaking, this operation is denoted by $f(t)*g(t)$ and the result is a function of time, t, defined by the following integrals:

$$(f*g)(t) = \int_{-\infty}^{+\infty} f(t-z)g(z)\,dz \tag{3.3}$$

or

$$(g*f)(t) = \int_{-\infty}^{+\infty} f(z)g(t-z)\,dz. \tag{3.4}$$

It can be shown that these two integrals are equal. When functions $f(t)$ and $g(t)$ are defined between 0 and $+\infty$ and zero for $t<0$, then the limits of the integrals can be restricted to the range 0 and t, which leads to

$$(f*g)(t)=\int_0^t f(t-z)g(z)dz \tag{3.5}$$

or

$$(g*f)(t)=\int_0^t f(z)g(t-z)dz. \tag{3.6}$$

3.11.1 Visual Explanation

The advantage of using a visual explanation for concepts like convolution integral lies in its ability to make complex integration more accessible and understandable. Convolution integral can be challenging to understand purely through equations. Visualisation breaks down the integration operation into simpler systematic processes and helps with understanding the process of combining two functions.

Let us consider two arbitrary functions $f(t)$ and $g(t)$ and divide the integration described in Equation 3.3 into the following intuitive steps:

Step 1: Sketch the graphs of both functions and express them in terms of the variable z; that is $f(t)\to f(z)$ and $g(t)\to g(z)$. See Figure 3.9 (step 1). The shape of the functions we have chosen is only for simple visualisation of the process.

Step 2: Reflect one of the functions about the vertical axis. Say, for calculating $f*g=\int_{-\infty}^{+\infty} f(t-z)g(z)dz$, reflect $f(z)$; that is $f(z)\to f(-z)$. see Figure 3.9 (step 2).

Step 3: Shift $f(-z)$ by t units to the right i.e. $f(-z)\to f(t-z)$, see Figure 3.9 (step 3). In this figure, the vertical dotted line is not vertical axis, and it represents the position of the shift.

Step 4: Start moving the variable t from $-\infty$ to the right, and whenever two functions overlap, integrate the product $f(t-z)g(z)$; see Figure 3.9 (step 4).

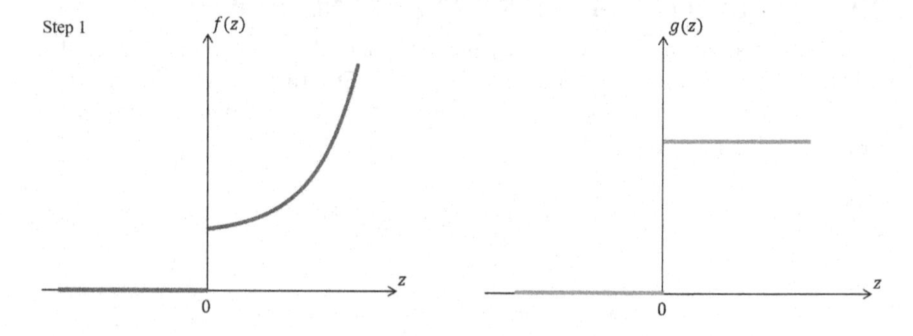

FIGURE 3.9 Visual representation of the convolution integral.

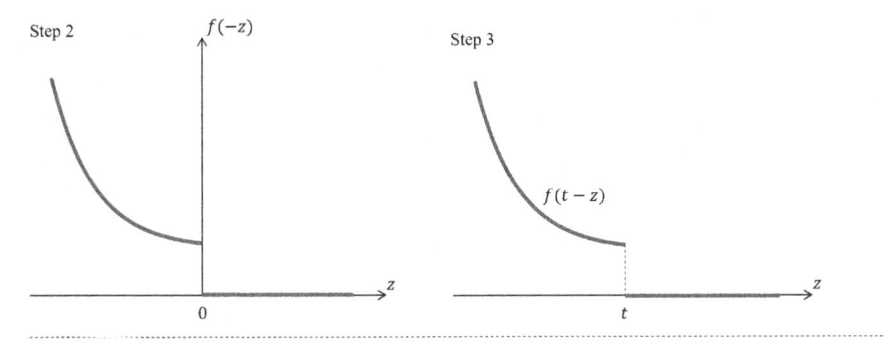

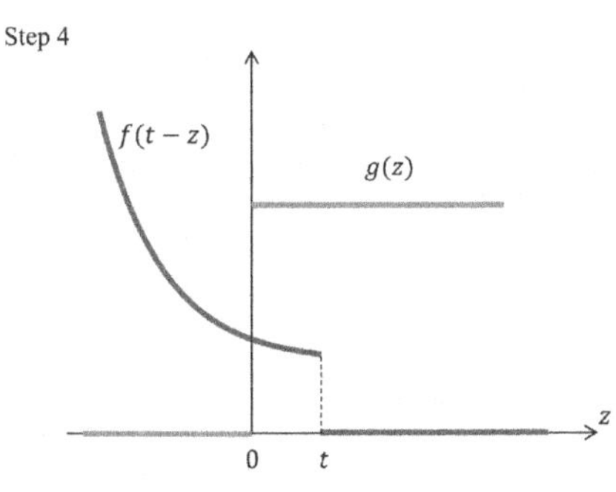

FIGURE 3.9 *(Continued)*

Example 3.15 Evaluate the convolution of the functions $f(t) = e^{-3t}u(t)$ and $g(t) = u(t-1)$, using Equation 3.3.

Solution

Step 1: Sketch the graph of both functions and express in terms of the variable z; see Figure 3.10 (step 1).

Step 2: For calculating $f * g = \int_{-\infty}^{+\infty} f(t-z)g(z)dz$, reflect the function $f(t)$ about the vertical axis; that is $f(z) \to f(-z)$; see Figure 3.10 (step 2).

Step 3: Shift $f(-z)$ by amount of t; that is $f(-z) \to f(t-z)$; see Figure 3.10 (step 3).

Note that the line joining the point t to 1 in Figure 3.10 (step 3) is not the vertical axis and it is only for the representation of the shift.

Step 4: Start t at $-\infty$ and slide it to the right, whenever the two functions overlap integrate the product $f(t-z) \times g(z)$. In this step, for this example, we consider two cases.

Case 1: For $t < 1$ as shown in Figure 3.11 (case 1), $g(z) = 0$, and thus, $f(t-z) \times g(z) = 0$, which implies $f * g = 0$.

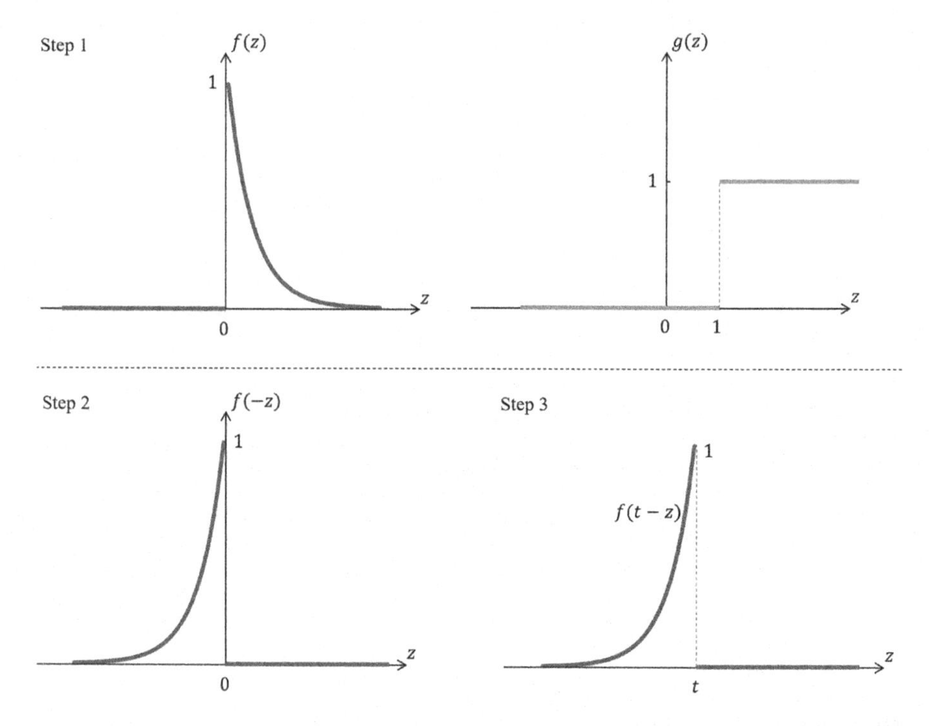

FIGURE 3.10 Visual representation of the convolution integral.

Case 2: We move t to the right until $t > 1$. For which the functions $f(t-z)$ and $g(z)$ overlap at non-zero values $f(t-z) = e^{-3(t-z)}$ and $g(z) = 1$ as shown in Figure 3.11 (case 2). Note that the variable z varies between $-\infty$ and $+\infty$, but $f(t-z) \times g(z)$ is non-zero, only for $1 < z < t$. Hence, for $t > 1$, the convolution integral is

$$(f*g)(t) = \int_1^t f(t-z)g(z)dz$$

$$= \int_1^t e^{-3(t-z)} \times 1dz$$

$$= \frac{1}{3}\left(1 - e^{-3t+3}\right).$$

Collecting the results obtained in cases 1 and 2, we write

$$(f*g)(t) = \begin{cases} 0, & t < 1 \\ \frac{1}{3}\left(1 - e^{-3t+3}\right). & t > 1 \end{cases}.$$

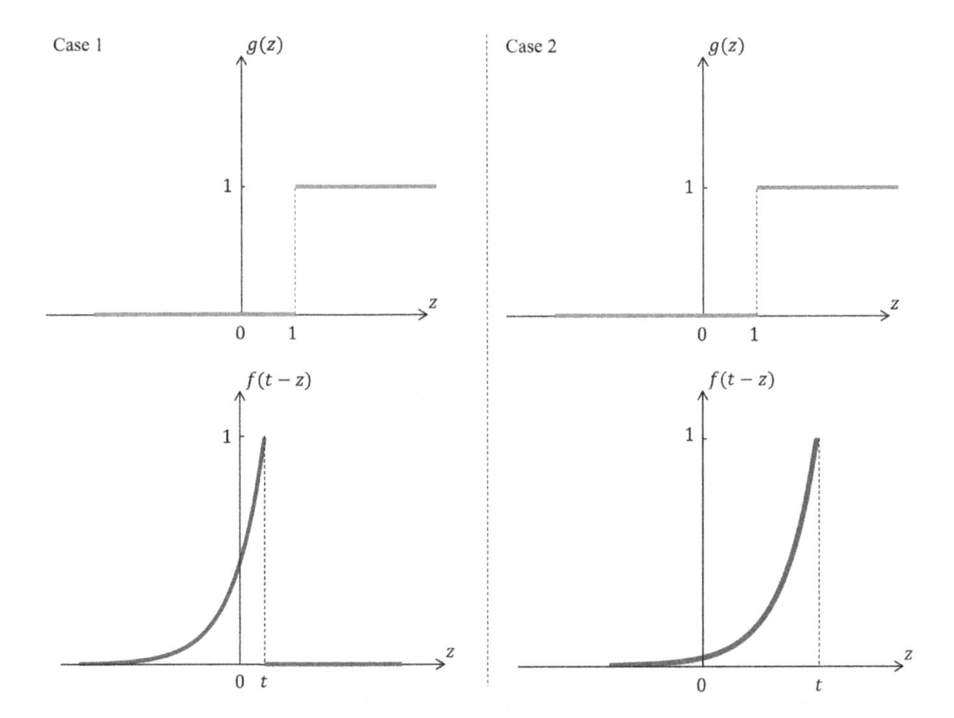

FIGURE 3.11 Visual representation of case 1 and case 2 in step 4.

Example 3.16 Evaluate the convolution of the functions $f(t)=e^{-3t}u(t)$ and $g(t)=u(t-1)$, which are also defined in the previous Example 3.15, using Equation 3.4.

Solution

Step 1: Sketch the graph of both functions and express them in terms of the variable z. See Figure 3.12 (step 1).

Step 2: For calculating $g*f=\int_{-\infty}^{+\infty} f(z)g(t-z)dz$, reflect $g(z)$ about the vertical axis; that is $g(z)\to g(-z)$. See Figure 3.12 (step 2).

Step 3: Shift $g(-z)$ by t units to the right or left; that is $g(-z)\to g(t-z)$. See Figure 3.12 (step 3).

Note that the line joining the point $t-1$ to 1 in Figure 3.12 (step 3) is not the vertical axis and that it is only for the illustration of the shift.

Step 4: We move t to the right, whenever the two functions overlap, we integrate the product $g(t-z)\times f(z)$ in an appropriate region. At this stage, we consider two cases for this example.

Case 1: For $t-1<0$ as shown in Figure 3.13 (case 1), $f(z)=0$, and thus, $g(t-z)f(z)=0$, which implies $f*g=0$.

Case 2: Moving t to the right until $t-1>0$ in which the functions $f(t-z)$ and $g(z)$ overlap at non-zero values $g(t-z)=1$

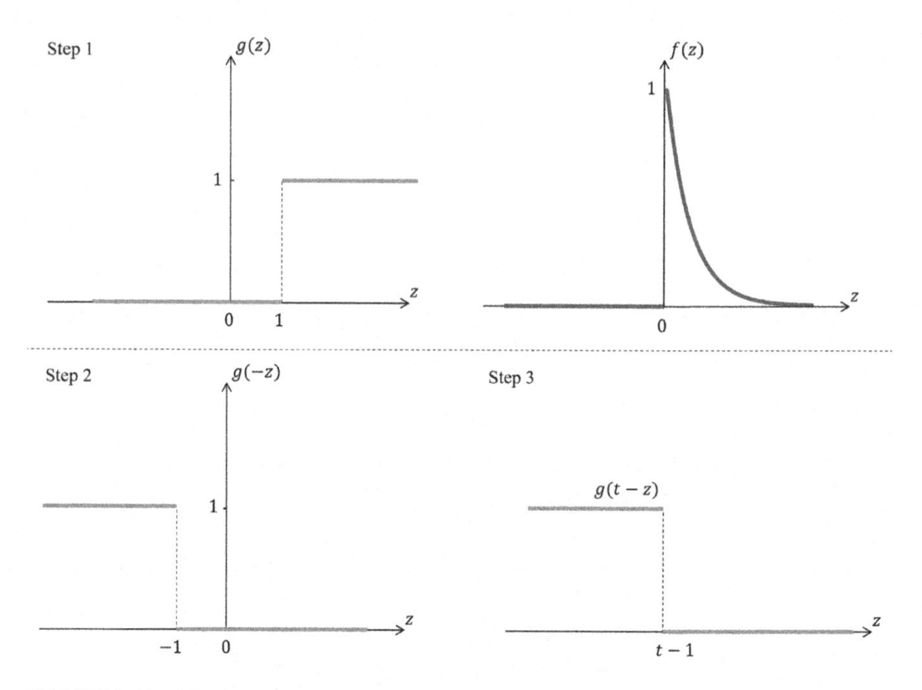

FIGURE 3.12 Visual representation of the convolution integral.

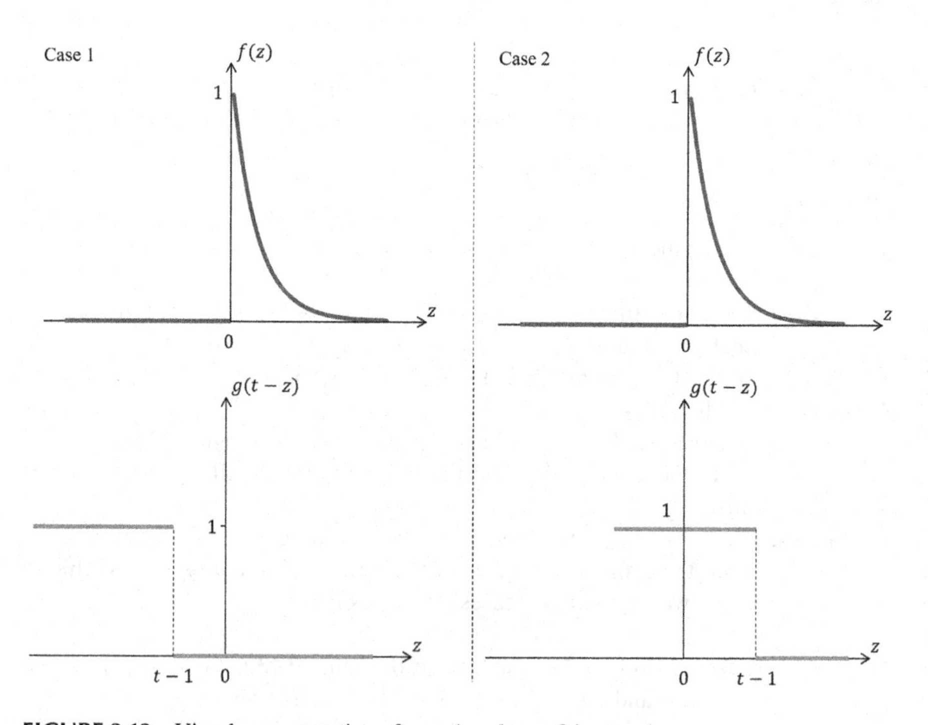

FIGURE 3.13 Visual representation of case 1 and case 2 in step 4.

and $f(z)=e^{-3t}$ as shown in Figure 3.13 (case 2). Note the variable z varies between $-\infty$ and $+\infty$, but $f(z)\times g(t-z)$ is non-zero for $0<z<t-1$. Hence, for $t>1$ the convolution integral is

$$(f*g)(t)=\int_0^{t-1} f(z)g(t-z)dz$$

$$=\int_0^{t-1} e^{-3z}\times 1\,dz$$

$$=\frac{1}{3}\left(1-e^{-3t+3}\right).$$

Collecting the results obtained in cases 1 and 2, we write

$$(g*f)(t)=\begin{cases}0, & t<1\\ \frac{1}{3}\left(1-e^{-3t+3}\right), & t>1\end{cases}.$$

Example 3.17 Evaluate the convolution of the functions $f(t)=1$ and $g(t)=te^{-4t}$, assuming $t>0$.

Solution

Since both functions are defined for $t>0$, we simply use Equation 3.5 to calculate the convolution *f* and *g*:

$$f(t-z)=1, \qquad g(z)=ze^{-4z}.$$

Hence,

$$f*g=\int_0^t f(t-z)g(z)dz=\int_0^t 1\times ze^{-4z}dz.$$

This can be integrated using the method of integration by parts. The integration details have been omitted, and we only present the result:

$$f*g=-\frac{1}{4}te^{-4t}-\frac{1}{16}e^{-4t}+\frac{1}{16}.$$

Example 3.18 In this example, we first describe the convolution of an arbitrary function with the Dirac function and then we express it for an exponential function.

(a) Determine the convolution of an arbitrary function $f(t)$ with the Dirac function $g(t)=\delta(t-\alpha)$.

(b) Determine the convolution of $f(t)=e^{-2t}$ with the Dirac function $g(t)=\delta(t-3)$.

Solution

(a) In part (a), the function $f(t)$ is any arbitrary function, and we use Property 3.1 of the Dirac function.

$$(f*g)(t)=\int_{-\infty}^{+\infty} f(t-z)g(z)dz$$

$$=\int_{-\infty}^{+\infty} f(t-z)\delta(z-\alpha)dz$$

$$=f(t-\alpha).$$

This result means that if an arbitrary function *f*(*t*) is convoluted with a Dirac function $\delta(t-\alpha)$, it causes the function *f*(*t*) to shift by *a* units; that is $f(t)\to f(t-\alpha)$.

(b) Using the result in part (a):

$$(f*g)(t)=e^{-2(t-3)}.$$

3.11.2 Some Useful Property of Convolution

Commutative property

$$f*g=g*f.$$

Proof follows the change of variable as described next:

$$(f*g)(t)=\int_{-\infty}^{+\infty} f(t-z)g(z)dz.$$

Assume $u=t-z\Rightarrow dz=-du$. Hence,

$$f*g=\int_{-\infty}^{+\infty} f(t-z)g(z)dz$$

$$=\int_{+\infty}^{-\infty} g(t-u)f(u)(-du)$$

$$=\int_{-\infty}^{+\infty} g(t-u)f(u)du=g*f.$$

Distributive property

$$f*(g+h)=f*g+f*h.$$

This property can be proved using the linearity of integration operation and it is very useful in the modelling and analysis of linear time-invariant systems.

3.11.3 Laplace Transform of $f * g$

If $F(s)$ and $G(s)$ are the Laplace transform of $f(t)$ and $g(t)$, respectively, then

$$\mathcal{L}\{f*g\} = F(s)\times G(s)$$

$$\mathcal{L}^{-1}\{F(s)\times G(s)\} = f*g.$$

For example: Invert the Laplace transform

$$\frac{1}{s^2(s^2-1)^2}.$$

The Laplace transform of this can be found using the partial fraction decomposition, but we use convolution integral.

Assume $F(s)=\frac{1}{s^2}$ and $G(s)=\frac{1}{(s-1)^2}$ with their inverses of $f(t)=t$ and $g(t)=te^t$ (using the table of the Laplace transform).

$$\frac{1}{s^2}\frac{1}{(s-1)^2} = F(s)G(s).$$

Hence,

$$\mathcal{L}^{-1}\left\{\frac{1}{s^2}\frac{1}{(s-1)^2}\right\} = \mathcal{L}^{-1}\{F(s)\times G(s)\} = (f*g)(t) = \int_0^t (t-z)\times ze^z dz.$$

This integral can be calculated using the method of integration by parts, but we omit the details and present the result:

$$\mathcal{L}^{-1}\left\{\frac{1}{s^2}\frac{1}{(s-1)^2}\right\} = 2+t+e^t(t-2).$$

3.11.4 Exercises

1. Assume $f(t)=u(t)$ and $(t)=2u(t-1)$, where u is the step function. Evaluate $f*g=\int_{-\infty}^{+\infty} f(z)g(t-z)dz$ and sketch its graph.
2. Assume

$$f(t)=2e^{-t}u(t) \quad \text{and} \quad g(t)=\begin{cases}0 & t<0\\ t & 0\le t<1.\\ 1 & t\ge 1\end{cases}$$

Evaluate $f*g$.

3. Assume $f(t)=2e^{-t}u(t)$ and $g(t)=\delta(t-4)$ is a shifted Dirac function. Evaluate $f*g$ and sketch its graph.
4. Assume

$$f(t)=\begin{cases} t+1, & -1<t<0 \\ -t+1, & 0<t<1 \end{cases}$$

$$g(t)=\delta(t-2)+\frac{1}{2}\delta(t-4).$$

(a) Draw the graph of each of the functions f and g.
(b) Evaluate $f*g$ and draw its graph.

5. Assume

$$f(t)=\begin{cases} t+1, -1<t<0 \\ -t+1, 0<t<1 \end{cases}$$

$$g(t)=\delta(t-2)+\delta(t-3).$$

(a) Draw the graph of each of the functions f and g.
(b) Evaluate $f*g$ and draw its graph.

6. Assume $f(t)=u(t)$ and $g(t)=2[u(t-2)-u(t-4)]$, where u is the unit step function.

(a) Evaluate $f*g$ and present it using step function.
(b) Draw the graph of $f*g$.

7. Assume $f(t)=2[u(t)-u(t-2)]$ and $g(t)=u(t)-u(t-4)$, where u is the unit step function. Evaluate $f*g$ and draw its graph.

Answers

1. $(f*g)(t)=\begin{cases} 0, & t<1 \\ 2t-2, & t\geq 1 \end{cases} = 2(t-1)u(t-1).$

Draw $y=2t-2$, truncate it for $t<1$ and set it to zero.

2. $(f*g)(t)=\begin{cases} 0, & t<0 \\ 2(e^{-t}+t-1), & 0<t<1. \\ 2(e^{-t}-e^{-t+1}+1), & t>1 \end{cases}$

3. $(f*g)(t)=2e^{-(t-4)}u(t-4)$. The graph is the exponential function $2e^{-t}$ but shifted 4 units to the right. Before the shift, it becomes zero due to the product with step function.

4.

(a) The graphs of f and g are shown in Figure 3.14i.

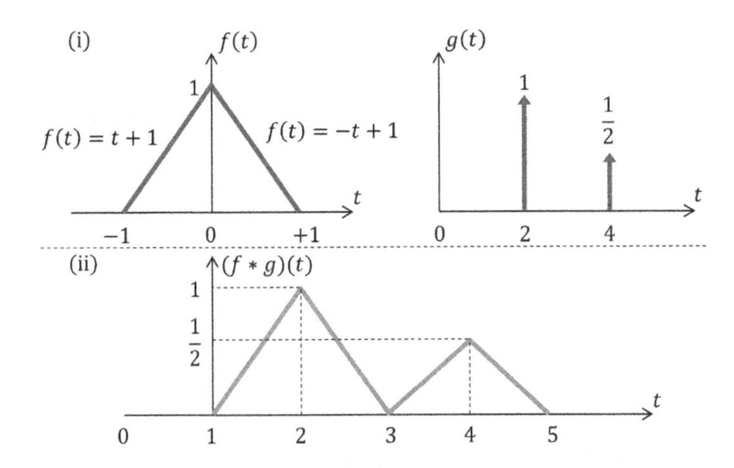

FIGURE 3.14 Convolution with the Dirac function.

(b)

$$(f*g)(t) = f(t-2) + \frac{1}{2} f(t-4)$$

$$= \begin{cases} t-1, & 1 < t < 2 \\ -t+3, & 2 < t < 3 \end{cases} + \frac{1}{2} \begin{cases} t-3, & 3 < t < 4 \\ -t+5, & 4 < t < 5 \end{cases}$$

The graph of $f * g$ is shown in Figure 3.14ii.

5. The graph of functions f and g is shown in Figure 3.15i.

(a)

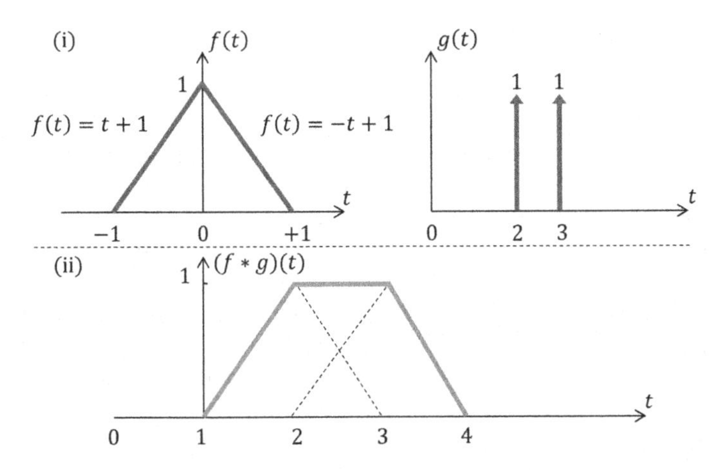

FIGURE 3.15 Convolution with the Dirac function.

(b) The graph is shown in Figure 3.15ii.

$$(f*g)(t) = f(t-2) + f(t-3)$$

$$= \begin{cases} t-1, & 1<t<2 \\ -t+3, & 2<t<3 \end{cases} + \begin{cases} t-2, & 2<t<3 \\ -t+4, & 3<t<4 \end{cases}$$

$$= \begin{cases} t-1, & 1<t<2 \\ 1, & 2<t<3. \\ -t+4, & 3<t<4 \end{cases}$$

6. It is a combination of step functions.

$$(f*g)(t) = \begin{cases} 0, & t<2 \\ 2t-4, & 2 \le t < 4 \\ 4, & t \ge 4 \end{cases}$$

$$= (2t-4)\left[u(t-2) - u(t-4)\right] + 4u(t-4).$$

7. The result is a combination of step functions

$$(f*g)(t) = \begin{cases} 0, & t<0 \\ 2t, & 0<t<2 \\ 4, & 2<t<4. \\ 12-2t, & 4<t<6 \\ 0, & t>6 \end{cases}$$

3.12 APPLICATION IN CONTROL ENGINEERING

Laplace transform is widely used in control engineering systems to describe the relationship between the input and the output of the systems. In a control system, the relationship between the input and output of the system is described by a transfer function. In this section, we explain the use of Laplace transform in describing the transfer function. In its simplest description, the transfer function is the Laplace transform of the impulse response of a linear time-invariant system when the system initial conditions are set to zero.

Consider a linear time-invariant control system which can be described by its impulse response, say, $g(t)$, at time $t = 0$. If the system is subjected to an arbitrary input signal $f(t)$, the output $x(t)$ can be represented by the convolution of $f(t)$ and $g(t)$:

$$x(t) = f(t) * g(t).$$

Calculation of the convolution integral is described in Section 3.11 and can be quite complex when the control system consists of many components. However, using the Laplace transform, the preceding equation can be converted into a simple multiplication:

$$X(s) = F(s) \times G(s)$$

where $X(s) = \mathcal{L}\{x(t)\}, F(s) = \mathcal{L}\{f(t)\}$ and $, G(s) = \mathcal{L}\{g(t)\}$. Rearranging gives

$$G(s) = \frac{X(s)}{F(s)}.$$

The function $G(s)$ is called the **transfer function** which is the ratio of the output function (signal) to the input function (signal) in the Laplace domain.

For example, take a simple harmonic oscillator, the spring-mass system without any damping term, which is considered in Section 2.11 in Chapter 2 (see also Example 3.14b), can be written as

$$m\frac{d^2x}{dt^2} + kx = f(t),\ x(0) = x'(0) = 0.$$

To find the impulse response of the system we set $f(t) = \delta(t)$ and solve for $x(t)$ by applying the Laplace transform to both sides of the equation:

$$ms^2X(s) + kX(s) = \mathcal{L}\{\delta(\mathrm{t})\} = 1$$

$$X(s) = \frac{1}{ms^2 + k} := G(s)$$

$$x(t) = \frac{1}{\sqrt{km}}\sin\left(\sqrt{\frac{k}{m}}t\right) := g(t).$$

Hence, in the time domain for the input $f(t) = \delta(t)$, the impulse response of the system is $g(t) = \frac{1}{\sqrt{km}}\sin\left(\sqrt{\frac{k}{m}}t\right)$ with its Laplace transform $G(s) = \frac{1}{ms^2+k}$ (Transfer Function).

Now, if we assume the input $f(t) = t$ (the ram function), then what is the output of the system? In the time domain,

$$x(t) = f(t) * g(t) = t * \frac{1}{\sqrt{km}}\sin\left(\sqrt{\frac{k}{m}}t\right).$$

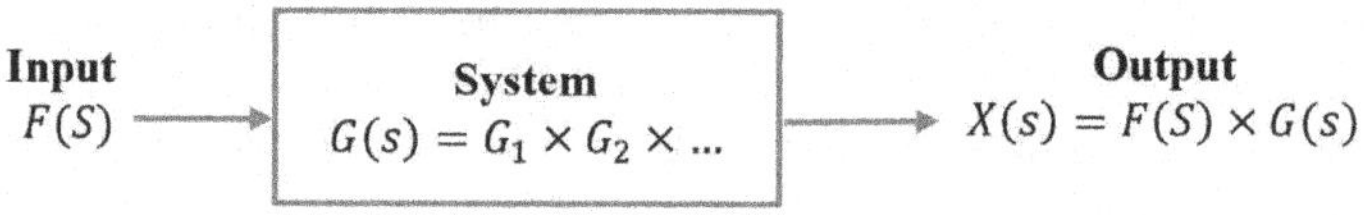

Input $F(S)$ → **System** $G(s) = G_1 \times G_2 \times \ldots$ → **Output** $X(s) = F(S) \times G(s)$

FIGURE 3.16 Block diagram of a linear time-invariant control system.

However, the convolution integral in the Laplace domain is the following multiplication:

$$X(s) = \frac{1}{s^2} \times \frac{1}{ms^2 + k}.$$

Control systems, in reality, can be very complex. They can be a combination of several components and each component can be represented as a transfer function in a block diagram. Then, using block diagram reduction techniques, the control system is simplified and reduced to the form shown in Figure 3.16, and an overall transfer function is determined. For example, where all components in a system are in series, the overall transfer function can be determined by multiplying the transfer function, G_i, of each component as shown in Figure 3.16.

3.13 TABLE OF LAPLACE TRANSFORMS (TABLE 3.1)

TABLE 3.1
Table of Laplace Transforms

$f(t)$	$F(s) = \mathcal{L}\{f(t)\}$
1	$\frac{1}{s}$
t	$\frac{1}{s^2}$
t^2	$2s^3$
t^n	$\frac{n!}{s^{n+1}}$
$e^{-\alpha t}$	$\frac{1}{s+\alpha}$
$1 - e^{-\alpha t}$	$\frac{\alpha}{s(s+\alpha)}$
$te^{-\alpha t}$	$\frac{1}{(s+\alpha)^2}$
$\sin(\omega t)$	$\frac{\omega}{s^2+\omega^2}$

(Continued)

TABLE 3.1 (*Continued*)
Table of Laplace Transforms

$\cos(\omega t)$	$\dfrac{s}{s^2+\omega^2}$
$1-\cos(\omega t)$	$\dfrac{\omega^2}{s\left(s^2+\omega^2\right)}$
$\omega t\sin(\omega t)$	$\dfrac{2\omega^2 s}{\left(s^2+\omega^2\right)^2}$
$\sin(\omega t)-\omega t\cos(\omega t)$	$\dfrac{2\omega^2}{(s^2+\omega^2)^2}$
$\sin(\omega t+\phi)$	$\dfrac{s\sin\phi+\omega\cos\phi}{s^2+\omega^2}$
$e^{-\alpha t}\sin(\omega t)$	$\dfrac{\omega}{(s+\alpha)^2+\omega^2}$
$e^{-\alpha t}\cos(\omega t)$	$\dfrac{s+\alpha}{(s+\alpha)^2+\omega^2}$
$e^{-\alpha t}\left[\cos(\omega t)-\alpha\omega\sin(\omega t)\right]$	$\dfrac{s}{(s+\alpha)^2+\omega^2}$
$e^{-\alpha t}+\alpha\omega\sin(\omega t)-\cos(\omega t)$	$\dfrac{\alpha^2+\omega^2}{(s+\alpha)\left(s^2+\omega^2\right)}$
$\sinh(\beta t)$	$\dfrac{\beta}{s^2-\omega^2}$
$e^{-\alpha t}\sinh(\beta t)$	$\dfrac{\beta}{(s+\alpha)^2-\beta^2}$
$\cosh(\beta t)$	$\dfrac{s}{s^2-\omega^2}$
$e^{-\alpha t}\cosh(\beta t)$	$\dfrac{s+\alpha}{(s+\alpha)^2-\beta^2}$
$e^{-\alpha t}f(t)$	$F(s+\alpha)$
$t\,f(t)$	$-F'(s)$
x	$\bar{x}$
$\dot{x}$	$s\bar{x}-x(0)$
$\ddot{x}$	$s^2\bar{x}-s\,x(0)-\dot{x}(0)$
$u(t-\alpha)=\begin{cases}0, & t<\alpha\\ 1, & t>\alpha\end{cases}$	$e^{-\alpha s}s$
$f(t-\alpha)u(t-\alpha)$	$e^{-\alpha s}F(s)$

(Continued)

TABLE 3.1 (*Continued*)
Table of Laplace Transforms

Dirac Delta Function	
$\delta(t-\alpha)$	$e^{-\alpha s}$
Convolution Integral	
$f * g$	$F(s)\times G(s)$

4 Linear Systems of Differential Equations

4.1 INTRODUCTION

Many real-life problems can be modelled by the systems of differential equations, for example electrical circuits having multiple loops can be modelled by systems of differential equations, in which each equation describes the current in each loop or one popular application occurs in predator prey systems, among others.

In mechanical engineering, it is impossible to describe the dynamics of complex vibrating systems unless they are modelled and solved using differential equations with multiple state variables. For example, consider a coupled spring-mass system shown in Figure 4.1, where two equal masses m_1 and m_2 connected by a spring of spring constant k and attached to the wall by two other springs of the same spring constant k. Assume that the masses are free to oscillate on a frictionless surface and are in equilibrium (rest).

We now pull the mass m_2 from its equilibrium state and release it so that in the absence of friction the springs will start to oscillate. Assume that at any instant of time the displacement of the mass m_1 is described by $x_1(t)$ and the mass on m_2 is described by $x_2(t)$. Assuming that $x_1 = 0$ and $x_2 = 0$ correspond to the equilibrium state of the system, that is when the springs are not stretched and the masses are free of any force.

When the system is not in equilibrium state, that is when the springs are stretched or compresses, then according to the Hooke's law the forces applied to the mass m_1, in Figure 4.1 , are $-kx_1$ by the spring on the left and $-kx_1 + kx_2$ by the middle spring. The forces applied to the mass m_2, in Figure 4.1, are $-kx_2 + kx_1$ by the middle spring and $-kx_2$ by the spring on the right.

Using Newton's second law of motion, the equations of motion of the two masses can be written in the following form:

$$\begin{aligned} m_1\ddot{x}_1 &= -kx_1 - kx_1 + kx_2 = -2kx_1 + kx_2 \\ m_2\ddot{x}_2 &= -kx_2 + kx_1 - kx_2 = kx_1 - 2kx_2. \end{aligned} \tag{4.1}$$

This is a pair of coupled second order equations and the method of solving these equations is described in Section 4.7.3.

In this chapter, although we explain how to solve linear systems of differential equations, most of this topic is related to finding eigenvalues and eigenvectors of matrices. We do not intend to expand on the topic of matrices and cover a full course in linear algebra. However, in order to familiarise you with these topics and be able to solve differential equations, we provide a substantial review of eigenvalues and eigenvectors.

DOI: 10.1201/9781032630694-4

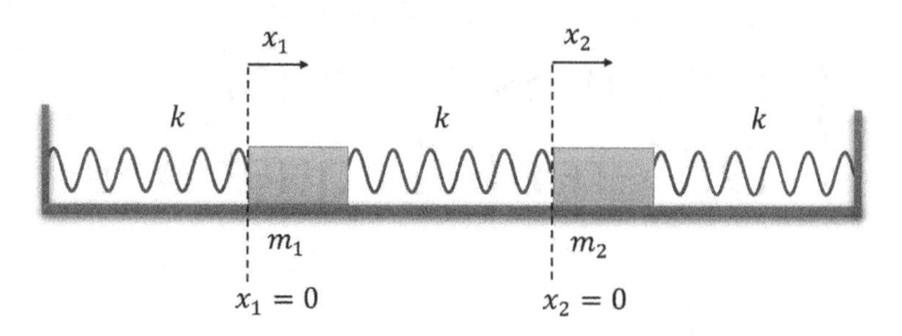

FIGURE 4.1 Two states coupled spring-mass system.

The outline of this chapter is as follows:

- Review of eigenvalues and eigenvectors
- Diagonalisation of matrices
- Solution of linear systems

4.2 EIGENVALUES

Given an $n \times n$ square matrix A, we can form a new matrix $A - \lambda I$ where λ is (yet) an unknown number and I is the identity matrix. The eigenvalues, λ, of the matrix A are the roots of its characteristic equation. The characteristic equation is

$$|A - \lambda I| = 0.$$

For example, if we start with the 2×2 matrix

$$A = \begin{bmatrix} 3 & 6 \\ 4 & 5 \end{bmatrix} \Rightarrow$$

$$(A - \lambda I) = \begin{bmatrix} 3-\lambda & 6 \\ 4 & 5-\lambda \end{bmatrix},$$

the evaluation of the determinant implies the characteristic equation

$$|A - \lambda I| = (3-\lambda)(5-\lambda) - 24 = 0.$$

Expand and simplify to give:

$$15 - 3\lambda - 5\lambda + \lambda^2 - 24 = 0 \Rightarrow$$

$$\lambda^2 - 8\lambda - 9 = 0.$$

The eigenvalues of A defined above are found by solving

$$\lambda^2 - 8\lambda - 9 = 0.$$

Factorising the left-hand side (LHS):

$$(\lambda-9)(\lambda+1)=0.$$

The eigenvalues are $\lambda_1=-1$ and $\lambda_2=9$.

The value of λ may be real or complex, we first consider real distinct values of λ or some real values of λ which are repeated.

4.3 EIGENVECTORS

Given an $n\times n$ matrix A and having found its eigenvalues, its eigenvectors are found as follows: For each eigenvalue separately, we need to solve the system of simultaneous equations

$$A\underline{x}=\lambda\underline{x},\ \underline{x}\neq\underline{0},$$

where $\underline{x}$ is a non-zero column vector of size n. For example, in the case of 2×2 or 3×3 matrices, we have

$$\underline{x}=\begin{bmatrix} v_1 \\ v_2 \end{bmatrix} \quad \text{and} \quad \underline{x}=\begin{bmatrix} v_1 \\ v_2 \\ v_3 \end{bmatrix}.$$

The equation $A\underline{x}=\lambda\underline{x}\Rightarrow(A-\lambda I)\underline{x}=\underline{0}$. Seeking the non-zero solutions implies that $(A-\lambda I)$ must be singular; that is $|A-\lambda I|=0$.

You will see that the simultaneous equations formed in this way always have an infinite number of solutions for each eigenvalue. Each solution $\underline{x}\neq\underline{0}$ is called an **eigenvector** of A.

Example 4.1 The eigenvalues of the following matrix A are $\lambda_1=-1$ and $\lambda_2=9$. Find the eigenvectors associated with these eigenvalues. The matrix A is defined as

$$A=\begin{bmatrix} 3 & 6 \\ 4 & 5 \end{bmatrix}.$$

Solution

Eigenvector $\underline{x}_1$ associated with $\lambda_1=-1$ can be determined as follows. In the equation $(A-\lambda I)\underline{x}_1=\underline{0}$ replace λ with -1 to give

$$\begin{bmatrix} 4 & 6 \\ 4 & 6 \end{bmatrix}\begin{bmatrix} v_1 \\ v_2 \end{bmatrix}=\begin{bmatrix} 0 \\ 0 \end{bmatrix}.$$

Multiplying the system of equations out gives $4v_1 + 6v_2 = 0$ and $4v_1 + 6v_2 = 0$. Since the two equations are the same, unique values of v_1 and v_2 cannot be obtained. The relationship between the components of the eigenvector is $4v_1 = -6v_2$ or $v_1 = -\frac{3}{2}v_2$. Hence,

$$\underline{x}_1 = \begin{bmatrix} v_1 \\ v_2 \end{bmatrix} = \begin{bmatrix} -\frac{3}{2}v_2 \\ v_2 \end{bmatrix} = v_2 \begin{bmatrix} -\frac{3}{2} \\ 1 \end{bmatrix}.$$

The value of v_2 is arbitrary. Changing its value merely alters the magnitude of the vector. Avoiding fractions (setting $v_2 = -2$), we can quote the eigenvector as

$$\underline{x}_1 = \begin{bmatrix} 3 \\ -2 \end{bmatrix} \text{ (or any multiple of it).}$$

Eigenvector $\underline{x}_2$ associated with $\lambda_2 = 9$ can be determined as follows. In the equation

$$(A - \lambda I)\underline{x}_2 = \underline{0},$$

replace λ with 9 to give

$$\begin{bmatrix} -6 & 6 \\ 4 & -4 \end{bmatrix} \begin{bmatrix} v_1 \\ v_2 \end{bmatrix} = \begin{bmatrix} 0 \\ 0 \end{bmatrix}.$$

Multiplying the system of equations out gives

$$-6v_1 + 6v_2 = 0$$
$$4v_1 - 4v_2 = 0.$$

Both of the two equations state that $v_1 = v_2$. Hence,

$$\underline{x}_2 = \begin{bmatrix} v_1 \\ v_2 \end{bmatrix} = \begin{bmatrix} v_2 \\ v_2 \end{bmatrix} = v_2 \begin{bmatrix} 1 \\ 1 \end{bmatrix}.$$

The value of v_2 is arbitrary. Changing its value merely alters the magnitude of the vector. We can quote the eigenvector as

$$\underline{x}_2 = \begin{bmatrix} 1 \\ 1 \end{bmatrix}.$$

Summary: To find the eigenvalues and the eigenvectors of a square matrix, say, A, the following steps must be considered:

1. Construct the characteristic equation $|A-\lambda I|=0$.
2. Solve the characteristic equation for λ.
3. Find the eigenvectors associated with each eigenvalue by solving $(A-\lambda I)\underline{x}=\underline{0}$ for the non-zero vector $\underline{x}$.

For the matrix A defined above the eigenvalues and the associated eigenvectors are

$$\left(\lambda_1=-1,\quad \underline{x}_1=\begin{bmatrix}3\\-2\end{bmatrix}\right) \text{ and } \left(\lambda_2=9,\quad \underline{x}_2=\begin{bmatrix}1\\1\end{bmatrix}\right).$$

Example 4.2 For the following matrix A, find its eigenvalues and eigenvectors:

$$A=\begin{bmatrix}-5 & 2 & 1\\-8 & 3 & 2\\-16 & 2 & 6\end{bmatrix}.$$

Solution

Find the characteristic equation.

$$|A-\lambda I|=\begin{vmatrix}-5-\lambda & 2 & 1\\-8 & 3-\lambda & 2\\-16 & 2 & 6-\lambda\end{vmatrix}=0.$$

Expand the determinant along the first row to give

$$|A-\lambda I|=$$

$$=(-5-\lambda)\begin{vmatrix}3-\lambda & 2\\2 & 6-\lambda\end{vmatrix}-2\begin{vmatrix}-8 & 2\\-16 & 6-\lambda\end{vmatrix}+1\begin{vmatrix}-8 & 3-\lambda\\-16 & 2\end{vmatrix}$$

$$=(-5-\lambda)\left[(3-\lambda)(6-\lambda)-4\right]-2\left[-8(6\lambda)+32\right]+$$

$$\left[-16+16(3-\lambda)\right]$$

$$=(-5-\lambda)\left[18-9\lambda+\lambda^2-4\right]-2[8\lambda-16]+[-16\lambda+32]$$

$$=(-5-\lambda)\left[\lambda^2-9\lambda+14\right]-16\lambda+32-16\lambda+32$$

$$=(-5-\lambda)\left[\lambda^2-9\lambda+14\right]-32\lambda+64=0.$$

Rather than multiplying out the first bracket to give a cubic equation, we factorise it and hope that a common factor can be taken out.

$$|A-\lambda I|=(-5-\lambda)(\lambda-7)(\lambda-2)-32(\lambda-2)=0.$$

We can now take out a common factor of $(\lambda-2)$.

$$|A-\lambda I|=(\lambda-2)\left[(-5-\lambda)(\lambda-7)-32\right]$$

$$=(\lambda-2)\left[(-5\lambda-\lambda^2+35+7\lambda-32\right]$$

$$=(\lambda-2)\left[(-\lambda^2+2\lambda+3\right]=-(\lambda-2)\left(\lambda^2-2\lambda-3\right)$$

$$=-(\lambda-2)(\lambda+1)(\lambda-3)=0.$$

Eigenvalues are

$$\lambda_1=-1, \qquad \lambda_2=2, \quad \text{and} \qquad \lambda_3=3.$$

Now we find the eigenvectors associated with each eigenvalue.

Eigenvector $\underline{x}_1$ associated with $\lambda_2=2$ can be determined as follows. In the equation $(A-\lambda I)\underline{x}=\underline{0}$, replace λ with 2 to give

$$\begin{bmatrix} -7 & 2 & 1 \\ -8 & 1 & 2 \\ -16 & 2 & 4 \end{bmatrix}\begin{bmatrix} v_1 \\ v_2 \\ v_3 \end{bmatrix} = \begin{bmatrix} 0 \\ 0 \\ 0 \end{bmatrix}.$$

This gives the system of equations:

$$-7v_1+2v_2+v_3=0 \qquad \text{(i)}$$

$$-8v_1+v_2+2v_3=0 \qquad \text{(ii)}$$

$$-16v_1+2v_2+4v_3=0 \qquad \text{(iii)}.$$

Equations (ii) and (iii) are the same since 2 × (ii) = (iii). Suppose we eliminate v_2:

$$(\text{i})-(\text{iii}): 9v_1-3v_3=0 \Rightarrow 9v_1=3v_3$$

$$\Rightarrow v_3=3v_1.$$

We can now express v_2 in terms of v_1. Replace v_3 with $3v_1$ in Equation (i).

$$-7v_1 + 2v_2 + 3v_1 = 0 \Rightarrow 2v_2 = 4v_1$$
$$\Rightarrow \quad v_2 = 2v_1.$$

The eigenvector can now be expressed in terms of the arbitrary value v_1:

$$\underline{x}_2 = \begin{bmatrix} v_1 \\ v_2 \\ v_3 \end{bmatrix} = \begin{bmatrix} v_1 \\ 2v_1 \\ 3v_1 \end{bmatrix} = v_1 \begin{bmatrix} 1 \\ 2 \\ 3 \end{bmatrix}.$$

Assuming $v_1 = 1$, we will write the eigenvector as

$$\left(\lambda_2 = 2,\ \underline{x}_2 = \begin{bmatrix} 1 \\ 2 \\ 3 \end{bmatrix} \right).$$

The other eigenvectors are

$$\left(\lambda_1 = -1,\ \underline{x}_1 = \begin{bmatrix} 1 \\ 1 \\ 2 \end{bmatrix} \right) \text{ and } \left(\lambda_3 = 3,\ \underline{x}_3 = \begin{bmatrix} 1 \\ 2 \\ 4 \end{bmatrix} \right).$$

Summary: The eigenvalues and eigenvectors of matrix A are shown in Table 4.1.

TABLE 4.1
Eigenvalues and Associated Eigenvectors

$\lambda_1 = -1$	$\underline{x}_1 = \begin{bmatrix} 1 \\ 1 \\ 2 \end{bmatrix}$,
$\lambda_2 = 2$	$\underline{x}_2 = \begin{bmatrix} 1 \\ 2 \\ 3 \end{bmatrix}$,
$\lambda_3 = 3$	$\underline{x}_3 = \begin{bmatrix} 1 \\ 2 \\ 4 \end{bmatrix}$

4.3.1 Exercises

For each of the following matrices, determine its eigenvalues and eigenvectors.

(a) $\begin{bmatrix} 1 & 2 \\ 4 & 3 \end{bmatrix}$ (b) $\begin{bmatrix} -4 & 2 \\ -2.5 & 2 \end{bmatrix}$

(c) $\begin{bmatrix} -1 & 1 & 0 \\ 1 & 2 & 1 \\ 0 & 3 & -1 \end{bmatrix}$ (d) $\begin{bmatrix} -1 & -1 & 0 \\ 0.75 & -1.5 & 3 \\ 0.125 & 0.25 & -0.5 \end{bmatrix}$

Answers

(a) $\left(\lambda_1 = 5, \quad \underline{x}_1 = c\begin{bmatrix} 1 \\ 2 \end{bmatrix}\right), \quad \left(\lambda_2 = -1, \quad \underline{x}_2 = c\begin{bmatrix} 1 \\ -1 \end{bmatrix}\right)$

(b) $\left(\lambda_1 = -3, \quad \underline{x}_1 = c\begin{bmatrix} 2 \\ 1 \end{bmatrix}\right), \quad \left(\lambda_2 = 1, \quad \underline{x}_2 = c\begin{bmatrix} 2 \\ 5 \end{bmatrix}\right)$

(c) $\left(\lambda_1 = -1, \ \underline{x}_1 = c\begin{bmatrix} 1 \\ 0 \\ -1 \end{bmatrix}\right), \left(\lambda_2 = 3, \ \underline{x}_2 = c\begin{bmatrix} 1 \\ 4 \\ 3 \end{bmatrix}\right), \left(\lambda_3 = -2, \ \underline{x}_3 = c\begin{bmatrix} 1 \\ -1 \\ 3 \end{bmatrix}\right)$

(d) $\left(\lambda_1 = -1, \underline{x}_1 = c\begin{bmatrix} 4 \\ 0 \\ -1 \end{bmatrix}\right), \quad \left(\lambda_2 = -0.5, \underline{x}_2 = c\begin{bmatrix} -12 \\ 6 \\ 5 \end{bmatrix}\right),$

$\left(\lambda_3 = -1.5, \underline{x}_3 = c\begin{bmatrix} 4 \\ 2 \\ -1 \end{bmatrix}\right).$

DEFINITION 4.1: LINEARLY INDEPENDENT VECTORS

Assume the matrix A has n eigenvectors $\underline{x}_1, \underline{x}_2, \underline{x}_3, \cdots, \underline{x}_n$. These vectors are said to be linearly independent if $c_1\underline{x}_1 + c_2\underline{x}_2 + c_2\ \underline{x}_3 + \cdots + c_n\underline{x}_n = 0$, then $c_1 = c_2 = c_3 = \cdots = c_n = 0$, where $c_1, c_2, \cdots, c_n$ are constants. This means that for linearly independent eigenvectors any of the eigenvectors cannot be written as a linear combination of the remaining vectors, otherwise they are linearly dependent.

In most practical cases, the eigenvalues of matrices are distinct (as in Examples 4.1 and 4.2). If the $n \times n$ matrix A has n distinct eigenvalues $\lambda_1, \lambda_2, \lambda_3, \ldots \lambda_n$, then the corresponding eigenvectors $\underline{x}_1$, $\underline{x}_2$, $\underline{x}_3, \ldots, x_n$ are linearly independent.

4.4 REPEATED EIGENVALUES

The characteristic equation

$$\lambda(\lambda-1)^2=0$$

has eigenvalues $\lambda=0,\ \lambda=1$ (twice). The root $\lambda=1$ is a repeated root or root of algebraic multiplicity 2.

The characteristic equation $(\lambda+1)(\lambda-2)^3=0$ has the eigenvalue $\lambda=2$ of algebraic multiplicity 3.

The multiplicity of the eigenvalue corresponds to the number of times λ is a root of the characteristic equation.

When a repeated eigenvalue occurs, there may be more than one linearly independent eigenvector corresponding to the repeated eigenvalue.

Example 4.3 Show that the matrix

$$A=\begin{bmatrix}5 & -2 & 0\\ 6 & -2 & 0\\ 8 & -4 & 1\end{bmatrix}$$

has repeated eigenvalues and find the associated eigenvector.

Solution

Characteristic equation is $-(\lambda-2)(\lambda-1)^2=0$.

Consider the eigenvector for $\lambda=1$.

We replace λ in the equation $(A-\lambda I)\underline{x}=\underline{0}$ with the value of 1. This produces

$$\begin{bmatrix}4 & -2 & 0\\ 6 & -3 & 0\\ 8 & -4 & 0\end{bmatrix}\begin{bmatrix}v_1\\ v_2\\ v_3\end{bmatrix}=\begin{bmatrix}0\\ 0\\ 0\end{bmatrix},$$

which gives the system of equations:

$$4v_1-2v_2=0\ (1)$$

$$6v_1-3v_2=0\ (2)$$

$$8v_1-4v_2=0\ (3).$$

These three equations all supply the same information, which is $v_2 = 2v_1$. This means that the component v_3 in the eigenvector is arbitrary. Because of this additional freedom of choice, we can generate two linearly independent eigenvectors.

$$\underline{x} = \begin{bmatrix} v_1 \\ v_2 \\ v_3 \end{bmatrix} = \begin{bmatrix} v_1 \\ 2v_1 \\ 0 \end{bmatrix} + \begin{bmatrix} 0 \\ 0 \\ v_3 \end{bmatrix} = v_1 \begin{bmatrix} 1 \\ 2 \\ 0 \end{bmatrix} + v_3 \begin{bmatrix} 0 \\ 0 \\ 1 \end{bmatrix}.$$

Hence, the eigenvalue $\lambda = 1$ of algebraic multiplicity 2 has 2 linearly independent eigenvectors

$$\begin{bmatrix} 1 \\ 2 \\ 0 \end{bmatrix} \quad \text{and} \quad \begin{bmatrix} 0 \\ 0 \\ 1 \end{bmatrix}.$$

In this case, we say that the matrix A has an algebraic multiplicity 2, which is associated to a **geometrical** multiplicity 2. An eigenvalue of multiplicity 2 does not always produce 2 linearly independent eigenvectors. For example, the matrix.

$$A = \begin{bmatrix} 1 & 2 & 2 \\ 0 & 2 & 1 \\ -1 & 2 & 2 \end{bmatrix}$$

has characteristic equation

$$-(\lambda - 1)(\lambda - 2)^2 = 0.$$

The eigenvalue $\lambda = 2$ of multiplicity 2 has only one eigenvector:

$$\underline{x} = \begin{bmatrix} 2 \\ 1 \\ 0 \end{bmatrix}.$$

In this case, matrix A has an algebraic multiplicity of 2, which is associated with a geometric multiplicity of 1. These types of matrices are called "defective"; that is their algebraic multiplicity and geometric multiplicities do not match.

4.5 COMPLEX EIGENVALUES AND ASSOCIATED EIGENVECTORS

In this section, we consider 2×2 or 3×3 real matrices that have complex eigenvalues. The complex eigenvalues and eigenvectors can be evaluated in the same way as we explained about real eigenvalues and eigenvectors in sections 4.2 and 4.3.

For a 2×2 matrix A with the real elements, the characteristic equation $|A - \lambda I| = 0$ can be written as the following quadratic equation

$$a\lambda^2 + b\lambda + c = 0,$$

where a, b and c are constant parameters. Suppose $b^2 - 4ac < 0$, so the roots of the characteristic equation are the following complex conjugate eigenvalues

$$\lambda = \alpha \pm \beta j, \quad j = \sqrt{-1}$$

where α and β are real numbers. The associated eigenvectors can be found by substituting $\lambda = \alpha \pm \beta j$ in the equation $(A - \lambda I)\underline{x} = \underline{0}$ and solving for the vector $\underline{x}$. We provide examples to clarify this topic.

Example 4.4 Determine the eigenvalues and eigenvectors of the matrix A:

$$A = \begin{bmatrix} -4 & -2 \\ 1 & -2 \end{bmatrix}.$$

Solution

Construct the characteristic equation and obtain the eigenvalues:

$$|A - \lambda I| = \begin{vmatrix} -4-\lambda & -2 \\ 1 & -2-\lambda \end{vmatrix} = 0.$$

Thus, we obtain the following quadratic equation, which has complex conjugate roots, and they are the complex eigenvalues:

$$\lambda^2 + 6\lambda + 10 = 0$$

$$\lambda = -3 \pm j, \quad j = \sqrt{-1}.$$

To obtain eigenvectors, we substitute one of the eigenvalues, say, $\lambda_p = -3 + j$, in $(A - \lambda I)\underline{x} = \underline{0}$ to get

$$\begin{bmatrix} -1-j & -2 \\ 1 & 1-j \end{bmatrix} \begin{bmatrix} v_1 \\ v_2 \end{bmatrix} = \begin{bmatrix} 0 \\ 0 \end{bmatrix}.$$

By expanding the above matrix-form, we get the following system of equations:

$$(-1-j)v_1 - 2v_2 = 0$$

$$v_1 + (1-j)v_2 = 0.$$

The two preceding equations are the same, Equation 2 is obtained by multiplying Equation 1 by $(1-j)$ and dividing by -2. Since the two equations are the same, unique values of v_1 and v_2 cannot be obtained. Both of the equations state that $v_1 = -(1-j)v_2$. Hence,

$$\underline{x}_p = \begin{bmatrix} v_1 \\ v_2 \end{bmatrix} = \begin{bmatrix} -(1-j)v_2 \\ v_2 \end{bmatrix}.$$

The value of v_2 is arbitrary, changing its value merely alters the magnitude of the vector. By setting $v_2 = 1$, we can quote the eigenvector associated with λ_p as

$$\underline{x}_p = \begin{bmatrix} -1+j \\ 1 \end{bmatrix} = \begin{bmatrix} -1 \\ 1 \end{bmatrix} + j\begin{bmatrix} 1 \\ 0 \end{bmatrix} := \underline{x}_1 + j\underline{x}_2,$$

where

$$\underline{x}_1 = \begin{bmatrix} -1 \\ 1 \end{bmatrix} \text{ and } \underline{x}_2 = \begin{bmatrix} 1 \\ 0 \end{bmatrix}.$$

So, for $\lambda_p = -3+j$, the associated eigenvector is $\underline{x}_p = \underline{x}_1 + j\underline{x}_2$.

For $\lambda_m = -3-j$, the associated eigenvector can be found by substituting $\lambda_m = -3-j$ in $(A-\lambda I)\underline{x} = \underline{0}$ and following the process described earlier, which gives

$$\underline{x}_m = \underline{x}_1 - j\underline{x}_2.$$

Summary: The eigenvalues and eigenvectors of matrix A are shown in Table 4.2.

The following general rule is true for any $n \times n$ matrix A with real elements and with complex eigenvalues and eigenvectors.

TABLE 4.2
Eigenvalues and Associated Eigenvectors

Eigenvalue	Associated eigenvector
$\lambda = -3 \pm j$	$\underline{x} = \underline{x}_1 \pm j\underline{x}_2$
	where $\underline{x}_1 = \begin{bmatrix} -1 \\ 1 \end{bmatrix}$, $\underline{x}_2 = \begin{bmatrix} 1 \\ 0 \end{bmatrix}$

If $Ax = \lambda x$ (where A is a real $n \times n$ matrix and $\boldsymbol{x}$ is an n-dimensional complex eigenvector associated with the complex eigenvalue λ) then

$$\overline{A\boldsymbol{x}} = \overline{\lambda \boldsymbol{x}}$$

$$\bar{A}\bar{\boldsymbol{x}} = \bar{\lambda}\bar{\boldsymbol{x}}$$

$$A\bar{\boldsymbol{x}} = \bar{\lambda}\bar{\boldsymbol{x}}.$$

Example 4.5 Determine the eigenvalues and eigenvectors of matrix A:

$$A = \begin{bmatrix} -2 & -4 \\ 1 & -2 \end{bmatrix}.$$

Solution

Construct the characteristic equation and obtain the eigenvalues:

$$|A - \lambda I| = \begin{vmatrix} -2-\lambda & -4 \\ 1 & -2-\lambda \end{vmatrix} = 0.$$

Hence, we obtain the following quadratic equation, which has complex conjugate roots, and they are the complex eigenvalues:

$$\lambda^2 + 4\lambda + 8 = 0$$

$$\lambda = -2 \pm 2j, \ j = \sqrt{-1}.$$

We just need to find an eigenvector associated with one of the eigenvalues. Therefore, we determine the eigenvector associated with $\lambda_p = -2 + 2j$ as follows. Substitute $\lambda_p = -2 + 2j$ in $(A - \lambda I)\underline{x} = \underline{0}$ to get

$$\begin{bmatrix} -2j & -4 \\ 1 & -2j \end{bmatrix} \begin{bmatrix} v_1 \\ v_2 \end{bmatrix} = \begin{bmatrix} 0 \\ 0 \end{bmatrix}.$$

Multiplying the system of equations out gives

$$-2jv_1 - 4v_2 = 0$$

$$v_1 - 2jv_2 = 0.$$

The preceding two equations are the same; Equation 2 is obtained by multiplying Equation 1 by j and dividing by 2. Since the two equations are the same, unique values of v_1 and v_2 cannot be obtained. Both equations state that $v_1 = 2jv_2$. Hence,

$$\underline{x} = \begin{bmatrix} v_1 \\ v_2 \end{bmatrix} = \begin{bmatrix} 2jv_2 \\ v_2 \end{bmatrix}.$$

The value of v_2 is arbitrary, changing its value merely alters the magnitude of the vector. By setting $v_2 = 1$, we can quote the eigenvector as

$$\underline{x}_p = \begin{bmatrix} 2j \\ 1 \end{bmatrix} = \begin{bmatrix} 0 \\ 1 \end{bmatrix} + j\begin{bmatrix} 2 \\ 0 \end{bmatrix}.$$

For $\lambda_m = -2 - 2j$, the associated eigenvector is:

$$\underline{x}_m = \begin{bmatrix} -2j \\ 1 \end{bmatrix} = \begin{bmatrix} 0 \\ 1 \end{bmatrix} - j\begin{bmatrix} 2 \\ 0 \end{bmatrix}.$$

Summary

The eigenvalues and eigenvectors of matrix A are shown in Table 4.3.

TABLE 4.3
Eigenvalues and Associated Eigenvectors

Eigenvalue	Associated Eigenvector
$\lambda_{p,m} = -2 \pm 2j$	$\underline{x}_{p,m} = \underline{x}_1 \pm j\underline{x}_2$
	where $\underline{x}_1 = \begin{bmatrix} 0 \\ 1 \end{bmatrix}$, $\underline{x}_2 = \begin{bmatrix} 2 \\ 0 \end{bmatrix}$

Example 4.6 Determine the eigenvalues and eigenvectors of matrix A:

$$A = \begin{bmatrix} 4 & -3 & 0 \\ 3 & 4 & 0 \\ 1 & 2 & 2 \end{bmatrix}.$$

Solution

Construct the characteristic equation and obtain the eigenvalues:

$$|A - \lambda I| = \begin{vmatrix} 4-\lambda & -3 & 0 \\ 3 & 4-\lambda & 0 \\ 1 & 2 & 2-\lambda \end{vmatrix} = 0.$$

The roots of the characteristic equation are two complex conjugate eigenvalues and a real eigenvalue:

$$-(\lambda^2 - 8\lambda + 25)(\lambda - 2) = 0$$

$$\lambda_{p,m} = 4 \pm 3j, \quad j = \sqrt{-1} \quad \text{and}$$

$$\lambda_r = 2.$$

Substitute $\lambda_p = 4+3j$ in $(A-\lambda I)\underline{x} = \underline{0}$ to get

$$\begin{bmatrix} -3j & -3 & 0 \\ 3 & -3j & 0 \\ 1 & 2 & -2-3j \end{bmatrix}\begin{bmatrix} v_1 \\ v_2 \\ v_3 \end{bmatrix} = \begin{bmatrix} 0 \\ 0 \\ 0 \end{bmatrix}.$$

Multiplying the system of equations out gives

$$\begin{aligned} -3jv_1 - 3v_2 &= 0 \quad \text{(i)} \\ 3v_1 - 3jv_2 &= 0 \quad \text{(ii)} \\ v_1 + 2v_2 - (2+3j)v_3 &= 0 \quad \text{(iii)}. \end{aligned}$$

Equations (i) and (ii) are the same; Equation (ii) is obtained by multiplying Equation (i) by j. Since the two equations are the same, unique values of v_1 and v_2 cannot be obtained. Both Equations (i) and (ii) give

$$v_2 = -jv_1.$$

Substituting this in Equation 3 and simplifying implies

$$\begin{aligned} (1-2j)v_1 &= (2+3j)v_3 \Rightarrow \\ v_1 &= \frac{1}{5}(-4+7j)v_3 \quad \text{and} \\ v_2 &= -jv_1 = \frac{-j}{5}(-4+7j)v_3 = \frac{1}{5}(7+4j)v_3. \end{aligned}$$

Hence, the eigenvector associated with $\lambda_p = (-4+7j)$ is

$$\underline{x}_p = \begin{bmatrix} v_1 \\ v_2 \\ v_3 \end{bmatrix} = \begin{bmatrix} \frac{1}{5}(-4+7j)v_3 \\ \frac{1}{5}(7+4j)v_3 \\ v_3 \end{bmatrix}.$$

Set $v_3 = 5$ gives

$$\underline{x}_p = \begin{bmatrix} -4+7j \\ 7+4j \\ 5 \end{bmatrix} = \begin{bmatrix} -4 \\ 7 \\ 5 \end{bmatrix} + j\begin{bmatrix} 7 \\ 4 \\ 0 \end{bmatrix}.$$

For $\lambda_m = 4 - 3j$,

$$\underline{x}_m = \begin{bmatrix} -4 \\ 7 \\ 5 \end{bmatrix} - j \begin{bmatrix} 7 \\ 4 \\ 0 \end{bmatrix}.$$

For $\lambda_r = 2$, the eigenvector is

$$\underline{x}_r = \begin{bmatrix} 0 \\ 0 \\ 1 \end{bmatrix}.$$

TABLE 4.4
Eigenvalues and Associated Eigenvectors

Eigenvalue	Associated Eigenvector
$\lambda_{p,m} = 4 \pm 3j$	$\underline{x}_{p,m} = \underline{x}_1 \pm j\underline{x}_2$ $\underline{x}_1 = \begin{bmatrix} -4 \\ 7 \\ 5 \end{bmatrix}, \underline{x}_2 = \begin{bmatrix} 7 \\ 4 \\ 0 \end{bmatrix}$
$\lambda_r = 2$	$\underline{x}_r = \begin{bmatrix} 0 \\ 0 \\ 1 \end{bmatrix}$

4.5.1 Exercises

For each of the matrices shown, determine its eigenvalues and eigenvectors.

(a) $\begin{bmatrix} 4 & 5 \\ -2 & 6 \end{bmatrix}$ (b) $\begin{bmatrix} 3 & -13 \\ 5 & 1 \end{bmatrix}$

(c) $\begin{bmatrix} 1 & 2 \\ -2 & 1 \end{bmatrix}$ (d) $\begin{bmatrix} 3 & 1 \\ -2 & 1 \end{bmatrix}$

(e) $\begin{bmatrix} 1 & -1 & 0 \\ 1 & 1 & 0 \\ 1 & 2 & 1 \end{bmatrix}$

Answers

In all the following answers, c is an arbitrary constant.

(a) $\lambda = 5 \pm 3j, \quad x = c\begin{bmatrix} \frac{1 \mp 3j}{2} \\ 1 \end{bmatrix}$

(b) $\lambda = 2 \pm 8j, \quad x = c\begin{bmatrix} \frac{1 \pm 8j}{5} \\ 1 \end{bmatrix}$

(c) $\lambda = 1 \pm 2j, \quad x = c\begin{bmatrix} \mp j \\ 1 \end{bmatrix}$

(d) $\lambda = 2 \pm j, \quad x = c\begin{bmatrix} \frac{-1 \mp j}{2} \\ 1 \end{bmatrix}$

(e) $\lambda_{p,m} = 1 \pm j, \quad j = \sqrt{-1}$ and $\lambda_r = 1$

(f) $\underline{x}_{p,m} = \begin{bmatrix} -2 \\ 1 \\ 5 \end{bmatrix} + j\begin{bmatrix} 1 \\ 2 \\ 0 \end{bmatrix}$, and $\underline{x}_r = \begin{bmatrix} 0 \\ 0 \\ 1 \end{bmatrix}$

4.6 DIAGONALISATION OF A MATRIX

The process of converting a matrix into a diagonal form that shares the same properties of the original matrix is called diagonalisation.

An $n \times n$ matrix A can be diagonalised if

Step 1: It has n distinct real eigenvalues $\lambda_1, \lambda_2, \lambda_3, \ldots, \lambda_n$.
OR
Step 2: It has n linearly independent real eigenvectors.

Assume matrix A has n distinct real eigenvalues $\lambda_1, \lambda_2, \lambda_3, \ldots, \lambda_n$ with the associated eigenvectors $\underline{x}_1, \underline{x}_2, \underline{x}_3, \ldots, \underline{x}_n$. We construct a matrix P using the eigenvectors of A; that is

$$P = [\underline{x}_1, \underline{x}_2, \ldots, \underline{x}_n].$$

In general, it can be proved that

$$P^{-1}AP = \operatorname{diag}[\lambda_1, \lambda_2, \ldots, \lambda_n] := D.$$

Note that the order of the diagonal elements in matrix D depends on the order of eigenvectors when the matrix P is constructed. Now we describe the diagonalisation process using a couple of examples.

Example 4.7 Consider the matrix

$$A = \begin{bmatrix} -5 & 2 & 1 \\ -8 & 3 & 2 \\ -16 & 2 & 6 \end{bmatrix},$$

which has following eigenvalues and associated eigenvectors:

$$\left(\lambda_1 = -1,\ \underline{x}_1 = \begin{bmatrix} 1 \\ 1 \\ 2 \end{bmatrix}\right), \left(\lambda_2 = 2,\ \underline{x}_2 = \begin{bmatrix} 1 \\ 2 \\ 3 \end{bmatrix}\right) \text{and} \left(\lambda_3 = 3,\ \underline{x}_3 = \begin{bmatrix} 1 \\ 2 \\ 4 \end{bmatrix}\right).$$

Construct the matrix P and show that

$$P^{-1}AP = \text{diag}\left[-1,2,3\right].$$

Solution

We form a 3×3 matrix P whose columns are the three eigenvectors:

$$P = \begin{bmatrix} 1 & 1 & 1 \\ 1 & 2 & 2 \\ 2 & 3 & 4 \end{bmatrix}.$$

The inverse of matrix P is found to be

$$P^{-1} = \begin{bmatrix} 2 & -1 & 0 \\ 0 & 2 & -1 \\ -1 & -1 & 1 \end{bmatrix}.$$

If we now form the matrix product $P^{-1}AP$, we obtain

$$P^{-1}AP = \begin{bmatrix} 2 & -1 & 0 \\ 0 & 2 & -1 \\ -1 & -1 & 1 \end{bmatrix} \times \begin{bmatrix} -5 & 2 & 1 \\ -8 & 3 & 2 \\ -16 & 2 & 6 \end{bmatrix} \times \begin{bmatrix} 1 & 1 & 1 \\ 1 & 2 & 2 \\ 2 & 3 & 4 \end{bmatrix}$$

$$= \begin{bmatrix} -1 & 0 & 0 \\ 0 & 2 & 0 \\ 0 & 0 & 3 \end{bmatrix} := D.$$

This means that $P^{-1}AP$ produced a diagonal matrix D, whose diagonal elements are the eigenvalues of A. Note that, in practice, we never evaluate the inverse matrix P^{-1}.

Note: One point to note is that the order in which the eigenvectors are laid down to form P determines the position of the eigenvalues in D; the eigenvector in column 1 of matrix P will correspond to the eigenvalue in row 1, column 1 of matrix D and so on. For example, suppose

$$P = \left[\underline{x}_2, \ \underline{x}_1, \ \underline{x}_3\right],$$

then

$$D = \begin{bmatrix} \lambda_2 & 0 & 0 \\ 0 & \lambda_1 & 0 \\ 0 & 0 & \lambda_3 \end{bmatrix}.$$

Example 4.8 For the matrix

$$A = \begin{bmatrix} 5 & -2 & 0 \\ 6 & -2 & 0 \\ 8 & -4 & 1 \end{bmatrix},$$

write matrix P which diagonalises A.

Solution

This matrix is considered in Example 4.3 and has the following eigenvalues and eigenvectors:

$$\left(\lambda_1 = 1\ (\text{repeated}),\ \underline{x}_1 = \begin{bmatrix} 1 \\ 2 \\ 0 \end{bmatrix},\ \underline{x}_2 = \begin{bmatrix} 0 \\ 0 \\ 1 \end{bmatrix}\right) \text{ and } \left(\lambda_2 = 2,\ \underline{x}_3 = \begin{bmatrix} 2 \\ 3 \\ 4 \end{bmatrix}\right).$$

Since A has 3 linearly independent eigenvectors then it can be diagonalised. Forming

$$P = \begin{bmatrix} 1 & 0 & 2 \\ 2 & 0 & 3 \\ 0 & 1 & 4 \end{bmatrix}$$

will produce

$$P^{-1}AP = \begin{bmatrix} 1 & 0 & 0 \\ 0 & 1 & 0 \\ 0 & 0 & 2 \end{bmatrix} := D.$$

Note: "Not every matrix can be diagonalised". The matrix

$$A = \begin{bmatrix} 1 & 2 & 2 \\ 0 & 2 & 1 \\ -1 & 2 & 2 \end{bmatrix}$$

has eigenvalues $\lambda_1 = 1$ and $\lambda_2 = 2$ (twice). The associated eigenvectors are as follows:

$$\left(\lambda_1 = 1,\ \underline{x}_1 = \begin{bmatrix} 1 \\ 1 \\ -1 \end{bmatrix}\right) \text{ and } \left(\lambda_2 = 2\ (\text{repeated}),\ \underline{x}_2 = \begin{bmatrix} 2 \\ 1 \\ 0 \end{bmatrix}\right).$$

Since the 3×3 matrix A has only 2 linearly independent eigenvectors, it cannot be diagonalised. As mentioned in Section 4.4, matrices of this type are called defective matrices. In order to solve systems of differential equations associated with defective matrices, we introduce **generalised eigenvectors** which are described in Sections 4.7.2–4.7.8.

4.6.1 Exercises

For each of the matrices below, determine its eigenvalues and eigenvectors. Determine the matrix P that diagonalises A.

(a) $A = \begin{bmatrix} 1 & -3 \\ 1 & -3 \end{bmatrix}$

(b) $A = \begin{bmatrix} -1 & 1 & 1 \\ -12 & 6 & 4 \\ 6 & -2 & 0 \end{bmatrix}$

(b) $A = \begin{bmatrix} -7 & -6 & 1 \\ -2 & -3 & -1 \\ 2 & -6 & -8 \end{bmatrix}$

(d) $A = \begin{bmatrix} 1 & 2 & 2 \\ 0 & 2 & 1 \\ -1 & 2 & 2 \end{bmatrix}$

Answers

(a) $\left(\lambda_1 = 0, \underline{x}_1 = c\begin{bmatrix} 3 \\ 1 \end{bmatrix}\right), \left(\lambda_2 = -2, \underline{x}_2 = c\begin{bmatrix} 1 \\ 1 \end{bmatrix}\right), P = \begin{bmatrix} 3 & 1 \\ 1 & 1 \end{bmatrix}$

(b) $\left(\lambda_1 = 1, \underline{x}_1 = c\begin{bmatrix} 1 \\ 4 \\ -2 \end{bmatrix}\right), \left(\lambda_2 = 2, \underline{x}_2 = c\begin{bmatrix} 1 \\ 0 \\ 3 \end{bmatrix} + d\begin{bmatrix} 0 \\ 1 \\ -1 \end{bmatrix}\right),$

$$P = \begin{bmatrix} 1 & 1 & 0 \\ 4 & 0 & 1 \\ -2 & 3 & -1 \end{bmatrix}$$

$$(\text{or } \underline{x}_2 = c\begin{bmatrix} 1 \\ 3 \\ 0 \end{bmatrix} + d\begin{bmatrix} 1 \\ 0 \\ 3 \end{bmatrix}, P = \begin{bmatrix} 1 & 1 & 1 \\ 4 & 3 & 0 \\ -2 & 0 & 3 \end{bmatrix}$$

$$(\text{or } \underline{x}_2 = c\begin{bmatrix} 1 \\ 0 \\ 3 \end{bmatrix} + d\begin{bmatrix} 0 \\ -1 \\ 1 \end{bmatrix}, P = \begin{bmatrix} 1 & 1 & 0 \\ 4 & 0 & -1 \\ -2 & 3 & 1 \end{bmatrix}$$

(c) $\left(\lambda_1 = 0, \underline{x}_1 = c\begin{bmatrix} 1 \\ -1 \\ 1 \end{bmatrix}\right), \left(\lambda_2 = -9, \underline{x}_2 = c\begin{bmatrix} 1 \\ 0 \\ -2 \end{bmatrix} + d\begin{bmatrix} 0 \\ 1 \\ 6 \end{bmatrix}\right),$

$$P = \begin{bmatrix} 1 & 1 & 0 \\ -1 & 0 & 1 \\ 1 & -2 & 6 \end{bmatrix}$$

$$\left(\text{or } \underline{x}_2 = c\begin{bmatrix} 3 \\ 1 \\ 0 \end{bmatrix} + d\begin{bmatrix} 1 \\ 0 \\ -2 \end{bmatrix}, P = \begin{bmatrix} 1 & 3 & 1 \\ -1 & 1 & 0 \\ 1 & 0 & -2 \end{bmatrix}\right)$$

$$\left(\text{or } \underline{x}_2 = c\begin{bmatrix} 3 \\ 1 \\ 0 \end{bmatrix} + d\begin{bmatrix} 0 \\ 1 \\ 6 \end{bmatrix}, P = \begin{bmatrix} 1 & 3 & 0 \\ -1 & 1 & 1 \\ 1 & 0 & 6 \end{bmatrix}\right)$$

(d) $\left(\lambda_1 = 1, \underline{x}_1 = c\begin{bmatrix} 1 \\ 1 \\ -1 \end{bmatrix}\right), \left(\lambda_2 = 2, \underline{x}_2 = c\begin{bmatrix} 2 \\ 1 \\ 0 \end{bmatrix}\right)$, A cannot be diagonalised.

4.7 HOMOGENEOUS LINEAR SYSTEMS

In Section 4.1, we described a coupled spring-mass system and its mathematical model led us to a second-order linear system of differential Equations 4.1. The system (Equation 4.1) can be solved in its present form or rewritten as four first-order equations by setting $x_3 = \dot{x}_1$ and $x_4 = \dot{x}_2$.

We now write the coupled-linear systems in matrix form and describe methods for finding their solutions. Suppose $x_1(t), x_2(t), \ldots, x_n(t)$ are unknown functions of t that are satisfied in the following equations:

$$\frac{dx_1}{dt} = a_{11}x_1(t) + a_{12}x_2(t) + \cdots + a_{1n}\,x_n(t)$$

$$\frac{dx_2}{dt} = a_{21}x_1(t) + a_{22}x_2(t) + \cdots + a_{2n}\,x_n(t)$$

$$\vdots$$

$$\frac{dx_n}{dt} = a_{n1}x_1(t) + a_{n2}\,x_2(t) + \cdots + a_{nn}\,x_n(t).$$

As can be seen, each of these equations has an unknown function of the other equations. This type of system is known as a coupled system of differential equations and cannot be solved by the direct integration of each equation. The system can be presented in matrix form as

$$\dot{\underline{x}}(t) = A\underline{x}, \tag{4.2}$$

where

$$\underline{x}(t) = \begin{bmatrix} x_1(t) \\ x_2(t) \\ \vdots \\ x_n(t) \end{bmatrix}, \quad \dot{\underline{x}}(t) = \begin{bmatrix} \dot{x}_1(t) \\ \dot{x}_2(t) \\ \vdots \\ \dot{x}_n(t) \end{bmatrix} \text{ and } A = \begin{bmatrix} a_{11} & a_{12} & \cdots & a_{1n} \\ a_{21} & a_{22} & \cdots & a_{2n} \\ \vdots & \vdots & \vdots & \vdots \\ a_{n1} & a_{n2} & \cdots & a_{nn} \end{bmatrix}.$$

The aim is to introduce the well-known process of solving Equation 4.2, which is called diagonalisation method. By using this method, the system of coupled equations is decoupled and solved by integrating each equation separately.

In this book, we only present methods of solving 2×2 and 3×3 systems. There are several cases for 2×2 and 3×3 systems that we consider in Sections 4.7.1–4.7.9.

4.7.1 Solving 2 × 2 Systems (Distinct Real Eigenvalues)

In Equation 4.2, let

$$\underline{x}(t) = \begin{bmatrix} x_1(tvn) \\ x_2(t) \end{bmatrix}, \dot{\underline{x}}(t) = \begin{bmatrix} \dot{x}_1(t) \\ \dot{x}_2(t) \end{bmatrix}, \text{ and } A = \begin{bmatrix} a_{11} & a_{12} \\ a_{21} & a_{22} \end{bmatrix}.$$

The solution of Equation 4.2 depends on the eigenvalues and the associated eigenvectors of the matrix A. Assume the matrix A has 2 distinct real eigenvalues λ_1 and λ_2 associated with the 2 linearly independent eigenvectors $\underline{u}_1$ and $\underline{u}_2$. Note that these vectors are not time-dependent, and their entries are constant numbers. The process of solving Equation 4.2 in this case is as follows:

Construct the matrix $P = [\underline{u}_1, \underline{u}_2]$, using the eigenvectors of A as the column, such that $P^{-1}AP = \text{diag}(\lambda_1, \lambda_2) := D$. Now, pre-multiply Equation 4.2 by the matrix P^{-1}:

$$\begin{aligned} &P^{-1}\underline{\dot{x}}(t) = P^{-1}APP^{-1}\underline{x}(t), \text{where } PP^{-1} = I \\ &\therefore P^{-1}\underline{\dot{x}}(t) = DP^{-1}\underline{x}(t). \end{aligned} \tag{4.3}$$

Assume

$$P^{-1}\underline{x}(t) := \underline{w}(t) = \begin{bmatrix} w_1(t) \\ w_2(t) \end{bmatrix} \Rightarrow \underline{x}(t) = P\underline{w}(t).$$

Note that since the elements of the matrix P are constants, we can conclude that

$$\underline{w}(t) = P^{-1}\underline{x}(t) \Rightarrow \underline{\dot{w}}(t) = P^{-1}\underline{\dot{x}}(t).$$

Hence, Equation 4.3 can be written as

$$\underline{\dot{w}}(t) = D\underline{w}(t). \tag{4.4}$$

Equation 4.4 can be expanded as follows:

$$\begin{bmatrix} \dot{w}_1(t) \\ \dot{w}_2(t) \end{bmatrix} = \begin{bmatrix} \lambda_1 & 0 \\ 0 & \lambda_2 \end{bmatrix} \begin{bmatrix} w_1(t) \\ w_2(t) \end{bmatrix},$$

which gives

$$\frac{dw_1(t)}{dt} = \lambda_1 w_1(t)$$

$$\frac{dw_2(t)}{dt} = \lambda_2 w_2(t).$$

As mentioned earlier, the process of diagonalisation transforms the coupled equations into uncoupled systems of equations so that each equation in the new system depends on only one variable, and hence, the system can be solved by direct integration of each equation separately:

$$w_1(t) = C_1 e^{\lambda_1 t}$$

$$w_2(t) = C_2 e^{\lambda_2 t},$$

where C_1 and C_2 are arbitrary constants. We write this in vector form as

$$\underline{w}(t) = \begin{bmatrix} w_1(t) \\ w_2(t) \end{bmatrix} = \begin{bmatrix} C_1 e^{\lambda_1 t} \\ C_2 e^{\lambda_2 t} \end{bmatrix}.$$

The original unknown solution $\underline{x}(t)$ is determined using $\underline{x}(t) = P\underline{w}(t)$:

$$\begin{aligned} \underline{x}(t) &= P\underline{w}(t) \\ &= [\underline{u}_1, \underline{u}_2] \cdot \begin{bmatrix} C_1 e^{\lambda_1 t} \\ C_2 e^{\lambda_2 t} \end{bmatrix} \\ &= C_1 \underline{u}_1 e^{\lambda_1 t} + C_2 \underline{u}_2 e^{\lambda_2 t} \end{aligned}$$

This is the general solution of the coupled system of linear ordinary differential equations for two distinct real eigenvalues associated with two linearly independent eigenvectors. The arbitrary constants of integration can be determined if the initial conditions are provided.

Note that in the diagonalisation process, as is clear from the earlier analysis, at any point of the process, the computation of the inverse matrix P^{-1} is not required.

Summary: To summarise the steps that we take to solve $\underline{x} = A\,\underline{x}$:

Step 1: Find the eigenvalues of the 2×2 matrix A.
Step 2: Find the corresponding eigenvectors of the matrix A.
Step 3: If the matrix A has 2 linearly independent eigenvectors $\underline{u}_1$ and $\underline{u}_2$ associated with the 2 distinct real eigenvalues λ_1 and λ_2, the general solution of the linear system $\dot{\underline{x}} = A\underline{x}$ is given as follows:

$$\underline{x}(t) = C_1 \underline{u}_1 e^{\lambda_1 t} + C_2 \underline{u}_2 e^{\lambda_2 t}, \tag{4.5}$$

where C_1 and C_2 are arbitrary constants.

Example 4.9 Find the general solution for the following second-order coupled system:

$$\dot{x}_1(t) = 3x_1(t) + 6x_2(t)$$
$$\dot{x}_2(t) = 4x_1(t) + 5x_2(t).$$

SOLUTION

We first write the system in matrix form as $\dot{\underline{x}}(t) = A\,\underline{x}(t)$, where

$$\underline{x}(t) = \begin{bmatrix} x_1(t) \\ x_2(t) \end{bmatrix}, \ \dot{\underline{x}}(t) = \begin{bmatrix} \dot{x}_1(t) \\ \dot{x}_2(t) \end{bmatrix} \text{ and } A = \begin{bmatrix} 3 & 6 \\ 4 & 5 \end{bmatrix}.$$

The eigenvalues and their associated eigenvectors are

$$\left(\lambda_1 = -1,\ \underline{u}_1 = \begin{bmatrix} 3 \\ -2 \end{bmatrix}\right) \text{ and } \left(\lambda_2 = 9,\ \underline{u}_2 = \begin{bmatrix} 1 \\ 1 \end{bmatrix}\right).$$

The matrix A has 2 linearly independent eigenvectors associated with 2 distinct real eigenvalues; hence, using the Equation 4.5, the general solution can be determined as follows:

$$\begin{bmatrix} x_1(t) \\ x_2(t) \end{bmatrix} = C_1 \begin{bmatrix} 3 \\ -2 \end{bmatrix} e^{-t} + C_2 \begin{bmatrix} 1 \\ 1 \end{bmatrix} e^{9t},$$

or we write it explicitly in terms of the unknown functions:

$$x_1(t) = 3C_1e^{-t} + C_2e^{9t}$$
$$x_2(t) = -2C_1e^{-t} + C_2e^{9t}.$$

4.7.2 Solving 2 × 2 Systems (Algebraic Multiplicity 2, Geometric Multiplicity 1)

In Equation 4.2, let

$$\underline{x}(t) = \begin{bmatrix} x_1(t) \\ x_2(t) \end{bmatrix}, \ \dot{\underline{x}}(t) = \begin{bmatrix} \dot{x}_1(t) \\ \dot{x}_2(t) \end{bmatrix} \text{ and } A = \begin{bmatrix} a_{11} & a_{12} \\ a_{21} & a_{22} \end{bmatrix}.$$

Moreover, we consider a case where λ_1 is a repeated eigenvalue of the matrix A and $\underline{u}_1$ is the only associated eigenvector. In this case, A is a "defective matrix", and the matrix P cannot be constructed; hence, A cannot be diagonalised. The only solution of Equation 4.2 is

$$\underline{x}_1(t) = \underline{u}_1 e^{\lambda_1 t}.$$

To get the general solution, we need another linearly independent solution. The process applied to second-order scalar equations works approximately in finding the second solution. Therefore, we introduce a second solution as

$$\underline{x}_2(t) = \left(\underline{v}_1 + t\underline{v}_0\right)e^{\lambda_1 t}, \tag{4.6}$$

where $\underline{v}_0$ is an integer multiple of the vector $\underline{u}_1$ ($\underline{v}_0 = k\underline{u}_1, k$ is a constant value). The vector $\underline{v}_1$ is called the "**generalised eigenvector**" associated with the eigenvalue λ_1. We determine $\underline{v}_1$ and $\underline{v}_0$ by substituting $\underline{x}_2(t)$ and its derivative $\dot{\underline{x}}_2(t)$ in Equation 4.2. So,

$$\dot{\underline{x}}_2(t) = \underline{v}_0\, e^{\lambda_1 t} + \lambda_1\left(\underline{v}_1 + t\underline{v}_0\right)e^{\lambda_1 t}$$

$$\Rightarrow$$

$$\underline{v}_0\, e^{\lambda_1 t} + \lambda_1\left(\underline{v}_1 + t\underline{v}_0\right)e^{\lambda_1 t} = A\left(\underline{v}_1 + t\underline{v}_0\right)e^{\lambda_1 t}$$

$$\Rightarrow$$

$$A\underline{v}_1 = \lambda_1\underline{v}_1 + \underline{v}_0 \quad \text{OR} \quad \left(A - \lambda_1 I\right)\underline{v}_1 = \underline{v}_0 \tag{4.7}$$

$$A\underline{v}_0 = \lambda_1\underline{v}_0 \quad \text{OR} \quad \left(A - \lambda_1 I\right)\underline{v}_0 = \underline{0}.$$

Multiplying Equation 4.7 by $\left(A - \lambda_1 I\right)$ gives

$$\left(A - \lambda_1 I\right)^2 \underline{v}_1 = \left(A - \lambda_1 I\right)\underline{v}_0 = 0.$$

Hence, to obtain the second solution, we need to determine the vectors $\underline{v}_1$ and $\underline{v}_0$ such that they satisfy the following conditions:

$$\left(A - \lambda_1 I\right)^2 \underline{v}_1 = \underline{0}$$

$$\left(A - \lambda_1 I\right)\underline{v}_1 = \underline{v}_0,$$

where $\underline{v}_1$ and $\underline{u}_1$ must be linearly independent and $\underline{v}_0 = k\underline{u}_1$ (k is a constant value). After finding the second solution, the general solution of the system can be written as

$$\underline{x}(t) = C_1\underline{u}_1 e^{\lambda_1 t} + C_2\left(\underline{v}_1 + t\underline{v}_0\right)e^{\lambda_1 t},$$

where C_1 and C_2 are arbitrary constants.

Summary: To solve the 2×2 systems $\dot{\underline{x}} = A\,\underline{x}$ when the matrix A has a repeated eigenvalue λ_1 with the associated eigenvector $\underline{u}_1$, we complete the following steps:

Step 1: $\underline{x}_1(t) = \underline{u}_1 e^{\lambda_1 t}$.

Step 2: We find the vectors $\underline{v}_1$ and $\underline{v}_0$ such that they satisfy the following conditions:

$$(A-\lambda I)^2\,\underline{v}_1 = \underline{0}$$

$$(A-\lambda I)\underline{v}_1 = \underline{v}_0.$$

We find the vectors $\underline{v}_1$ and $\underline{v}_0$ such that they satisfy the following conditions:

$$(A-\lambda I)^2\,\underline{v}_1 = \underline{0}$$

$$(A-\lambda I)\underline{v}_1 = \underline{v}_0,$$

where $\underline{v}_1$ is the generalised eigenvector associated with λ_1 that is linearly independent of $\underline{u}_1$ and $\underline{v}_0 = k\underline{u}_1$ (k is an integer constant).

Step 3: $\underline{x}_2(t) = \left(\underline{v}_1 + t\underline{v}_0\right)e^{\lambda_1 t}$.

Step 4: The general solution is

$$\underline{x}(t) = C_1\underline{u}_1 e^{\lambda_1 t} + C_2\left(\underline{v}_1 + t\underline{v}_0\right)e^{\lambda_1 t}, \tag{4.8}$$

where C_1 and C_2 are arbitrary constants.

Example 4.10 Find the general solution for the following coupled system:

$$\dot{x}_1(t) = x_1(t) - x_2(t)$$

$$\dot{x}_2(t) = x_1(t) + 3x_2(t).$$

SOLUTION

We first write the system in matrix form as $\dot{\underline{x}}(t) = A\,\underline{x}(t)$, where

$$\underline{x}(t) = \begin{bmatrix} x_1(t) \\ x_2(t) \end{bmatrix},\ \dot{\underline{x}}(t) = \begin{bmatrix} \dot{x}_1(t) \\ \dot{x}_2(t) \end{bmatrix} \text{ and } A = \begin{bmatrix} 1 & -1 \\ 1 & 3 \end{bmatrix}.$$

The characteristic equation has repeated roots, as shown:

$$|A-\lambda I|=0$$

$$\lambda^2-4\lambda+4=0$$

$$(\lambda-2)^2=0\ (\lambda_1=2\,\text{repeated}).$$

The only eigenvector of matrix A associated with the repeated eigenvalue $\lambda_1=2$ is

$$\underline{u}_1=\begin{bmatrix}-1\\1\end{bmatrix}$$

So, a solution of the system is $\underline{x}_1(t)=\underline{u}_1e^{2t}$. To find the second linearly independent solution, we find the vectors $\underline{v}_1$ and $\underline{v}_0$ such that they satisfy the following conditions:

$$(A-2I)^2\,\underline{v}_1=\underline{0}$$

$$(A-2I)\underline{v}_1=\underline{v}_0.$$

The matrix $(A-2I)^2$ is a matrix with zero elements:

$$(A-2I)^2=\begin{bmatrix}-1&-1\\1&1\end{bmatrix}\times\begin{bmatrix}-1&-1\\1&1\end{bmatrix}=\begin{bmatrix}0&0\\0&0\end{bmatrix}.$$

Hence, given that $\underline{v}_1$ and $\underline{u}_1$ must be linearly independent vectors, v_1 can be identified. We choose

$$\underline{v}_1=\begin{bmatrix}0\\1\end{bmatrix}$$

$$\Rightarrow$$

$$(A-2I)\underline{v}_1=\underline{v}_0$$

$$\Rightarrow$$

$$\underline{v}_0=\begin{bmatrix}-1&-1\\1&1\end{bmatrix}\begin{bmatrix}0\\1\end{bmatrix}=\begin{bmatrix}-1\\1\end{bmatrix}.$$

Note that $\underline{v}_0=\underline{u}_1$. Hence, the second solution of the system is

$$\underline{x}_2(t)=(\underline{v}_1+t\underline{v}_0)e^{2t}=\left(\begin{bmatrix}0\\1\end{bmatrix}+t\begin{bmatrix}-1\\1\end{bmatrix}\right)e^{2t}.$$

Now, we can write the general solution of the system using Equation 4.8 as follows:

$$\begin{bmatrix} x_1(t) \\ x_2(t) \end{bmatrix} = C_1 \begin{bmatrix} -1 \\ 1 \end{bmatrix} e^{2t} + C_2 \left(\begin{bmatrix} 0 \\ 1 \end{bmatrix} + t \begin{bmatrix} -1 \\ 1 \end{bmatrix} \right) e^{2t},$$

where C_1 and C_2 are arbitrary constants. Alternatively, writing it explicitly in terms of the unknown functions, we get

$$x_1(t) = -C_1 e^{2t} - C_2 t e^{2t}$$

$$x_2(t) = C_1 e^{2t} + C_2 (1+t) e^{2t}.$$

4.7.3 Solving 2 × 2 Second-Order Systems

The diagonalisation process can be extended to second-order linear systems of ordinary differential equations

$$\ddot{x}(t) = Ax(t).$$

In this book, we only consider the case where matrix A has two distinct real eigenvalues.

We omit the details of the diagonalisation process and limit ourselves to two examples in this context.

Example 4.11 Find the general solution for the following second-order coupled system:

$$\ddot{x}_1(t) = -6x_1(t) + 4x_2(t)$$

$$\ddot{x}_2(t) = -5x_1(t) + 3x_2(t).$$

Solution

In this example, the diagonalisation method explained in Section 4.7.1 is used to solve a second-order system. We write the system in matrix form as $\ddot{x} = A\underline{x}$, where

$$\underline{x}(t) = \begin{bmatrix} x_1(t) \\ x_2(t) \end{bmatrix}, \ \underline{\ddot{x}}(t) = \begin{bmatrix} \ddot{x}_1(t) \\ \ddot{x}_2(t) \end{bmatrix} \text{ and } A = \begin{bmatrix} -6 & 4 \\ -5 & 3 \end{bmatrix}.$$

The eigenvalues and their associated eigenvectors of the matrix A are

$$\left(\lambda_1 = -1, \ \underline{u}_1 = \begin{bmatrix} 4 \\ 5 \end{bmatrix} \right) \text{ and } \left(\lambda_2 = -2, \ \underline{u}_2 = \begin{bmatrix} 1 \\ 1 \end{bmatrix} \right).$$

Since A has distinct eigenvalues and linearly independent eigenvectors, it can be diagonalised.

Forming $P=\begin{bmatrix}4 & 1\\ 5 & 1\end{bmatrix}$ will produce $P^{-1}AP=\begin{bmatrix}-1 & 0\\ 0 & -2\end{bmatrix}=D$. Now we can uncouple the system as follows:

$$\ddot{\underline{x}}(t)=A\,\underline{x}(t)\Rightarrow$$

$$P^{-1}\ddot{\underline{x}}(t)=P^{-1}APP^{-1}\underline{x}(t) \text{ OR}$$

$$\ddot{\underline{w}}(t)=D\underline{w}(t),$$

where $\underline{w}(t)=P^{-1}\underline{x}(t)$. The system $\ddot{\underline{w}}(t)=D\underline{w}(t)$ can be written out in full as

$$\begin{bmatrix}\ddot{w}_1(t)\\ \ddot{w}_2(t)\end{bmatrix}=\begin{bmatrix}-1 & 0\\ 0 & -2\end{bmatrix}\begin{bmatrix}w_1(t)\\ w_2(t)\end{bmatrix},$$

which when multiplied out gives the uncoupled differential equations

$$\frac{d^2w_1(t)}{dt^2}=-w_1(t) \text{ and } \frac{d^2w_2(t)}{dt^2}=-2w_2(t).$$

Rearranging the equations gives

$$\frac{d^2w_1(t)}{dt^2}+w_1(t)=0 \text{ and } \frac{d^2w_2(t)}{dt^2(t)}+2w_2(t)=0.$$

The solution for these equations is fully described in Chapter 2. The roots of characteristic equation for both equations are complex numbers with the real parts zero and the imaginary parts 1 and $\sqrt{2}$, respectively. So, the solutions are as follows:

$$w_1(t)=A_1\sin t+B_1\cos t \text{ and } w_2(t)=A_2\sin\sqrt{2}t+B_2\cos\sqrt{2}t,$$

where A_1, B_1, A_2 and B_2 are arbitrary constants. The solution in terms of the original functions $x_1(t)$ and $x_2(t)$ is now found using $\underline{x}(t)=P\underline{w}(t)$ as follows:

$$\begin{bmatrix}x_1(t)\\ x_2(t)\end{bmatrix}=\begin{bmatrix}4 & 1\\ 5 & 1\end{bmatrix}\begin{bmatrix}w_1(t)\\ w_2(t)\end{bmatrix}.$$

Hence, multiplying out the matrix gives

$$x_1(t)=4w_1(t)+w_2(t)=4A_1\sin 2t+4B_1\cos 2t+A_2\sin\sqrt{2}t+B_2\cos\sqrt{2}\,t$$

$$x_2(t)=5\,w_1(t)+w_2(t)=5A_1\sin 2t+5B_1\cos 2t+A_2\sin\sqrt{2}\,t+B_2\cos\sqrt{2}\,t.$$

A more precise way to write the general solution is as follows:

$$\begin{bmatrix} x_1(t) \\ x_2(t) \end{bmatrix} = (A_1 \sin t + B_1 \cos t)\begin{bmatrix} 4 \\ 5 \end{bmatrix} + (A_2 \sin\sqrt{2}\,t + B_2 \cos\sqrt{2}t)\begin{bmatrix} 1 \\ 1 \end{bmatrix}.$$

As you can see from the solution, the general solution depends on the eigenvalues and eigenvectors of the matrix A, so we can write in matrix form as

$$\underline{x}(t) = (A_1 \sin t + B_1 \cos t)\underline{u}_1 + (A_2 \sin\sqrt{2}\,t + B_2 \cos\sqrt{2}\,t)\underline{u}_2.$$

Note: The solution obtained in Example 4.11 can be generalised and extended to any $n \times n$ matrix with n distinct real negative eigenvalues $-\omega_1^2, -\omega_2^2, -\omega_3^2, \cdots, -\omega_n^2$ associated with n linearly independent eigenvectors $\underline{u}_1, \underline{u}_2, \underline{u}_3, \cdots, \underline{u}_n$. The general solution of such systems can be written as

$$\underline{x}(t) = \sum_{i=1}^{n} (A_i \sin\omega_i t + B_i \cos\omega_i t)\underline{u}_i,$$

where A_i and B_i are arbitrary constants.

Example 4.12 Find the solution of the following coupled system subject to the initial conditions $x_1(0) = x_2(0) = 1$ and $\dot{x}_1(0) = \dot{x}_2(0) = 0$:

$$\ddot{x}_1(t) = -6x_1(t) + 4x_2(t)$$

$$\ddot{x}_2(t) = -5x_1(t) + 3x_2(t).$$

Solution

We obtained the following general solution of this system in Example 4.11:

$$\begin{bmatrix} x_1(t) \\ x_2(t) \end{bmatrix} = (A_1 \sin t + B_1 \cos t)\begin{bmatrix} 4 \\ 5 \end{bmatrix} + (A_2 \sin\sqrt{2}\,t + B_2 \cos\sqrt{2}\,t)\begin{bmatrix} 1 \\ 1 \end{bmatrix}.$$

At $t = 0$, $x_1(0) = x_2(0) = 1$, so

$$\begin{bmatrix} 1 \\ 1 \end{bmatrix} = (A_1 \sin 0 + B_1 \cos 0)\begin{bmatrix} 4 \\ 5 \end{bmatrix} + (A_2 \sin 0 + B_2 \cos 0)\begin{bmatrix} 1 \\ 1 \end{bmatrix}$$

$$\begin{bmatrix} 1 \\ 1 \end{bmatrix} = B_1\begin{bmatrix} 4 \\ 5 \end{bmatrix} + B_2\begin{bmatrix} 1 \\ 1 \end{bmatrix} \Rightarrow \begin{matrix} 1 = 4B_1 + B_2 \\ 1 = 5B_1 + B_2 \end{matrix}.$$

Subtracting the two equations gives $B_1 = 0$ and hence $B_2 = 1$. Before using the other two conditions, we have to differentiate the general solution:

$$\begin{bmatrix} \dot{x}_1(t) \\ x_2(t) \end{bmatrix} = \left(A_1 \cos t - B_1 \sin t\right)\begin{bmatrix} 4 \\ 5 \end{bmatrix} + \left(\sqrt{2}A_2 \cos\sqrt{2}\,t - \sqrt{2}B_2 \sin\sqrt{2}\,t\right)\begin{bmatrix} 1 \\ 1 \end{bmatrix}.$$

At $t = 0$, $\dot{x}_1(0) = \dot{x}_2(0) = 0$, so

$$\begin{bmatrix} 0 \\ 0 \end{bmatrix} = \left(A_1 \cos 0 - B_1 \sin 0\right)\begin{bmatrix} 4 \\ 5 \end{bmatrix} + \left(\sqrt{2}A_2 \cos 0 - \sqrt{2}B_2 \sin 0\right)\begin{bmatrix} 1 \\ 1 \end{bmatrix}$$

$$\begin{bmatrix} 0 \\ 0 \end{bmatrix} = A_1\begin{bmatrix} 4 \\ 5 \end{bmatrix} + \sqrt{2}A_2\begin{bmatrix} 1 \\ 1 \end{bmatrix} \Rightarrow \begin{matrix} 0 = 4A_1 + \sqrt{2}A_2 \\ 0 = 5A_1 + \sqrt{2}A_2. \end{matrix}$$

Subtracting the two equations gives $A_1 = 0 \Rightarrow A_2 = 0$. Hence, the particular solution is

$$\begin{bmatrix} x_1(t) \\ x_2(t) \end{bmatrix} = \left(0\sin 2t + 0\cos 2t\right)\begin{bmatrix} 4 \\ 5 \end{bmatrix} + \left(0\sin\sqrt{2}\,t + \cos\sqrt{2}\,t\right)\begin{bmatrix} 1 \\ 1 \end{bmatrix}$$

$$x_1(t) = \cos\sqrt{2}\,t \text{ and } x_2(t) = \cos\sqrt{2}\,t.$$

4.7.4 Solving 3 × 3 Systems (3 Distinct Real Eigenvalues)

In this section, we reconsider Equation 4.2 with

$$\underline{x}(t) = \begin{bmatrix} x_1(t) \\ x_2(t) \\ x_3(t) \end{bmatrix},\ \dot{\underline{x}}(t) = \begin{bmatrix} \dot{x}_1(t) \\ \dot{x}_2(t) \\ \dot{x}_3(t) \end{bmatrix} \text{ and } A = \begin{bmatrix} a_{11} & a_{12} & a_{13} \\ a_{21} & a_{22} & a_{23} \\ a_{31} & a_{32} & a_{33} \end{bmatrix}.$$

The solution of Equation 4.2, as explained for 2×2 systems, depends on the eigenvalues and eigenvectors of the matrix A. Assume the characteristic equation of the matrix A is

$$|A - \lambda I| = (\lambda - \lambda_1)(\lambda - \lambda_2)(\lambda - \lambda_3) = 0,$$

and hence, λ_1, λ_2 and λ_3 are 3 distinct real eigenvalues. Let $\underline{u}_1$, $\underline{u}_2$ and $\underline{u}_3$ be 3 associated linearly independent eigenvectors. In this case, the matrix $P = \left[\underline{u}_1, \underline{u}_2, \underline{u}_3\right]^T$ diagonalises the matrix A, so we can use the same process as described in Section 4.7.1 and obtain the general solution as

$$\underline{x}(t) = C_1\underline{u_1}e^{\lambda_1 t} + C_2\underline{u_2}e^{\lambda_2 t} + C_3\underline{u_3}e^{\lambda_3 t}, \tag{4.9}$$

where C_1, C_2 and C_3 are arbitrary constants.

Example 4.13 Find the general solution for the following coupled system:

$$\begin{aligned}\dot{x}_1 &= -5x_1 + 2x_2 + x_3\\ \dot{x}_2 &= -8x_1 + 3x_2 + 2x_3\\ \dot{x}_3 &= -16x_1 + 2x_2 + 6x_3.\end{aligned}$$

Solution

In matrix form $\underline{\dot{x}} = A\underline{x}$, where

$$\underline{x}(t) = \begin{bmatrix} x_1(t) \\ x_2(t) \\ x_3(t) \end{bmatrix}, \underline{\dot{x}}(t) = \begin{bmatrix} \dot{x}_1(t) \\ \dot{x}_2(t) \\ \dot{x}_3(t) \end{bmatrix} \text{ and } A = \begin{bmatrix} -5 & 2 & 1 \\ -8 & 3 & 2 \\ -16 & 2 & 6 \end{bmatrix}.$$

In Example 4.2, we have already calculated the eigenvalues and eigenvectors of matrix A. The eigenvalues and their associated eigenvectors are

$$\left(\lambda_1 = -1,\ \underline{u_1} = \begin{bmatrix} 1 \\ 1 \\ 2 \end{bmatrix}\right), \left(\lambda_2 = 2,\ \underline{u_2} = \begin{bmatrix} 1 \\ 2 \\ 3 \end{bmatrix}\right), \text{ and } \left(\lambda_3 = 3,\ \underline{u_3} = \begin{bmatrix} 1 \\ 2 \\ 4 \end{bmatrix}\right).$$

Since A has distinct eigenvalues, using Equation 4.9, the general solution is

$$\underline{x}(t) = C_1\underline{u_1}e^{-t} + C_2\underline{u_2}e^{2t} + +C_3\underline{u_3}e^{3t},$$

where C_1, C_2 and C_3 are arbitrary constants.

4.7.5 Solving 3×3 Systems (Algebraic Multiplicity 2 and Geometric Multiplicity 2)

Assume in Equation 4.2 that matrix A is a 3×3 matrix and the characteristic equation of the matrix A is

$$|A - \lambda I| = (\lambda - \lambda_1)^2(\lambda - \lambda_2) = 0.$$

Let $\underline{u}_1$ and $\underline{u}_1'$ be two linearly independent eigenvectors associated with the repeated eigenvalue λ_1 and $\underline{u}_2$ linearly independent eigenvector associated with the eigenvalue λ_2. In this case, the matrix $P = \left[\underline{u}_1, \underline{u}_1', \underline{u}_2\right]^T$ diagonalises the matrix A as $P^{-1}AP = \text{diag}(\lambda_1, \lambda_1, \lambda_2)$. So, we can use the same process as described in Section 4.7.1 and obtain the general solution

$$\underline{x}(t) = \left(C_1\underline{u}_1 + C_2\underline{u}_1'\right)e^{\lambda_1 t} + C_3\underline{u}_2 e^{\lambda_2 t}, \tag{4.10}$$

where C_1, C_2 and C_3 are arbitrary constants.

Example 4.14

Find the general solution for the following coupled system:

$$\dot{x}_1(t) = 5x_1(t) - 2x_2(t)$$
$$\dot{x}_2(t) = 6x_1(t) - 2x_2(t)$$
$$\dot{x}_3(t) = 8x_1(t) - 4x_2(t) + x_3(t).$$

Solution

In matrix form the system can be written as $\underline{\dot{x}} = A\underline{x}$, where

$$\underline{x}(t) = \begin{bmatrix} x_1(t) \\ x_2(t) \\ x_3(t) \end{bmatrix}, \underline{\dot{x}}(t) = \begin{bmatrix} \dot{x}_1(t) \\ \dot{x}_2(t) \\ \dot{x}_3(t) \end{bmatrix} \text{and } A = \begin{bmatrix} 5 & -2 & 0 \\ 6 & -2 & 0 \\ 8 & -4 & 1 \end{bmatrix}.$$

The matrix A is considered in Example 4.3 and has a repeated eigenvalue $\lambda_1 = 1$ and an eigenvalue $\lambda_2 = 2$. The associated eigenvectors are

$$\left(\lambda_1 = 1, \underline{u}_1 = \begin{bmatrix} 1 \\ 2 \\ 0 \end{bmatrix}, \underline{u}_1' = \begin{bmatrix} 0 \\ 0 \\ 1 \end{bmatrix}\right) \text{and} \left(\lambda_2 = 2, \underline{u}_2 = \begin{bmatrix} 2 \\ 3 \\ 4 \end{bmatrix}\right).$$

The general solution for the equation, using Equation 4.10, is given:

$$\underline{x}(t) = \left(C_1\underline{u}_1 + C_2\underline{u}_1'\right)e^{t} + C_3\underline{u}_2 e^{2t}.$$

A more precise way to write the general solution is as follows:

$$x_1(t) = C_1e^t + 2C_3e^{2t}$$
$$x_2(t) = 2C_1e^t + 3C_3e^{2t}$$
$$x_3(t) = C_2e^t + 4C_3e^{2t}.$$

where C_1, C_2 and C_3 are arbitrary constants.

4.7.6 Solving 3 × 3 Systems (Algebraic Multiplicity 2 and Geometric Multiplicity 1)

Assume in Equation 4.2 that matrix A is a 3 × 3 matrix and the characteristic equation of the matrix A is

$$|A - \lambda I| = (\lambda - \lambda_1)^2(\lambda - \lambda_2) = 0.$$

Let $\underline{u}_1$ be the only eigenvector associated with the eigenvalue λ_1 and $\underline{u}_2$ be the eigenvector associated with the eigenvalue λ_2.

In this case, since the matrix A is defective (with the algebraic multiplicity 2 but geometric multiplicity 1) and the matrix P cannot be constructed, A cannot be diagonalised. The two linearly independent solutions can be written as

$$\underline{x}_1(t) = \underline{u}_1e^{\lambda_1 t} \text{ and } \underline{x}_2(t) = \underline{u}_2e^{\lambda_2 t}.$$

In order to write the general solution, we need the third linearly independent solution. We apply the same process as described in Section 4.7.2 and write the third solution as

$$\underline{x}_3(t) = (\underline{v}_1 + t\underline{v}_0)e^{\lambda_1 t}.$$

We find the vectors $\underline{v}_1$ and $\underline{v}_0$ such that they satisfy the following conditions:

$$(A - \lambda_1 I)^2 \underline{v}_1 = 0$$
$$(A - \lambda_1 I)\underline{v}_1 = \underline{v}_0,$$

where $\underline{v}_1$ is the generalised eigenvector associated with λ_1; the vectors $\underline{v}_1$, $\underline{u}_1$ and $\underline{u}_2$ must be linearly independent; and $\underline{v}_0 = k\underline{u}_1$ (k is an integer constant). The general solution can be written as a linear combination of $\underline{x}_1(t), \underline{x}_2(t)$ and $\underline{x}_3(t)$:

$$\underline{x}(t)=C_1\underline{u}_1e^{\lambda_1 t}+C_2\underline{u}_2e^{\lambda_2 t}+C_3\left(\underline{v}_1+t\underline{v}_0\right)e^{\lambda_1 t}, \tag{4.11}$$

where C_1, C_2 and C_3 are arbitrary constants.

Example 4.15 Find the general and particular solution for the following coupled system:

$$\dot{x}_1(t)=x_1(t)+2x_2(t)+2x_3(t)$$

$$\dot{x}_2(t)=2x_2(t)+x_3(t)$$

$$\dot{x}_3(t)=-x_1(t)+2x_2(t)+2x_3(t)$$

$$x_1(0)=1, x_2(0)=-1, x_3(0)=3.$$

Solution

In the matrix form, $\underline{\dot{x}}=A\underline{x}$, where

$$\underline{x}(t)=\begin{bmatrix} x_1(t) \\ x_2(t) \\ x_3(t) \end{bmatrix}, \underline{\dot{x}}(t)=\begin{bmatrix} \dot{x}_1(t) \\ \dot{x}_2(t) \\ \dot{x}_3(t) \end{bmatrix} \text{ and } A=\begin{bmatrix} 1 & 2 & 2 \\ 0 & 2 & 1 \\ -1 & 2 & 2 \end{bmatrix}.$$

The characteristic equation of the matrix A is

$$|A-\lambda I|=(\lambda-2)^2(\lambda-1)=0.$$

Hence, the eigenvalues and the associated eigenvectors of the matrix A are as follows:

$$\left(\lambda_1=2\ (\text{repeated}),\ \underline{u}_1=\begin{bmatrix} 2 \\ 1 \\ 0 \end{bmatrix}\right) \text{ and } \left(\lambda_2=1,\ \underline{u}_2=\begin{bmatrix} 1 \\ 1 \\ -1 \end{bmatrix}\right).$$

Matrix A with algebraic multiplicity 2 and geometrical multiplicity 1 is a defective matrix. Using the eigenvalues and the eigenvectors presented earlier, we can write two linearly independent solutions of the given system as $\underline{x}_1(t)=\underline{u}_1e^{2t}$ and $\underline{x}_2(t)=\underline{u}_2e^{t}$. In order to write the general solution, we need the third linearly independent solution. We write the third solution as $\underline{x}_3(t)=(\underline{v}_1+t\underline{v}_0)e^{2t}$ and obtain $\underline{v}_1$ and $\underline{v}_0$ using the following conditions:

$$(A-2I)^2\underline{v}_1=0$$

$$(A-2I)\underline{v}_1=\underline{v}_0,$$

where $\underline{v}_1,\underline{u}_1$ and $\underline{u}_2$ must be linearly independent vectors, and $\underline{v}_0=k\underline{u}_1$ (k is a constant integer).

First, we determine v_1:

$$(A-2I)^2\underline{v}_1=\begin{bmatrix}-1&2&2\\0&0&1\\-1&2&0\end{bmatrix}\times\begin{bmatrix}-1&2&2\\0&0&1\\-1&2&0\end{bmatrix}\begin{bmatrix}v_1\\v_2\\v_3\end{bmatrix}$$

$$=\begin{bmatrix}-1&2&0\\-1&2&0\\1&-2&0\end{bmatrix}\begin{bmatrix}v_1\\v_2\\v_3\end{bmatrix}=\begin{bmatrix}0\\0\\0\end{bmatrix}.$$

The expansion of the preceding equation provides only one equation:

$$-v_1+2v_2=0\Rightarrow v_1=2v_2.$$

We can assign any number to v_2 and v_3 given that $\underline{v}_1,\underline{u}_1$ and $\underline{u}_2$ are linearly independent vectors. One possible option is to choose $v_2=0$ and $v_3=1$, which gives

$$\underline{v}_1=\begin{bmatrix}v_1\\v_2\\v_3\end{bmatrix}=\begin{bmatrix}0\\0\\1\end{bmatrix},$$

or the other option is to choose $v_2=1$ and $v_3=1$, which gives

$$\underline{v}_1=\begin{bmatrix}v_1\\v_2\\v_3\end{bmatrix}=\begin{bmatrix}2\\1\\1\end{bmatrix}.$$

Now, using $\underline{v}_1$, we determine $\underline{v}_0$

$$(A-2I)\underline{v}_1=\underline{v}_0\Rightarrow\underline{v}_0=\begin{bmatrix}-1&2&2\\0&0&1\\-1&2&0\end{bmatrix}\begin{bmatrix}0\\0\\1\end{bmatrix}=\begin{bmatrix}2\\1\\0\end{bmatrix}.$$

Note that $\underline{v}_0 = \underline{u}_1$. Therefore, using $\underline{v}_0$ and $\underline{v}_1$, we can write the third solution $\underline{x}_3(t)$. Referring to Equation 4.11, the general solution is the linear combination of $\underline{x}_1(t)$, $\underline{x}_2(t)$ and $\underline{x}_3(t)$:

$$\underline{x}(t) = C_1\underline{u}_1 e^{2t} + C_2\underline{u}_2 e^{t} + C_3\left(\underline{v}_1 + t\underline{v}_0\right)e^{2t}$$

$$= C_1\begin{bmatrix}2\\1\\0\end{bmatrix}e^{2t} + C_2\begin{bmatrix}1\\1\\-1\end{bmatrix}e^{t} + C_3\left(\begin{bmatrix}0\\0\\1\end{bmatrix} + t\begin{bmatrix}2\\1\\0\end{bmatrix}\right)e^{2t},$$

where C_1, C_2 and C_3 are arbitrary constants. A more precise way to write the general solution is as follows:

$$x_1(t) = 2C_1e^{2t} + C_2e^{t} + 2C_3te^{2t}$$

$$x_2(t) = C_1e^{2t} + C_2e^{t} + C_3te^{2t}$$

$$x_3(t) = -C_2e^{t} + C_3e^{2t}.$$

Using the initial conditions, we can eliminate the arbitrary constants C_1, C_2 and C_3:

$$x_1(0) = 2C_1 + C_2 = 1$$

$$x_2(0) = C_1 + C_2 = -1$$

$$x_3(t) = -C_2 + C_3 = 3$$

$$\Rightarrow C_1 = 2,\ C_2 = -3,\ C_3 = 0.$$

Hence, the particular solution is

$$x_1(t) = 4e^{2t} - 3e^{t},\ x_2(t) = 2e^{2t} - 3e^{t},\ x_3(t) = 3e^{t}$$

4.7.7 Solving 3×3 Systems (Algebraic Multiplicity 3 and Geometric Multiplicity 2)

Assume in Equation 4.2 that matrix A is a 3×3 matrix and the characteristic equation of the matrix A is

$$|A - \lambda I| = (\lambda - \lambda_1)^3 = 0.$$

Assume the repeated eigenvalue λ_1 is associated with two linearly independent eigenvectors $\underline{u}_1$ and $\underline{u}_1'$ (i.e., matrix A is a defective matrix with algebraic multiplicity

3 and geometric multiplicity 2). The two linearly independent solutions of the system can be written as

$$\underline{x}_1(t) = \underline{u}_1 e^{\lambda_1 t} \text{ and } \underline{x}_2(t) = \underline{u}_1' e^{\lambda_1 t}.$$

In order to write the general solution, we need the third linearly independent solution. We write the third solution as

$$\underline{x}_3(t) = \left(\underline{v}_1 + t\underline{v}_0\right) e^{\lambda_1 t}$$

and seek to find the vectors $\underline{v}_1$ and $\underline{v}_0$, such that they satisfy the following conditions:

$$(A - \lambda I)^2 \underline{v}_1 = 0$$

$$(A - \lambda I)\underline{v}_1 = \underline{v}_0,$$

where $\underline{v}_1$ is the generalised eigenvector associated with λ_1; the vectors $\underline{v}_1$, $\underline{u}_1$, $\underline{u}_1'$ must be linearly independent and $\underline{v}_0 = k_0\underline{u}_1 + k_1\underline{u}_1'$ (k_0 and k_1 are integer constants).

The general solution can be written as a linear combination of $\underline{x}_1(t), \underline{x}_2(t)$ and $\underline{x}_3(t)$:

$$x(t) = (C_1\underline{u}_1 + C_2\underline{u}_1')e^{\lambda_1 t} + C_3\left(\underline{v}_1 + t\underline{v}_0\right)e^{\lambda_1 t}, \tag{4.12}$$

where C_1, C_2 and C_3 are arbitrary constants

Example 4.16 Find the general solution for the following coupled system:

$$\dot{x}_1(t) = 5x_1(t) - 3x_2(t) - 2x_3(t)$$

$$\dot{x}_2(t) = 8x_1(t) - 5x_2(t) - 4x_3(t)$$

$$\dot{x}_3(t) = -4x_1(t) + 3x_2(t) + 3x_3(t).$$

Solution

In matrix form, the system can be written as $\dot{\underline{x}} = A\underline{x}$, where

$$\underline{x} = \begin{bmatrix} x_1 \\ x_2 \\ x_3 \end{bmatrix}, \dot{\underline{x}} = \begin{bmatrix} \dot{x}_1 \\ \dot{x}_2 \\ \dot{x}_3 \end{bmatrix} \text{ and } A = \begin{bmatrix} 5 & -3 & -2 \\ 8 & -5 & -4 \\ -4 & 3 & 3 \end{bmatrix}.$$

The characteristic equation is

$$|A-\lambda I| = -(\lambda-1)^3 = 0 \Rightarrow \lambda_1 = 1.$$

Hence, the matrix A has an eigenvalue $\lambda_1 = 1$ with algebraic multiplicity 3. The eigenvectors associated with the repeated eigenvalue λ_1 are as follows:

$$\left(\lambda_1 = 1,\ \underline{u}_1 = \begin{bmatrix}1\\0\\2\end{bmatrix}\right) \text{and} \left(\lambda_1 = 1,\ \underline{u}_1' = \begin{bmatrix}3\\4\\0\end{bmatrix}\right).$$

Hence, the two linearly independent solutions of the system can be written as

$$\underline{x}_1(t) = \underline{u}_1 e^t \text{ and } \underline{x}_2(t) = \underline{u}_1'.$$

We write the third solution as

$$\underline{x}_3(t) = (\underline{v}_1 + t\underline{v}_0)e^t$$

and obtain $\underline{v}_1$ and $\underline{v}_0$ such that they satisfy the following conditions:

$$(A-I)^2 \underline{v}_1 = 0$$

$$(A-I)\underline{v}_1 = \underline{v}_0$$

given that $\underline{v}_1, \underline{u}_1$ and $\underline{u}_1'$ must be linearly independent vectors and $\underline{v}_0 = k_0\underline{u}_1 + k_1\underline{u}_1'$ (k_0 and k_1. are integer constants).

First, we determine v_1:

$$(A-2I)^2 \underline{v}_1 = \begin{bmatrix}4 & -3 & -2\\8 & -6 & -4\\-4 & 3 & 2\end{bmatrix} \times \begin{bmatrix}4 & -3 & -2\\8 & -6 & -4\\-4 & 3 & 2\end{bmatrix}\begin{bmatrix}v_1\\v_2\\v_3\end{bmatrix}$$

$$= \begin{bmatrix}0 & 0 & 0\\0 & 0 & 0\\0 & 0 & 0\end{bmatrix}\begin{bmatrix}v_1\\v_2\\v_3\end{bmatrix} = \begin{bmatrix}0\\0\\0\end{bmatrix}.$$

We can assign any number to v_1, v_2 and v_3, given that $\underline{v}_1, \underline{u}_1$ and $\underline{u}_1'$ must be linearly independent vectors. One possible option is to choose $v_1 = v_2 = 0$ and $v_3 = 1$, which gives

$$\underline{v}_1 = \begin{bmatrix}v_1\\v_2\\v_3\end{bmatrix} = \begin{bmatrix}0\\0\\1\end{bmatrix}.$$

Now, using $\underline{v}_1$, we determine $\underline{v}_0$

$$\left(A-I\right)\underline{v}_1=\underline{v}_0 \Rightarrow \underline{v}_0=\begin{bmatrix} 4 & -3 & -2 \\ 8 & -6 & -4 \\ -4 & 3 & 2 \end{bmatrix}\begin{bmatrix} 0 \\ 0 \\ 1 \end{bmatrix}=\begin{bmatrix} -2 \\ -4 \\ 2 \end{bmatrix}.$$

Note that $\underline{v}_0=\underline{u}_1-\underline{u}_1'$. Therefore, using $\underline{v}_0$ and $\underline{v}_1$ we can write the third solution $\underline{x}_3\left(t\right)$. By referring to Equation 4.12, the general solution is the linear combination of $\underline{x}_1\left(t\right)$, $\underline{x}_2\left(t\right)$ and $\underline{x}_3\left(t\right)$:

$$x\left(t\right)=(C_1\underline{u}_1+C_2\underline{u}_1')e^t+C_3\left(\underline{v}_1+t\underline{v}_0\right)e^t$$

$$=\left(C_1\begin{bmatrix} 1 \\ 0 \\ 2 \end{bmatrix}+C_2\begin{bmatrix} 3 \\ 4 \\ 0 \end{bmatrix}\right)e^t+C_3\left(\begin{bmatrix} 0 \\ 0 \\ 1 \end{bmatrix}+t\begin{bmatrix} -2 \\ -4 \\ 2 \end{bmatrix}\right)e^t,$$

where C_1, C_2 and C_3 are arbitrary constants. A more precise way to write the general solution is as follows:

$$x_1\left(t\right)=\left(C_1+3C_2\right)e^t-2C_3te^t$$

$$x_2\left(t\right)=4C_2e^t-4C_3te^t$$

$$x_3\left(t\right)=2C_1e^t+\left(C_3+2C_3t\right)e^t.$$

Example 4.17 Find the particular solution for the following coupled system:

$$\dot{x}_1\left(t\right)=-3x_1\left(t\right)-x_3\left(t\right)$$

$$\dot{x}_2\left(t\right)=-2x_1\left(t\right)-x_2\left(t\right)-x_3\left(t\right)$$

$$\dot{x}_3\left(t\right)=+4x_1\left(t\right)+x_3\left(t\right)$$

$$x_1\left(0\right)=2,\ x_2\left(0\right)=3 \text{ and } x_3\left(0\right)=-4.$$

Solution

In matrix form, the system can be written as $\dot{\underline{x}}=A\underline{x}$, where

$$\underline{x}=\begin{bmatrix} x_1 \\ x_2 \\ x_3 \end{bmatrix},\ \dot{\underline{x}}=\begin{bmatrix} \dot{x}_1 \\ \dot{x}_2 \\ \dot{x}_3 \end{bmatrix} \text{ and } A=\begin{bmatrix} -3 & 0 & -1 \\ -2 & -1 & -1 \\ 4 & 0 & 1 \end{bmatrix}.$$

The characteristic equation is

$$|A-\lambda I| = -(\lambda+1)^3 = 0 \Rightarrow \lambda_1 = -1.$$

The eigenvectors associated with the repeated eigenvalue λ_1 are as follows:

$$\left(\lambda_1 = -1,\ \underline{u}_1 = \begin{bmatrix} -1 \\ 0 \\ 2 \end{bmatrix}\right) \text{ and } \left(\lambda_1 = -1,\ \underline{u}_1' = \begin{bmatrix} 0 \\ 1 \\ 0 \end{bmatrix}\right).$$

Hence, the two linearly independent solutions of the system can be written as

$$\underline{x}_1(t) = \underline{u}_1 e^{-t} \text{ and } \underline{x}_2(t) = \underline{u}_1' e^{-t}.$$

We write the third solution as

$$\underline{x}_3(t) = \left(\underline{v}_1 + t\underline{v}_0\right)e^{-t}$$

and seek to find $\underline{v}_1$ and $\underline{v}_0$ such that they satisfy the following conditions:

$$(A-I)^2 \underline{v}_1 = 0$$

$$(A-I)\underline{v}_1 = \underline{v}_0$$

given that $\underline{v}_1, \underline{u}_1$ and $\underline{u}_1'$ are linearly independent vectors and $\underline{v}_0 = k_0\underline{u}_1 + k_1\underline{u}_1'$ (k_0 and k_1 are integer constants).

First, we determine v_1 :

$$(A-2I)^2 \underline{v}_1 = \begin{bmatrix} -2 & 0 & -1 \\ -2 & 0 & -1 \\ 4 & 0 & 2 \end{bmatrix} \times \begin{bmatrix} -2 & 0 & -1 \\ -2 & 0 & -1 \\ 4 & 0 & 2 \end{bmatrix} \begin{bmatrix} v_1 \\ v_2 \\ v_3 \end{bmatrix}$$

$$= \begin{bmatrix} 0 & 0 & 0 \\ 0 & 0 & 0 \\ 0 & 0 & 0 \end{bmatrix} \begin{bmatrix} v_1 \\ v_2 \\ v_3 \end{bmatrix} = \begin{bmatrix} 0 \\ 0 \\ 0 \end{bmatrix}.$$

We can assign any number to v_1, v_2, and v_3, given that $\underline{v}_1, \underline{u}_1$ and $\underline{u}_1'$ are linearly independent vectors. One possible option is to choose $v_1 = v_2 = 0$ and $v_3 = 1$, which gives

$$\underline{v}_1 = \begin{bmatrix} v_1 \\ v_2 \\ v_3 \end{bmatrix} = \begin{bmatrix} 0 \\ 0 \\ 1 \end{bmatrix}.$$

Now, using $\underline{v}_1$, we determine $\underline{v}_0$

$$\left(A-I\right)\underline{v}_1=\underline{v}_0 \Rightarrow \underline{v}_0=\begin{bmatrix}-2 & 0 & -1\\ -2 & 0 & -1\\ 4 & 0 & 2\end{bmatrix}\begin{bmatrix}0\\0\\1\end{bmatrix}=\begin{bmatrix}-1\\-1\\2\end{bmatrix}.$$

Therefore, using $\underline{v}_0$ and $\underline{v}_1$ we can write the third solution $\underline{x}_3\left(t\right)$. By referring to Equation 4.12, the general solution is the linear combination of $\underline{x}_1\left(t\right)$, $\underline{x}_2\left(t\right)$ and $\underline{x}_3\left(t\right)$:

$$x\left(t\right)=(C_1\underline{u}_1+C_2\underline{u}_1')e^{-t}+C_3\left(\underline{v}_1+t\underline{v}_0\right)e^{-t}$$

$$=\left(C_1\begin{bmatrix}-1\\0\\2\end{bmatrix}+C_2\begin{bmatrix}0\\1\\0\end{bmatrix}\right)e^{-t}+C_3\left(\begin{bmatrix}0\\0\\1\end{bmatrix}+t\begin{bmatrix}-1\\-1\\2\end{bmatrix}\right)e^{-t},$$

where C_1, C_2 and C_3 are arbitrary constants. Note that $\underline{v}_0=\underline{u}_1-\underline{u}_1'$. A more precise way to write the general solution is as follows:

$$x_1\left(t\right)=-C_1e^{-t}-C_3te^{-t}$$

$$x_2\left(t\right)=C_2e^{-t}-C_3te^{-t}$$

$$x_3\left(t\right)=2C_1\mathrm{e}^{t}+\left(C_3+2C_3t\right)e^{-t}$$

To obtain the particular solution we use the initial condition $\underline{x}\left(0\right)=\left[2,3,-4\right]^T$.

$$\underline{x}\left(0\right)=\left(C_1\begin{bmatrix}-1\\0\\2\end{bmatrix}+C_2\begin{bmatrix}0\\1\\0\end{bmatrix}\right)e^{-0}+C_3\left(\begin{bmatrix}0\\0\\1\end{bmatrix}+0\times\begin{bmatrix}-1\\-1\\2\end{bmatrix}\right)e^{-0}=\begin{bmatrix}2\\3\\-4\end{bmatrix}$$

Multiplying out gives

$$C_1=-2, C_2=3 \text{ and } C_3=0.$$

Hence the particular solution is

$$\underline{x}\left(t\right)=(-2\underline{u}_1+3\underline{u}_1')e^{-t}$$

OR

$$x_1\left(t\right)=2e^{-t},\ x_2\left(t\right)=3e^{-t},\ x_3\left(t\right)=-4e^{-t}.$$

4.7.8 Solving 3×3 Systems (Algebraic Multiplicity 3 and Geometric Multiplicity 1)

Suppose that in Equation 4.2 the matrix A is a 3×3 matrix and the characteristic equation of the matrix A is

$$|A - \lambda I| = (\lambda - \lambda_1)^3 = 0.$$

Assume the repeated eigenvalue λ_1 is associated with only one eigenvector $\underline{u}_1$ (i.e., matrix A is a defective matrix with algebraic multiplicity 3 and geometric multiplicity 1). In this case, we can only identify one solution of the system as

$$\underline{x}_1(t) = \underline{u}_1 e^{\lambda_1 t}.$$

In order to write the general solution, we need two more linearly independent solutions. We write the second and the third solution as

$$\underline{x}_2(t) = \left(\underline{v}_1 + t\underline{v}_0\right) e^{\lambda_1 t}$$

$$\underline{x}_3(t) = \left(\underline{v}_2 + \underline{v}_1 t + \frac{\underline{v}_0}{2} t^2\right) e^{\lambda_1 t}$$

and seek to find vectors $\underline{v}_2$, $\underline{v}_1$ and $\underline{v}_0$ that satisfy the following conditions:

$$(A - \lambda_0 I)^3 \underline{v}_2 = 0$$

$$(A - \lambda_0 I)^2 \underline{v}_2 = \underline{v}_0$$

$$(A - \lambda I) \underline{v}_2 = \underline{v}_1.$$

Note that $\underline{v}_2$, $\underline{v}_1$ and $\underline{u}_1$ must be linearly independent vectors, and $\underline{v}_0 = k\underline{u}_1$ (k is a constant integer).

The general solution, in this case, can be written as

$$\underline{x}(t) = C_1 \underline{u}_1 e^{\lambda_1 t} + C_2 \left(\underline{v}_1 + t\underline{v}_0\right) e^{\lambda_1 t} + C_3 \left(\underline{v}_2 + \underline{v}_1 t + \frac{\underline{v}_0}{2} t^2\right) e^{\lambda_1 t}, \quad (4.13)$$

where C_1, C_2 and C_3 are arbitrary constants.

Example 4.18 Find the particular solution for the following coupled system:

$$\dot{x}_1(t) = -x_1(t) + x_2(t)$$
$$\dot{x}_2(t) = -2x_1(t) - 3x_2(t) + x_3(t)$$
$$\dot{x}_3(t) = x_1(t) + x_2(t) - 2x_3(t)$$
$$x_1(0) = 2, x_2(0) = 3 \text{ and } x_3(0) = -4.$$

Solution

In matrix form, the system can be written as $\underline{\dot{x}} = A\underline{x}$, where

$$\underline{x} = \begin{bmatrix} x_1 \\ x_2 \\ x_3 \end{bmatrix}, \underline{\dot{x}} = \begin{bmatrix} \dot{x}_1 \\ \dot{x}_2 \\ \dot{x}_3 \end{bmatrix} \text{ and } A = \begin{bmatrix} -1 & 1 & 0 \\ -2 & -3 & 1 \\ 1 & 1 & -2 \end{bmatrix}.$$

The characteristic equation is

$$|A - \lambda I| = -(\lambda + 2)^3 = 0 \Rightarrow \lambda_1 = -2 \text{ (multiplicity 3)}.$$

The eigenvector $\underline{u}_1$ associated with λ_1 must satisfy the following

$$(A + 2I)\underline{u}_1 = 0 \Rightarrow \begin{bmatrix} 1 & 1 & 0 \\ -2 & -1 & 1 \\ 1 & 1 & 0 \end{bmatrix} \begin{bmatrix} u_1 \\ u_2 \\ u_3 \end{bmatrix} = \begin{bmatrix} 0 \\ 0 \\ 0 \end{bmatrix}.$$

This equation can be expanded to give

$$u_1 + u_2 = 0 \quad \text{(i)}$$
$$-2u_1 - u_2 + u_3 = 0 \quad \text{(ii)}$$
$$u_1 + u_2 = 0. \quad \text{(iii)}$$

Equations (i) and (iii) supply the same information, which is $u_2 = -u_1$. Substituting it in Equation (ii) gives $u_3 = u_1$, and hence,

$$\underline{u}_1 = \begin{bmatrix} u_1 \\ u_2 \\ u_3 \end{bmatrix} = \begin{bmatrix} u_1 \\ -u_1 \\ u_1 \end{bmatrix} = u_1 \begin{bmatrix} 1 \\ -1 \\ 1 \end{bmatrix}.$$

Setting $u_1 = 1$ gives $\underline{u}_1 = [1, -1, 1]^T$, which is the only eigenvector associated with eigenvalue $\lambda_1 = -2$. One solution of the system is

$$\underline{x}_1(t) = \underline{u}_1 e^{-2t}.$$

In order to write the general solution, we need two more linearly independent solutions. We write the second and the third solution as

$$\underline{x}_2(t) = (\underline{v}_1 + t\underline{v}_0)e^{-2t}$$

$$\underline{x}_3(t) = \left(\underline{v}_2 + \underline{v}_1 t + \frac{\underline{v}_0}{2}t^2\right)e^{-2t}$$

and seek to find vectors $\underline{v}_2$, $\underline{v}_1$ and $\underline{v}_0$ that satisfy the following conditions:

$$(A + 2I)^3 \underline{v}_2 = 0$$

$$(A + 2I)^2 \underline{v}_2 = \underline{v}_0$$

$$(A + 2I)\underline{v}_2 = \underline{v}_1,$$

where $\underline{v}_2$, $\underline{v}_1$ and $\underline{u}_1$ must be linearly independent vectors.

First, we determine $\underline{v}_2$:

$$(A + 2I)^3 \underline{v}_2 = \begin{bmatrix} 1 & 1 & 0 \\ -2 & -1 & 1 \\ 1 & 1 & 0 \end{bmatrix} \times \begin{bmatrix} 1 & 1 & 0 \\ -2 & -1 & 1 \\ 1 & 1 & 0 \end{bmatrix} \times \begin{bmatrix} 1 & 1 & 0 \\ -2 & -1 & 1 \\ 1 & 1 & 0 \end{bmatrix} = \begin{bmatrix} v_1 \\ v_2 \\ v_3 \end{bmatrix}$$

$$= \begin{bmatrix} 0 & 0 & 0 \\ 0 & 0 & 0 \\ 0 & 0 & 0 \end{bmatrix} \begin{bmatrix} v_1 \\ v_2 \\ v_3 \end{bmatrix} = \begin{bmatrix} 0 \\ 0 \\ 0 \end{bmatrix}$$

We can assign any number to v_1, v_2 and v_3 given that $\underline{v}_2$ and $\underline{u}_1$ must be linearly independent vectors. One possible option is to choose $v_1 = v_2 = 0$ and $v_3 = 1$, which gives

$$\underline{v}_2 = \begin{bmatrix} v_1 \\ v_2 \\ v_3 \end{bmatrix} = \begin{bmatrix} 0 \\ 0 \\ 1 \end{bmatrix}.$$

We now determine $\underline{v}_1$ using the condition $(A + 2I)\underline{v}_2 = \underline{v}_1$:

$$\underline{v}_1 = \begin{bmatrix} 1 & 1 & 0 \\ -2 & -1 & 1 \\ 1 & 1 & 0 \end{bmatrix} \begin{bmatrix} 0 \\ 0 \\ 1 \end{bmatrix} = \begin{bmatrix} 0 \\ 1 \\ 0 \end{bmatrix}.$$

Finally, we determine $\underline{v}_0$ using the condition $(A+2I)^2\,\underline{v}_2 = \underline{v}_0$:

$$\underline{v}_0 = \begin{bmatrix} 1 & 1 & 0 \\ -2 & -1 & 1 \\ 1 & 1 & 0 \end{bmatrix} \times \begin{bmatrix} 1 & 1 & 0 \\ -2 & -1 & 1 \\ 1 & 1 & 0 \end{bmatrix} \begin{bmatrix} 0 \\ 0 \\ 1 \end{bmatrix} = \begin{bmatrix} 1 \\ -1 \\ 1 \end{bmatrix}.$$

So, the general solution is the linear combination of the three solutions $\underline{x}_1(t), \underline{x}_2(t)$ and $\underline{x}_3(t)$, which can be written as

$$\underline{x}(t) = C_1\underline{u}_1 e^{-2t} + C_2\left(\underline{v}_1 + t\underline{v}_0\right)e^{-2t} + C_3\left(\underline{v}_2 + \underline{v}_1 t + \frac{\underline{v}_0}{2}t^2\right)e^{-2t}$$

$$= C_1\begin{bmatrix} 1 \\ -1 \\ 1 \end{bmatrix}e^{-2t} + C_2\left(\begin{bmatrix} 0 \\ 1 \\ 0 \end{bmatrix} + t\begin{bmatrix} 1 \\ -1 \\ 1 \end{bmatrix}\right)e^{-2t} + C_3\left(\begin{bmatrix} 0 \\ 0 \\ 1 \end{bmatrix} + \begin{bmatrix} 0 \\ 1 \\ 0 \end{bmatrix}t + \frac{1}{2}\begin{bmatrix} 1 \\ -1 \\ 1 \end{bmatrix}t^2\right)e^{-2t}.$$

A more precise way to write the general solution is as follows:

$$x_1(t) = C_1 e^{-2t} + C_2 t e^{-2t} + \frac{1}{2}C_3 t^2 e^{-2t}$$

$$x_2(t) = -C_1 e^{-2t} + C_2(1-t)e^{-2t} + C_3(t - \frac{1}{2}t^2)e^{-2t}$$

$$x_3(t) = C_1 e^{-2t} + C_2 t e^{-2t} + C_3(1 + \frac{1}{2}t^2)e^{-2t}$$

Using the initial conditions gives

$$x_1(0) = C_1 = 2, x_2(0) = -C_1 + C_2 = 3, x_3(0) = C_1 + C_3 = -4$$
$$\Rightarrow C_1 = 2, C_2 = 5, C_3 = -6$$

Substitute to give

$$x_1(t) = (2 + 5t - 3t^2)e^{-2t}$$

$$x_2(t) = \left(3 - 11t + 3t^2\right)e^{-2t}$$

$$x_3(t) = \left(4 - 5t + 3t^2\right)e^{-2t}.$$

4.7.9 Exercises

1. Use the results of Exercises 4.3.1 to determine the general solutions to the following coupled systems of ordinary differential equations:

(a) $\begin{cases} \dot{x}_1 = x_1 + 2x_2 \\ \dot{x}_2 = 4x_1 + 3x_2 \end{cases}$

(b) $\begin{cases} \dot{x}_1 = -4x_1 + 2x_2 \\ \dot{x}_2 = -2.5x_1 + 2x_2 \end{cases}$

(c) $\begin{cases} \dot{x}_1 = -x_1 + x_2 \\ \dot{x}_2 = x_1 + 2x_2 + x_3 \\ \dot{x}_3 = 3x_2 + x_3 \end{cases}$

(d) $\begin{cases} \dot{x}_1 = -x_1 - x_2 \\ \dot{x}_2 = 0.75x_1 - 1.5x_2 + 3x_3 \\ \dot{x}_3 = 0.125\text{x}_1 + 0.25x_2 - 0.5x_3 \end{cases}$

2. Determine the particular solution of the following system of differential equations.

(a) $\begin{cases} \dot{x}_1 = x_1 + 2x_2 \\ \dot{x}_2 = 4x_1 + 3x_2 \\ (x_1(0) = 0 \text{ and } x_2(0) = 1) \end{cases}$

(b) $\begin{cases} \dot{x}_1 = 2x_1 - 3x_2 \\ \dot{x}_2 = 3x_1 + 8x_2 \\ (x_1(0) = 3, x_2(0) = 3) \end{cases}$

(c) $\begin{cases} \ddot{x}_1 = -4x_1 + 2x_2 \\ \ddot{x}_2 = -2.5x_1 + 2x_2 \\ (x_1(0) = x_2(0) = 0, \dot{x}_1(0) = 16\,\dot{x}_2(0) = 0) \end{cases}$

3. Determine the particular solution of the following system of differential equations.

(a) Three distinct real eigenvalues

$$\begin{cases} \dot{x}_1 = -x_2 - x_3 \\ \dot{x}_2 = -3x_1 + 2x_2 + x_3 \\ \dot{x}_3 = -x_1 + x_2 + 2x_3 \\ (x_1(0) = 2, x_2(0) = 1, x_3 = -4) \end{cases}$$

(b) Algebraic multiplicity 2 and geometric multiplicity 2

$$\begin{cases} \dot{x}_1 = x_2 \\ \dot{x}_2 = x_1 \\ \dot{x}_3 = -x_1 + x_2 + x_3 \\ (x_1(0) = 2, x_2(0) = 0, x_3 = -1) \end{cases}$$

(c) Algebraic multiplicity 2 and geometric multiplicity 2

$$\begin{cases} \dot{x}_1 = x_1 - 2x_2 + 2x_3 \\ \dot{x}_2 = -2x_1 + x_2 + 2x_3 \\ \dot{x}_3 = 2x_1 + 2x_2 + x_3 \\ \left(x_1(0) = 1, x_2(0) = -2, x_3 = 1\right) \end{cases}$$

(d) Algebraic multiplicity 2 and geometric multiplicity 1

$$\begin{cases} \dot{x}_1 = 3x_1 + x_3 \\ \dot{x}_2 = x_1 + x_2 - x_3 \\ \dot{x}_3 = -x_1 + x_3 \\ \left(x_1(0) = -1, x_2(0) = 0, x_3 = 1\right) \end{cases}$$

(e) Algebraic multiplicity 3 and geometric multiplicity 2

$$\begin{cases} \dot{x}_1 = -3x_1 + x_3 \\ \dot{x}_2 = -2x_1 - x_2 + x_3 \\ \dot{x}_3 = -4x_1 + x_3 \\ \left(x_1(0) = -1, x_2(0) = -1, x_3 = -2\right) \end{cases}$$

(f) Algebraic multiplicity 3 and geometric multiplicity 1"

$$\begin{cases} \dot{x}_1 = 3x_1 - x_3 \\ \dot{x}_2 = x_1 + x_2 - x_3 \\ \dot{x}_3 = 4x_1 - x_3 \\ \left(x_1(0) = 1, x_2(0) = -1, x_3 = 3\right) \end{cases}$$

Answers

1.

(a) $$\begin{cases} x_1(t) = Ae^{-t} + Be^{5t} \\ x_2(t) = -Ae^{-t} + 2Be^{5t} \end{cases}$$

(b) $$\begin{cases} x_1(t) = 2Ae^{-3t} + 2Be^{t} \\ x_2(t) = Ae^{-3t} + 5Be^{t} \end{cases}$$

(c) $$\begin{cases} x_1(t) = Ae^{-t} + Be^{3t} + Ce^{-2t} \\ x_2(t) = 4Be^{3t} - Ce^{-2t} \\ x_3(t) = -Ae^{-t} + 3Be^{3t} + 3Ce^{-2t} \end{cases}$$

(d) $\begin{cases} x_1(t) = 4Ae^{-t} - 12Be^{-0.5t} + 4Ce^{-1.5t} \\ x_2(t) = 6Be^{-0.5t} + 2Ce^{-1.5t} \\ x_3(t) = -Ae^{-t} + 5Be^{-0.5t} - Ce^{-1.5t} \end{cases}$

2. (a) $\begin{cases} x_1(t) = e^{5t} - e^{-t} \\ x_2(t) = 2e^{5t} + e^{-t} \end{cases}$

(b) $\begin{cases} x_1(t) = 3e^{5t}(-6t+1) \\ x_2(t) = 3e^{5t}(6t+1) \end{cases}$

(c) $\begin{cases} x_1(t) = -2e^{t} + 2e^{-t} + \dfrac{20}{\sqrt{3}}\sin\sqrt{3}t \\ x_2(t) = -5e^{t} + 5e^{-t} + \dfrac{10}{\sqrt{3}}\sin\sqrt{3}t \end{cases}$

3. (a) The particular solution is obtained using the initial conditions:

$$\begin{cases} x_1(t) = e^{-t} + e^{4t} \\ x_2(t) = -\dfrac{1}{3}e^{-t}\left(-3 - 7e^{2t} + 7e^{5t}\right). \\ x_3(t) = -\dfrac{1}{3}e^{t}\left(7 + 5e^{3t}\right) \end{cases}$$

(b) The particular solution is obtained using the initial conditions:

$$\begin{cases} x_1(t) = e^{-t} + e^{t} \\ x_2(t) = -e^{-t} + e^{t}. \\ x_3(t) = e^{-t} - 2e^{t} \end{cases}$$

(c) The particular solution is obtained using the initial conditions:

$$\begin{cases} x_1(t) = -\dfrac{2}{3}e^{-3t} + \dfrac{5}{3}e^{3t} \\ x_2(t) = -\dfrac{2}{3}e^{-3t} - \dfrac{4}{3}e^{3t}. \\ x_3(t) = \dfrac{2}{3}e^{-3t} + \dfrac{1}{3}e^{3t} \end{cases}$$

(d) $\begin{cases} x_1(t) = -e^{2t} \\ x_2(t) = 2e^{t} - 2e^{2t} \\ x_3(t) = e^{2t} \end{cases}$

(e) $\begin{cases} x_1(t) = -e^{-t} \\ x_2(t) = -e^{-t} \\ x_3(t) = -2e^{-t} \end{cases}$

(f) $\begin{cases} x_1(t) = e^t - te^t \\ x_2(t) = -2e^t + (-1 - 2t + \frac{1}{2}t^2)e^t \\ x_3(t) = (3 - 2t)e^t \end{cases}$

4.8 SYSTEMS WITH COMPLEX EIGENVALUES

In Section 4.7, we described the solution of linear systems of ordinary differential equations with real eigenvalues. Now we explain how to solve the system

$$\dot{\underline{x}}(t) = A\underline{x}(t) \tag{4.14}$$

when the roots of the characteristic equation $|A - \lambda I| = 0$ are complex numbers. We only describe the method for solving 2×2 systems, however it can be easily extended to higher dimensional systems.

For a 2×2 system, the characteristic equation is the following quadratic equation:

$$a\lambda^2 + b\lambda + c = 0,$$

where a, b and c are arbitrary constants. Suppose $b^2 - 4ac < 0$; then the roots of the characteristic equation are complex numbers. In this case the eigenvalues of the matrix A can be denoted as follows:

$$\lambda = \alpha \pm \beta j, j = \sqrt{-1},$$

where α and β are real numbers. Next step in the process of solving the system (Equation 4.14) is to find the associated eigenvectors. Since the elements of matrix A are constant, the associated eigenvectors are also complex.

Let $\underline{x}_1 \pm j\underline{x}_2$ be the eigenvectors associated with the complex eigenvalues $\lambda = \alpha \pm j\beta$. These eigenvalues and eigenvectors give the complex solution of the system (Equation 4.14) as follows:

$$\underline{x}(t) = e^{(\alpha \pm \beta j)t}\left(\underline{x}_1 \pm j\underline{x}_2\right). \tag{4.15}$$

Usually, the solutions that we want must be real. Below we show that the real solutions are the real and the imaginary parts of Equation 4.15.

Let us start with $\lambda = \alpha + \beta j$ and its associated eigenvector $\underline{x}_1 + j\underline{x}_2$:

$$\underline{x}(t) := \underline{x}_1(t) + j\underline{x}_2(t) = e^{(\alpha + j\beta)t}(\underline{x}_1 + j\underline{x}_2)$$

$$= e^{\alpha t}[\cos\beta t + j\sin\beta t)](\underline{x}_1 + j\underline{x}_2)$$

$$= e^{\alpha t}\left\{[\cos\beta t\underline{x}_1 - \sin\beta t\underline{x}_2] + j[\cos\beta t\underline{x}_2 + \sin\beta t\underline{x}_1]\right\}.$$

Equating the real parts and imaginary parts implies:

$$\underline{x}_1(t) = e^{\alpha t}[\cos\beta t\underline{x}_1 - \sin\beta t\underline{x}_2]$$

$$\underline{x}_2(t) = e^{\alpha t}[\cos\beta t\underline{x}_2 + \sin\beta t\underline{x}_1].$$

It is easy to show that the functions $\underline{x}_1(t)$ and $\underline{x}_2(t)$ are linearly independent and real solutions of the system (Equation 4.14; see Box 4.1). Therefore, the general real solution of Equation 4.14 is as follows:

$$\underline{x}(t) = C_1\underline{x}_1(t) + C_2\underline{x}_2(t)$$

$$= C_1 e^{\alpha t}[\cos\beta t\underline{x}_1 - \sin\beta t\underline{x}_2] + C_2 e^{\alpha t}[\cos\beta t\underline{x}_2 + \sin\beta t\underline{x}_1] \tag{4.16}$$

BOX 4.1: PROOF FOR THE REAL SOLUTION OF EQUATION 4.14

Since $\underline{x}(t) = \underline{x}_1(t) + j\underline{x}_2(t)$ is the solution of system (4.14), hence

$$\frac{d}{dt}[\underline{x}_1(t) + j\underline{x}_2(t)] = A[\underline{x}_1(t) + j\underline{x}_2(t)]$$

$$\dot{\underline{x}}_1(t) + j\dot{\underline{x}}_2(t) = A\underline{x}_1(t) + jA\underline{x}_2(t)$$

$$\dot{\underline{x}}_1(t) = A\underline{x}_1(t)$$

$$\dot{\underline{x}}_2(t) = A\underline{x}_2(t)$$

Remark: The eigenvalue of $\lambda = \alpha - \beta j$ leads to the same solution obtained as equation (4.16), so in practice, we only need to use one of the eigenvalues.

Example 4.19 Determine the general solution of the following coupled system of differential equations.

$$\begin{cases} \dot{x}_1 = -4x_1 - 2x_2 \\ \dot{x}_2 = x_1 - 2x_2 \end{cases}$$

Solution

We first write the system in matrix form:

$$\dot{\underline{x}}(t) = A\underline{x}(t),$$

where

$$A = \begin{bmatrix} -4 & -2 \\ 1 & -2 \end{bmatrix}, \underline{x}(t) = \begin{bmatrix} x_1(t) \\ x_2(t) \end{bmatrix} \text{ and } \dot{\underline{x}}(t) = \begin{bmatrix} \dot{x}_1(t) \\ \dot{x}_2(t) \end{bmatrix}.$$

In Section 4.5, we showed that the matrix A has the following complex conjugate eigenvalues:

$$\lambda = -3 \pm j, j = \sqrt{-1}.$$

The eigenvectors associated with these eigenvalues are

$$\underline{x} = \begin{bmatrix} -1 \pm j \\ 1 \end{bmatrix} = \begin{bmatrix} -1 \\ 1 \end{bmatrix} \pm j \begin{bmatrix} 1 \\ 0 \end{bmatrix} := \underline{x}_1 \pm j\underline{x}_2.$$

Now, we can write the general solution by replacing $\alpha = -3$ and $\beta = 1$ in Equation 4.16:

$$\underline{x}(t) = C_1 e^{-3t}\left[\cos(t)\underline{x}_1 - \sin(t)\underline{x}_2\right] + C_2 e^{-3t}\left[\cos(t)\underline{x}_2 + \sin(t)\underline{x}_1\right],$$

where C_1 and C_2 are arbitrary constants and

$$\underline{x}_1 = \begin{bmatrix} -1 \\ 1 \end{bmatrix} \text{ and } \underline{x}_2 = \begin{bmatrix} 1 \\ 0 \end{bmatrix}.$$

Example 4.20 Determine the solution of the following coupled system of differential equations.

$$\begin{cases} \dot{x}_1 = -2x_1 - 4x_2 \\ \dot{x}_2 = x_1 - 2x_2 \end{cases}$$

Solution

We first write the system in matrix form:

$$\dot{\underline{x}}(t) = A\underline{x}(t),$$

where

$$A = \begin{bmatrix} -2 & -4 \\ 1 & -2 \end{bmatrix}, \underline{x}(t) = \begin{bmatrix} x_1(t) \\ x_2(t) \end{bmatrix} \text{ and } \dot{\underline{x}}(t) = \begin{bmatrix} \dot{x}_1(t) \\ \dot{x}_2(t) \end{bmatrix}.$$

In Section 4.5, we showed the matrix A has the following complex conjugate eigenvalues:

$$\lambda = -2 \pm 2j, j = \sqrt{-1}$$

with the associated eigenvectors

$$\underline{x} = \begin{bmatrix} \pm 2j \\ 1 \end{bmatrix} = \begin{bmatrix} 0 \\ 1 \end{bmatrix} \pm j \begin{bmatrix} 2 \\ 0 \end{bmatrix} := \underline{x}_1 \pm j\underline{x}_2.$$

Now, we can write the general solution by replacing $\alpha = -2$ and $\beta = 2$ in Equation 4.16:

$$\underline{x}(t) = C_1 e^{-2t}\left[\cos(2t)\underline{x}_1 - \sin(2t)\underline{x}_2\right] + C_2 e^{-2t}\left[\cos(2t)\underline{x}_2 + \sin(2t)\underline{x}_1\right],$$

where C_1 and C_2 are arbitrary constants and

$$\underline{x}_1 = \begin{bmatrix} 0 \\ 1 \end{bmatrix} \text{ and } \underline{x}_1 = \begin{bmatrix} 2 \\ 0 \end{bmatrix}.$$

Example 4.21 Determine the solution of the following coupled system of differential equations.

$$\begin{cases} \dot{x}_1(t) = 4x_1(t) - 3x_2(t) \\ \dot{x}_2(t) = 3x_1(t) + 4x_2(t) \\ \dot{x}_3(t) = x_1(t) + 2x_2(t) + x_3(t) \end{cases}$$

Solution

We first write the system in matrix form:

$$\dot{\underline{x}}(t) = A\underline{x}(t),$$

where

$$A = \begin{bmatrix} 4 & -3 & 0 \\ 3 & 4 & 0 \\ 1 & 2 & 1 \end{bmatrix} \text{ and } \underline{x}(t) = \begin{bmatrix} x_1(t) \\ x_2(t) \\ x_3(t) \end{bmatrix}, \dot{\underline{x}}(t) = \begin{bmatrix} \dot{x}_1(t) \\ \dot{x}_2(t) \\ \dot{x}_3(t) \end{bmatrix}.$$

In Section 4.5, we showed that matrix A has the following complex conjugate eigenvalues:

$$\lambda_{p,m} = 4 \pm 3j,\ j = \sqrt{-1} \text{ and } \lambda_r = 1.$$

The associated eigenvectors for the complex conjugate eigenvalues $\lambda_{1,2} = 4 \pm 3j$ are

$$\underline{x}_{p,m} = \begin{bmatrix} -3 \\ 9 \\ 5 \end{bmatrix} \pm j \begin{bmatrix} 9 \\ 3 \\ 0 \end{bmatrix} := \underline{x}_1 \pm j\underline{x}_2,$$

and for the real eigenvalue $\lambda_r = 1$ is

$$\underline{x}_r = \begin{bmatrix} 0 \\ 0 \\ 1 \end{bmatrix}.$$

The general solution can be written by replacing $\alpha = 4$ and $\beta = 3$ in Equation 4.16 and using the real eigenvalue with its associated eigenvector $\underline{x}_3$:

$$\underline{x}(t) = C_1 e^{4t}\left[\cos(3t)\underline{x}_1 - \sin(3t)\underline{x}_2\right] + C_2 e^{4t}\left[\cos(3t)\underline{x}_2 + \sin(3t)\underline{x}_1\right] + C_3 e^{t}\underline{x}_r,$$

where C_1, C_2 and C_3 are arbitrary constants and

$$\underline{x}_1 = \begin{bmatrix} -3 \\ 9 \\ 5 \end{bmatrix}, \underline{x}_2 = \begin{bmatrix} 9 \\ 3 \\ 0 \end{bmatrix} \text{ and } x_r = \begin{bmatrix} 0 \\ 0 \\ 1 \end{bmatrix}.$$

4.8.1 Exercises

Using the results of exercises in Section 4.5.1, determine the solution of the following coupled systems of differential equations.

(a) $$\begin{cases} \dot{x}_1(t) = 4x_1(t) + 5x_2(t) \\ \dot{x}_2(t) = -2x_1(t) + 6x_2(t) \end{cases}$$

(b) $$\begin{cases} \dot{x}_1(t) = 4x_1(t) + 5x_2(t) \\ \dot{x}_2(t) = -2x_1(t) + 6x_2(t) \end{cases}$$

(c) $$\begin{cases} \dot{x}_1(t) = x_1(t) + 2x_2(t) \\ \dot{x}_2(t) = -2x_1(t) + x_2(t) \end{cases}$$

(d) $$\begin{cases} \dot{x}_1(t) = 3x_1(t) + x_2(t) \\ \dot{x}_2(t) = -2x_1(t) + x_2(t) \end{cases}$$

(e) $$\begin{cases} \dot{x}_1(t) = x_1(t) - x_2(t) \\ \dot{x}_2(t) = x_1(t) + x_2(t) \\ \dot{x}_3(t) = x_1(t) + 2x_2(t) + x_3(t) \end{cases}$$

Answers

(a) $$\begin{cases} x_1(t) = Ae^{5t}(\cos 3t + 3\sin 3t) + B\left(e^{5t}(-3\cos 3t + \sin 3t)\right) \\ x_2(t) = 2Ae^{5t}\cos 3t + 2Be^{5t}\sin 3t \end{cases}$$

(b) $$\begin{cases} x_1(t) = Ae^{2t}(\cos 8t - 8\sin 8t) + Be^{2t}(8\cos 8t + \sin 8t) \\ x_2(t) = 5Ae^{2t}\cos 8t + 5Be^{2t}\sin 8t \end{cases}$$

(c) $$\begin{cases} x_1(t) = e^{t}\left(A\sin 2t - B\cos 2t\right) \\ x_2(t) = e^{t}\left(A\cos 2t + B\sin 2t)\right) \end{cases}$$

(d) $$\begin{cases} x_1(t) = Ae^{2t}(-\cos t + \sin t) + Be^{2t}(-\cos t - \sin t) \\ x_2(t) = e^{2t}(2A\cos t + 2Be^{2t}\cos t) \end{cases}$$

(e) $$\begin{cases} x_1(t) = e^{t}(A\cos t + B\sin t) \\ x_2(t) = e^{t}(A\sin t - B\cos t) \\ x_3(t) = e^{t}(-2A\cos t + A\sin t - B\cos t - 2B\sin t + C) \end{cases}$$

4.9 ENGINEERING APPLICATIONS

Linear systems of differential equations represent a fundamental framework for modelling dynamic phenomena in various scientific and engineering disciplines. In this section, we consider a couple of such models including a mixture problem and a mechanical system with 2 degrees of freedom which is a spring-mass-damper system with two masses.

4.9.1 A Two-Tank Mixing Problem

Two large tanks A and B, each containing 24 litres of salt water, are connected through pipes as illustrated in Figure 4.2. Tank A receives a continuous influx of fresh water at a rate of 6 l/min, while an equivalent fluid is simultaneously drained from Tank B. Additionally, 8 l/min of fluid is transferred from Tank A to Tank B and 2 l/min flows from Tank B to Tank A. Effective mixing mechanisms ensure homogeneity of the liquid contents within each tank. Initially, Tank A contains a brine solution with 1 kg of salt, while Tank B holds a brine solution with 6 kg of salt. Determine the amount of salt present in each tank at a given time.

Solution

$x_1(t) =$ is the amount of salt in Tank A at time t.

$x_2(t) =$ is the amount of salt in Tank B at time t.

Recall from Chapter 2 that the rate of change of the salt in each tank is as follows:

$$\frac{dx_i}{dt} = \text{inflow} - \text{outflow} = C_1 \times Q_{in} - C_2 \times Q_{out}$$

$$\frac{dx_i}{dt} = (\text{concentration}) \times (\text{flow rate in}) - (\text{concentration}) \times (\text{flow rate out})$$

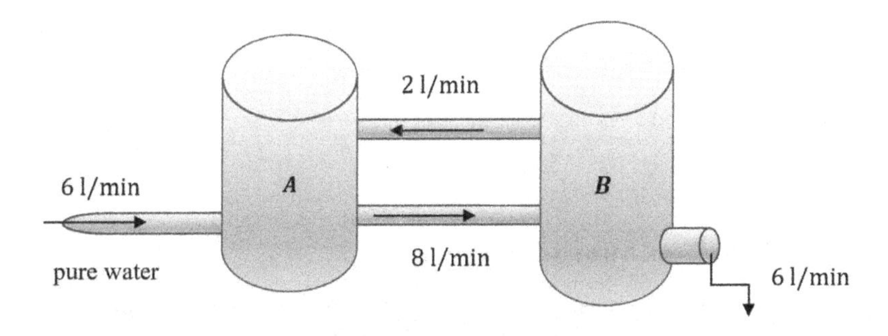

FIGURE 4.2 A two-tank mixing problem.

$$V_1 = 24 + t \times (6 + 2 - 8) = 24 \text{(Volume of Tank } A \text{ is constant at any } t\text{)}$$

$$V_2 = 24 + t \times (8 - 2 - 6) = 24 \text{(Volume of Tank } B \text{ is constant at any } t\text{)}$$

$$\frac{dx_1}{dt} = 0 \times 6 + \frac{x_2}{24} \times 2 - \frac{x_1}{24} \times 8$$

$$\frac{dx_2}{dt} = \frac{x_1}{24} \times 8 - \frac{x_2}{24} \times 6 - \frac{x_2}{24} \times 2,$$

which is simplified to give the following system of equations:

$$\frac{dx_1}{dt} = -\frac{1}{3} x_1 + \frac{1}{12} x_2$$

$$\frac{dx_2}{dt} = \frac{1}{3} x_1 - \frac{1}{3} x_2.$$

The eigenvalues and the associated eigenvectors are

$$\left(\lambda_1 = -\frac{1}{2},\ x_1 = \begin{bmatrix} -1 \\ 2 \end{bmatrix} \right) \text{ and}$$

$$\left(\lambda_2 = -\frac{1}{6},\ x_2 = \begin{bmatrix} 1 \\ 2 \end{bmatrix} \right).$$

The general solution is

$$\underline{x}(t) = C_1 e^{-\frac{1}{2}t} \underline{x_1} + C_2 e^{-\frac{1}{6}t} \underline{x_2}.$$

Using the initial condition $\underline{x} = \begin{bmatrix} 1 \\ 6 \end{bmatrix}$, we obtain

$$x_1(t) = -e^{-\frac{1}{2}t} + 2e^{-\frac{1}{6}t}$$

$$x_2(t) = 2e^{-\frac{1}{2}t} + 4e^{-\frac{1}{6}t}.$$

4.9.2 A Three-Tank Mixing Problem

(a) Given Figure 4.3, construct a mathematical model for the amount of salt $x_1(t), x_2(t)$, and $x_3(t)$ in Tanks A, B and C, respectively. Assume that

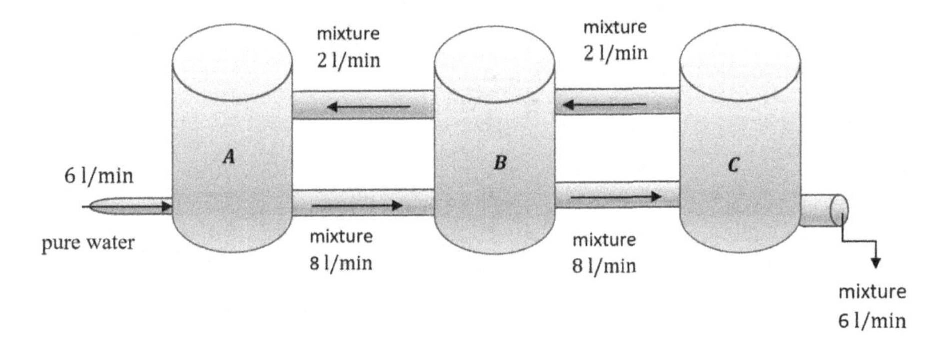

FIGURE 4.3 A three-tank mixing problem.

each tank contains 24 litres of solution (mixture of water and salt). Initially 1 kg salt is dissolved in Tank *A*, 6 kg of salt is dissolved in in Tank *B* and 1 kg of salt is dissolved in Tank *C*. Determine the amount of salt present in each tank at a given time t.

Solution

$x_1(t)$ = is the amount of salt in Tank A at time t.
$x_2(t)$ = is the amount of salt in Tank B at time t.
$x_3(t)$ = is the amount of salt in Tank C at time t.

$$V_1 = 24 + t \times (6 + 2 - 8) = 24 \quad \text{(Volume of Tank } A \text{ is constant at any } t)$$
$$V_2 = 24 + t \times (8 + 2 - 2 - 8) = 24 \quad \text{(Volume of Tank } B \text{ is constant at any } t)$$
$$V_3 = 24 + t \times (8 - 2 - 6) = 24 \quad \text{(Volume of Tank } C \text{ is constant at any } t).$$

So, the volume of each tank remains constant (24 litres) at time *t*. The amount of salt in each tank can be described by the following differential equations:

$$\frac{dx_1}{dt} = \left(0 \times 6 + \frac{x_2}{24} \times 2\right) - \frac{x_1}{24} \times 8$$

$$\frac{dx_2}{dt} = \left(\frac{x_1}{24} \times 8 + \frac{x_3}{24} \times 2\right) - \left(\frac{x_2}{24} \times 8 + \frac{x_2}{24} \times 2\right)$$

$$\frac{dx_3}{dt} = \left(\frac{x_2}{24} \times 8\right) - \left(\frac{x_3}{24} \times 2 + \frac{x_3}{24} \times 6\right),$$

which is simplified to give the following system of equations:

$$\frac{dx_1}{dt} = -\frac{1}{3}x_1 + \frac{1}{12}x_2$$

$$\frac{dx_2}{dt} = \frac{1}{3}x_1 - \frac{5}{12}x_2 + \frac{1}{12}x_3$$

$$\frac{dx_3}{dt} = \frac{1}{3}x_2 - \frac{1}{3}x_3.$$

The eigenvalues and the associated eigenvectors are

$$\left(\lambda_1 = -\frac{1}{3}, \underline{x}_1 = \begin{bmatrix} -1 \\ 0 \\ 4 \end{bmatrix}\right)$$

$$\left(\lambda_2 = \frac{1}{24}\left(-9+\sqrt{33}\right), \underline{x}_2 = \begin{bmatrix} 2 \\ -1+\sqrt{33} \\ 8 \end{bmatrix}\right)$$

$$\left(\lambda_3 = \frac{1}{24}\left(-9-\sqrt{33}\right), \underline{x}_3 = \begin{bmatrix} 2 \\ -1-\sqrt{33} \\ 8 \end{bmatrix}\right).$$

Hence the general solution is

$$\underline{x} = C_1 e^{-\frac{1}{3}t}\underline{x}_1 + C_2 e^{\frac{1}{24}\left(-9+\sqrt{33}\right)t}\underline{x}_2 + C_3 e^{\frac{1}{24}\left(-9-\sqrt{33}\right)t}\underline{x}_3.$$

Using the initial condition $\underline{x}(0) = \left[1,6,1\right]^T$, we obtain the approximate solutions as

$$x_1 \approx \frac{3}{8}e^{-\frac{1}{3}t} + 1.41e^{-0.14t} - 0.79e^{-0.61t}$$

$$x_2 \approx 2.55e^{-0.61t} + 3.35e^{-0.14t}$$

$$x_3 \approx -\frac{3}{2}e^{-\frac{1}{3}t} - 3.15e^{-0.61t} + 5.65e^{-0.14t}.$$

(b) In Figure 4.2, assume the solution (mixture of salt and water) in each tank is 100 litres. Pure water flows into Tank A at a rate of 3 l/min. The other flows are as follows:

- Tank A into Tank B at 4 l/min
- Tank A into Tank C at 3 l/min
- Tank B into Tank A at 2 l/min
- Tank B into Tank C at 2 l/min
- Tank C into Tank A at 2 l/min
- The solution leaves Tank C at 3 l/min.

(i) Construct a mathematical model for the amount of salt $x_1(t), x_2(t)$ and $x_3(t)$ in tanks A, B and C, respectively.
(ii) Initially 1 kg of salt is dissolved in Tank A, 6 kg of salt is dissolved in Tank B and 1 kg of salt is dissolved in Tank C. Determine the amount of salt in each tank at a given time t.
(iii) Determine the amount of salt in each tank after a very long time.

Solution

$$V_1 = 100 + t\times(+7-7) = 100 \quad (\text{Volume of tank } A \text{ is constant at any } t)$$
$$V_2 = 100 + t\times(+4-4) = 100 \quad (\text{Volume of tank } B \text{ is constant at any } t)$$
$$V_3 = 24 + t\times(+5-5) = 100 \quad (\text{Volume of tank } C \text{ is constant at any } t).$$

So, volume of each tank remains constant (100 litres) at time t. The amount of salt in each tank can be described by the following differential equations:

$$\frac{dx_1}{dt} = \left(0\times 3 + \frac{x_2}{100}\times 2 + \frac{x_3}{100}\times 2\right) - \frac{x_1}{100}\times 4 - \frac{x_1}{100}\times 3$$
$$= -0.07x_1 + 0.02x_2 + 0.02x_3$$

$$\frac{dx_2}{dt} = \left(\frac{x_1}{100}\times 4 - \frac{x_2}{100}\times 4\right) = 0.04x_1 - 0.04x_2$$

$$\frac{dx_3}{dt} = \left(\frac{x_1}{100}\times 3 + \frac{x_2}{100}\times 2\right) - \left(\frac{x_3}{100}\times 5\right) = 0.03x_1 + 0.02x_2 - 0.05x_3.$$

In matrix form, $\dot{\underline{x}}(t) = A\underline{x}(t)$, where

$$A = \begin{bmatrix} -0.07 & 0.02 & 0.02 \\ 0.04 & -0.04 & 0 \\ 0.03 & 0.02 & -0.05 \end{bmatrix} \text{ and } \underline{x}(t) = \begin{bmatrix} x_1(t) \\ x_2(t) \\ x_3(t) \end{bmatrix}, \dot{\underline{x}}(t) = \begin{bmatrix} \dot{x}_1(t) \\ \dot{x}_2(t) \\ \dot{x}_3(t) \end{bmatrix}.$$

The characteristic equation is

$$\lambda^3 + 0.16\lambda^2 + 0.0069\lambda + 0.00006 = 0.$$

This equation has no analytical solution. Therefore, we use the numerical method to find approximate solutions.

The eigenvalues and the associated eigenvectors are

$$\left(\lambda_1 \approx -0.092, x_1 \approx \begin{bmatrix} -0.77 \\ 0.59 \\ 0.27 \end{bmatrix}\right)$$

$$\left(\lambda_2 \approx -0.056, x_2 \approx \begin{bmatrix} 0.24 \\ -0.60 \\ 0.77 \end{bmatrix}\right)$$

$$\left(\lambda_3 \approx -0.012, x_3 \approx \begin{bmatrix} -0.44 \\ -0.61 \\ -0.66 \end{bmatrix}\right).$$

Hence the general solution is

$$\underline{x}(t) \approx C_1 e^{-0.092t} \underline{x_1} + C_2 e^{-0.056t} \underline{x_2} + C_3 e^{-0.012t} \underline{x_3}.$$

(ii)

Using the initial condition, $\underline{x}(0) = [4,6,5]^T$, we obtain

$$x_1(t) \approx 0.32\mathrm{e}^{-0.092t} - 0.27\mathrm{e}^{-0.056t} + 3.95\mathrm{e}^{-0.012t}$$

$$x_2(t) \approx -0.25\mathrm{e}^{-0.092t} + 0.68\mathrm{e}^{-0.056t} + 5.56\mathrm{e}^{-0.012t}$$

$$x_3(t) \approx 0.11\mathrm{e}^{-0.092t} - 0.87\mathrm{e}^{-0.056t} + 5.98\mathrm{e}^{-0.012t}.$$

(iii) When $t \to \infty$,

$$x_1 = x_2 = x_3 = 0.$$

4.9.3 Modelling 2 Degrees-of-Freedom Spring-Mass-Damper System

In this section, we extend the mechanical system in Section 4.1 by adding dampers to delay the oscillation of the springs, which is shown in Figure 4.4. In this figure in addition to spring and mass, we have introduced dashpots of damping coefficient c.

We now pull the mass on the right from its equilibrium state and release it so that the springs will start to oscillate. The forces applied by the springs are the same as described in Section 4.1, however the additional forces applied to the mass on the left, in Figure 4.4, are $-c\,x_1$ by the damper on the left and $-c\dot{x}_1 + c\dot{x}_2$ by the middle damper. The forces applied to the mass on the right, in Figure 4.4, are $-c\dot{x}_2 + c\dot{x}_1$ by the middle damper and $-c\dot{x}_2$ by the damper on the right.

Using Newton's second law of motion, the equations of motion of the two masses can be written in the following form:

$$m_1\ddot{x}_1 = -2c\dot{x}_1 + c\dot{x}_2 - 2kx_1 + kx_2$$

$$m_2\ddot{x}_2 = c\dot{x}_1 - 2c\dot{x}_2 + kx_1 - 2kx_2.$$

Now we can write this system of second-order equations into a coupled linear system with four first-order equations by assuming $x_3 = \dot{x}_1$ and $x_4 = \dot{x}_2$, which gives

$$x_1 = x_3$$

$$x_2 = x_4$$

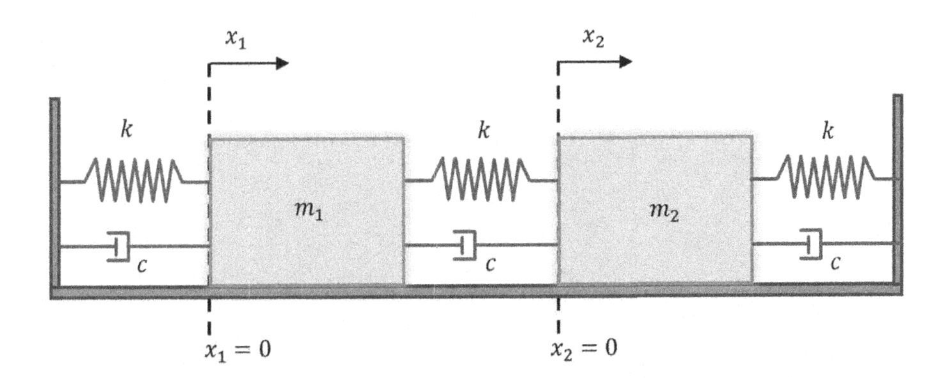

FIGURE 4.4 Two degrees-of-freedom spring-mass-damper system.

$$m_1\dot{x}_3 = -2cx_3 + cx_4 - 2kx_1 + kx_2$$

$$m_2\dot{x}_4 = cx_3 - 2cx_4 + kx_1 - 2kx_2.$$

This is a 4×4 coupled system of linear differential equations and can be solved using the matrix methods considered in this book; however, we do not cover the solution here.

5 Fourier Series

5.1 INTRODUCTION

The Fourier series is named after the French mathematician Joseph Fourier, who introduced a trigonometric series to solve the heat equation that had no analytical solution. The series he introduced was constructed by combining an infinite number of simple harmonics, which approximated periodic functions and the frequency of each harmonic in the Fourier series is an integer multiple of the frequency of the periodic functions.

Fourier series are more useful than McLaren or Taylor series; McLaren or Taylor series use polynomials to produce local approximations for a given function $f(t)$ and provide a close fit at a particular point. However, in some engineering applications, it is required to approximate $f(t)$ over a wide region. The finite Fourier series provides a somewhat reliable approximation for continuous periodic functions over the entire region. Furthermore, some periodic functions cannot be expanded to Maclaren or Taylor series, but the Fourier series can be constructed for any periodic function.

The outline of this chapter is as follows:

- Periodic functions
- Adding signals of unequal frequencies
- The beat phenomenon
- Frequency spectrum
- Constructing Fourier series
- Engineering applications

5.2 PERIODIC FUNCTIONS

In this section, we begin with a brief introduction of time periodic functions, whose graph repeats over a fixed time interval. Periodic functions are often used in signal processing and vibration analysis. Frequently used periodic functions are oscillatory trigonometric functions $f(t) = A\sin(\omega t)$ and $f(t) = A\cos(\omega t)$, which are shown in Figure 5.1.

The time period of a wave function (signal) is the time at which the graph of the function completes one full cycle, and it is the shortest time for the complete cycle to repeat. The time period is usually denoted by T.

For the sinusoidal wave functions defined earlier and shown in Figure 5.1, the period is

$$T = \frac{2\pi}{\omega},$$

DOI: 10.1201/9781032630694-5

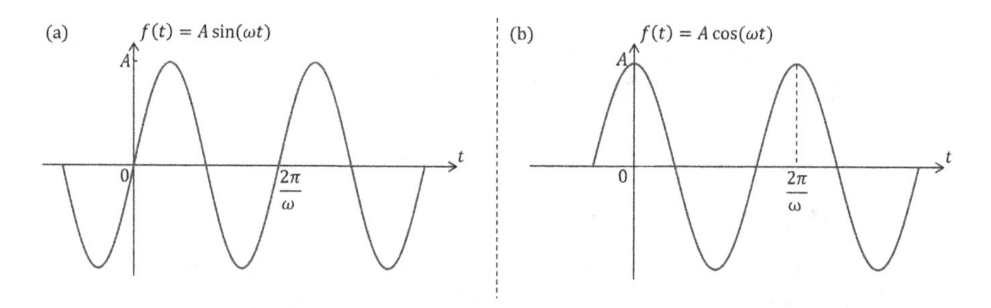

FIGURE 5.1 Periodic functions sine and cosine.

where the parameter ω is known as the angular frequency of the waveform. Note that when ω is large, then T will be small. The frequency of a periodic function, measured in Hertz (Hz), is the number of complete cycles in 1 second and is given by

$$f = \frac{1}{T}.$$

Hence, the frequency of both $\sin(\omega t)$ and $\cos(\omega t)$ functions is

$$f = \frac{\omega}{2\pi}.$$

The sinusoidal functions are called "simple harmonic" functions. In general, a function $f(t)$ will be periodic with time period T if the following condition is satisfied:

$$f(t+T) = f(t), \ (T \text{ is the smallest value}).$$

In Section 5.6, we consider non-simple periodic functions that can be expressed in terms of an infinite series of simple sine and cosine harmonics.

Now we introduce some trigonometric identities that will be used in the next sections.

$$\sin(-A) = -\sin(+A) \text{ and } \cos(-A) = \cos(+A)$$

$$\sin(A)\sin(B) = \frac{1}{2}\left[\cos(A-B) - \cos(A+B)\right]$$

$$\cos(A)\cos(B) = \frac{1}{2}\left[\cos(A+B) + \cos(A-B)\right]$$

$$\sin(A)\cos(B) = \frac{1}{2}\left[\sin(A+B) + \sin(A-B)\right].$$

For an integer n, the following conditions are true:

$$\sin(n\pi) = 0, \text{ and } \cos(n\pi) = (-1)^n = \begin{cases} +1, & n \text{ even} \\ -1, & n \text{ odd} \end{cases}.$$

5.3 ADDING PERIODIC SIGNALS WITH UNEQUAL FREQUENCIES

It is simple to add two periodic signals (waves) with the same frequency together, and the resulting signal has the same periodicity and frequency as the original signals. For example, it can easily be shown that

$$-2\sin(2t)+6\cos(2t)=\sqrt{40}\cos\left(2t+0.32^{c}\right).$$

When two periodic sinusoidal waves with equal frequencies are combined, the resulting signal will be a sinusoidal wave with the same frequency.

In this section, our focus is on adding periodic signals, having unequal frequencies. This topic will be an introduction to learning the basic principles of constructing Fourier series with the addition of simple harmonics.

Suppose there are two periodic signals $g_1(t)$ and $g_2(t)$ with the frequencies f_1 Hz and f_2 Hz and the time periods T_1 and T_2, respectively. Say, one could be a square wave and the other a sine wave. The question is, Will the combined signal, $g(t)=g_1(t)+g_2(t)$, be a periodic signal? If it is periodic what will be its frequency?

There is not a simple mathematical proof to these questions. However, the answer relies on whether $\frac{f_1}{f_2}$ is a rational number. In other words, are there two (smallest) integers a and b such that $\frac{f_1}{f_2}=\frac{a}{b}$?

The ratio $\frac{f_1}{f_2}=\frac{a}{b}$ can be easily checked and verified. But as we see in the following, the combined frequency depends on the link between the time periods, T_1 and T_2, of the signals g_1 and g_2:

$$\begin{aligned}
\frac{f_1}{f_2}&=\frac{a}{b}\Rightarrow\\
a\times f_2&=b\times f_1\\
a\times\frac{1}{f_1}&=b\times\frac{1}{f_2}\\
a\times T_1&=b\times T_2:=T_0.
\end{aligned}\tag{5.1}$$

The time $t=T_0$ is important. Since starting at $t=0$ the signal g_1 completes a cycles and the signal g_2 completes b cycles before returning to the same position at $t=T_0$. When the signals are added together, their values at each moment are added together to give a new value for the combined signal. However, the value of the combined signal at $t=0$ and $t=T_0$ will be the same. Hence, T_0 will be the time period of the combined signal $g=g_1+g_2$ and its frequency, $f_0=\frac{1}{T_0}$.

Example 5.1 Given two signals

$$g_1(t)=2\sin(2\pi t)\ \text{ and }\ g_2(t)=\sin\left(\frac{2\pi}{1.25}t\right),$$

(i) Determine whether the sum of these signals, $g=g_1+g_2$, is periodic?
(ii) If it is periodic what is the frequency of the combined signal?

Solution

(i) The answer for this part depends on whether $\frac{f_1}{f_2}$ is a rational number.

$$T_1 = \frac{2\pi}{2\pi} = 1 \quad \Rightarrow \quad f_1 = 1\,\text{Hz}$$

$$T_2 = \frac{2\pi}{\frac{2\pi}{1.25}} = 1.25 \quad \Rightarrow \quad f_2 = \frac{1}{1.25} = 0.8\,\text{Hz}.$$

Hence,

$$\frac{f_1}{f_2} = \frac{1}{0.8} = \frac{5}{4}$$

which is a rational number, so $g = g_1 + g_2$ is periodic. Note that the fraction is in its simplified form.

(ii) The answer for this part depends on the link between the time periods of the two signals, which we first explain visually and then we adopt a practical approach using $\gcd(f_1, f_2)$:

$$\frac{f_1}{f_2} = \frac{1}{0.8} = \frac{5}{4} \Rightarrow 5 \times T_1 = 4 \times T_2 = 5 := T_0.$$

This means that in the duration of $T_0 = 5$ signal g_1 completes five full cycles but signal g_2 completes four full cycles (see Figures 5.2a and 5.2b).

A plot of these two signals in the same coordinate system is given in Figure 5.2c. As it is shown in this figure, starting at $t = 0$ the signal $g_1(t)$ (solid curve) completes 5 cycles (5 × 1) in the range between $t = 0$ to $t = 5$, whereas the signal $g_2(t)$ (dashed curve) completes 4 cycles (4 ×1.25) in this range. The duration $t = 5$ is important, as both signals return to the same position again after integer number of cycles.

When two signals are added together, $g(t) = g_1(t) + g_2(t) = 2\sin(2\pi t) + \sin(2\pi / 1.25t)$, the values at every point on the curves are added together at any moment in time to provide a new value for the combined signal $g(t)$. The value of the combined signal at $t = 0$ and $t = 5$ will be the same (see Figure 5.2d). This process is repeated between $t = 5$ and $t = 10$ and again every 5 seconds. Thus, the combined signal $g(t)$ is a periodic signal with the time period $T_0 = 5$ and the frequency $f_0 = 0.2$ Hz.

An alternative (practical) approach can be adopted for determining the frequency, f_0, and the time period, T_0, of the combined signal g. Suppose $g = g_1 + g_2$ is a periodic signal with frequency f_0 and time period T_0. The reciprocal of both sides of Equation 5.1 implies

$$f_1 = af_0$$
$$f_2 = bf_0.$$

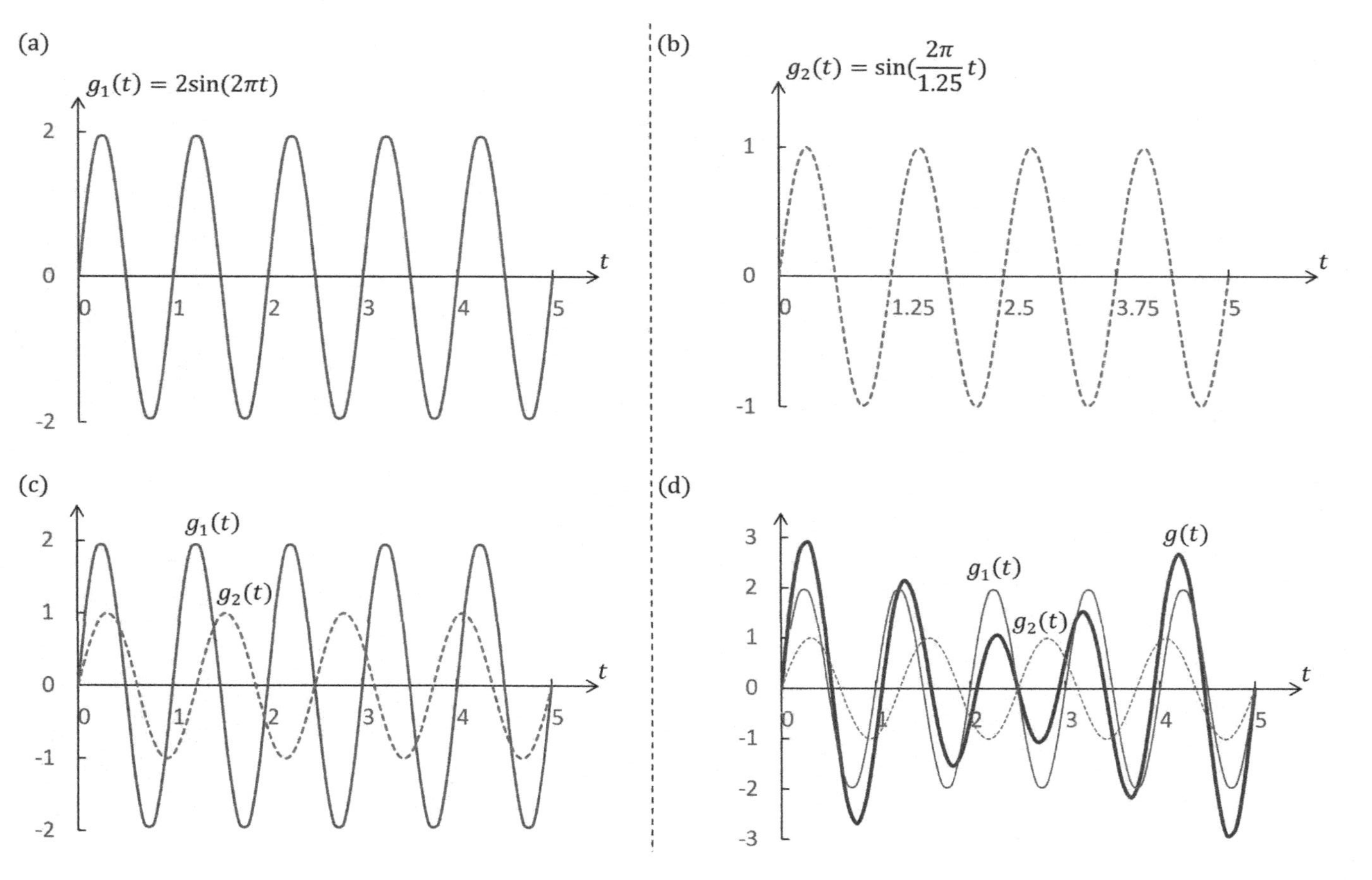

FIGURE 5.2 The visual representation of the time-period for the combined signal.

Hence f_0 is the greatest common divisor between the frequencies f_1 and f_2. This can be tricky when the frequencies are fractions. However, it can be solved by simplifying the fractions and making the denominators equal or using the following formula.

Note: The greatest common divisor between, say, the three fractions $\frac{a}{b}, \frac{c}{d}$ and $\frac{e}{f}$, can be determined using the following formula:

$$\gcd\left(\frac{a}{b}, \frac{c}{d}, \frac{e}{f}\right) = \frac{\gcd(a,c,e)}{\text{lcm}(b,d,f)}.$$

For example, the greatest common divisor between $\frac{5}{16}, \frac{25}{12}$ and $\frac{15}{8}$ is

$$\gcd\left(\frac{5}{16}, \frac{25}{12}, \frac{15}{8}\right) = \frac{\gcd(5,25,15)}{\text{lcm}(16,12,8)} = \frac{5}{48}.$$

The frequency f_0 is called the **fundamental frequency**. Now we use this alternative way to determine the fundamental frequency f_0 for the signals given in

Now we use this alternative way to determine the fundamental frequency f_0 for the signals given in Example 5.1. We find the greatest common divisor between $f_1 = 1$ and $f_2 = 0.8 = \frac{4}{5}$

$$f_0 = \gcd\left(1, \frac{4}{5}\right) = \frac{\gcd(1,4)}{\text{lcm}(1,5)} = \frac{1}{5}$$

and $\omega_0 = 2\pi f_0 = \frac{2\pi}{5}$ is the fundamental angular frequency. The signals can be represented in terms of ω_0 as

$$g(t) = 2\sin(2\pi t) + \sin(2\pi / 1.25t) = 2\sin(5\omega_0 t) + \sin(4\omega_0 t).$$

This is useful when combined signals are represented using the frequency spectrum diagram.

Note: Unlike sinusoidal wave functions with equal frequencies, the combined signal for signals with unequal frequencies will not be a sinusoidal signal as shown in Figure 5.2d.

Example 5.2 Given two sinusoidal signals $g_1(t)$ and $g_2(t)$ with frequencies $f_1 = 1.2$ Hz and $f_2 = 1.6$ Hz, respectively.

(i) Determine whether sum of these signals, $g = g_1 + g_2$, is periodic.
(ii) If it is periodic what is the frequency of the combined signal?

Solution

(i) The answer to this part depends on whether $\frac{f_1}{f_2}$ is a rational number.

$$\frac{f_1}{f_2} = \frac{1.2}{1.6} = \frac{3}{4}.$$

This is a rational number (must be in simplified form), hence the combined signal, $g(t)$, is a periodic signal.

(ii) To determine the frequency of $g = g_1 + g_2$, we try to find the relationship between the time period of these signals as follows:

$$\frac{f_1}{f_2} = \frac{3}{4} \Rightarrow$$

$$3 \times f_2 = 4 \times f_1$$

$$3 \times \frac{1}{f_1} = 4 \times \frac{1}{f_2}$$

$$3 \times T_1 = 4 \times T_2 = 2.5,$$

where $T_1 = \frac{5}{6}$ and $T_2 = \frac{5}{8}$. This means that the signal with the time period $T_1 = \frac{5}{.6}$ completes 3 cycles in the range $t = 0$ to $t = 2.5$, while the signal with the time period $T_2 = \frac{5}{8}$ completes 4 full cycles in this range (see Figure 5.3a). The duration $t = 2.5$ is important, as both signals return to the same position again after integer number of cycles. The process repeats every 2.5 seconds.

Hence as explained in Example 0.2, $g = g_1 + g_2$ is a signal with the time period $T_0 = \frac{5}{2}$ and the frequency $f_0 = \frac{2}{5}$ Hz. The graph of one full cycle of the signal $g(t)$ is represented in Figure 5.3b.

An alternative and a practical way to determine f_0 is to find the greatest common divisor between the frequencies $f_1 = 1.2 = \frac{6}{5}$ and $f_2 = 1.6 = \frac{8}{5}$:

$$\text{f}_0 \ \text{gcd}\left(\frac{6}{5}, \frac{8}{5}\right) = \frac{\text{gcd}(6,8)}{\text{lcm}(5,5)} = \frac{2}{5}.$$

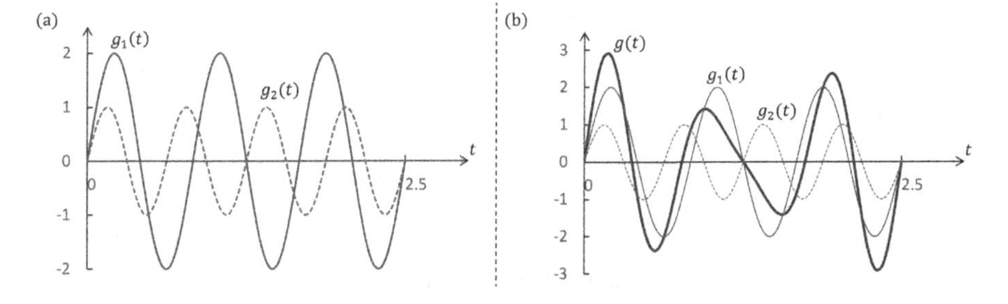

FIGURE 5.3 The visual representation of the time period for the combined signal.

Example 5.3 Given

$$g(t) = 2 + \sin\left(\frac{1}{2}t\right) + 3\sin\left(\frac{2}{3}t\right) + 5\sin\left(\frac{7}{6}t\right)$$

(i) Determine its fundamental frequency f_0.
(ii) Write down $g(t)$ in terms the fundamental angular frequency ω_0.

Solution

(i) The signal $g(t)$ consists of three simple harmonics $g_1(t) = \sin\left(\frac{1}{2}t\right)$, $g_2(t) = 3\sin\left(\frac{2}{3}t\right), g_3(t) = 5\sin\left(\frac{7}{6}t\right)$ and a constant non-periodic term. Their frequencies are

$$f_1 = \frac{1/2}{2\pi} = \frac{1}{4\pi}, f_2 = \frac{2/3}{2\pi} = \frac{1}{3\pi} \text{ and } f_3 = \frac{7/6}{2\pi} = \frac{7}{12\pi}.$$

The frequency of the constant term is zero and does not contribute to the frequency of the combined signal.

The ratios $\frac{f_1}{f_2} = \frac{3}{4}$, $\frac{f_1}{f_3} = \frac{3}{7}$ and $\frac{f_2}{f_3} = \frac{4}{7}$ are rational numbers; hence, the combined signal is periodic.

A practical way to obtain the fundamental frequency f_0, is to determine the greatest common divisor between the angular frequencies $\omega_1 = \frac{1}{2}$, $\omega_2 = \frac{2}{3}$ and $\omega_3 = \frac{7}{6}$:

$$\omega_0 = \gcd\left(\frac{1}{2}, \frac{2}{3}, \frac{7}{6}\right) = \frac{\gcd(1,2,7)}{\text{lcm}(2,3,6)} = \frac{1}{6}.$$

Hence, the fundamental frequency is $f_0 = \frac{\omega_0}{2\pi} = \frac{1}{12\pi}$, and the fundamental period is $T_0 = 12\pi$.

(ii) We can write the original signals in terms of $\omega_0 = 2\pi f_0 = \frac{1}{6}$:

$$g(t) = 2 + \sin\left(\frac{1}{2}t\right) + 3\sin\left(\frac{2}{3}t\right) + 5\sin\left(\frac{7}{6}t\right)$$

$$= 2 + \sin(3 \times \omega_0 t) + 3\sin(4 \times \omega_0 t) + 5\sin(7 \times \omega_0 t).$$

The graph of these three signals and the combined signal is given in Figure 5.4. Starting at $t = 0$ they return to the same position at $T_0 = 12\pi$.

Example 5.4 Assume $g_1(t) = sin(t)$ and $g_2(t) = sin(2\pi t)$. Determine whether $g_1 + g_2$ is periodic.

Solution

The frequency of the signals g_1 and g_2 are $f_1 = \frac{1}{2\pi}$ and $f_2 = 1$, respectively. This implies $\frac{f_2}{f_1} = 2\pi$, and it is an irrational number; hence, when they are added together, the resultant signal $g_1 + g_2$ will not be periodic.

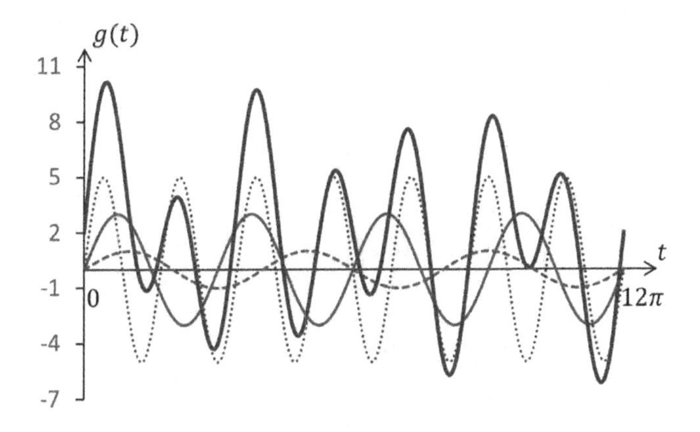

FIGURE 5.4 Graph of the three signals along with the combined signal.

5.4 ADDING SIGNALS WHEN FREQUENCIES ARE INTEGER MULTIPLE OF SMALLEST FREQUENCY

In this section, we consider combining signals whose frequencies are integer multiples of each other. Suppose the sinusoidal signals g_1 of frequency f_1 Hz (time period T_1) and g_2 of frequency f_2 (time period T_2) and $f_2 = nf_1$ (or $T_1 = nT_2$), where n is an integer number.

For example,

(i) $f_1 = 1.0$ Hz and $f_2 = 100$ Hz $\left(f_2 = 100 f_1 \Rightarrow T_1 = 100\,T_2\right)$

(ii) $f_1 = 0.4$ Hz and $f_2 = 1.2$ Hz $\left(f_2 = 3 f_1 \Rightarrow T_1 = 3T_2\right)$

(iii) $f_1 = 1.5$ Hz and $f_2 = 3$ Hz $\left(f_2 = 2 f_1 \Rightarrow T_1 = 2T_2\right)$.

The time period $T_1 = nT_2$ means that the signal with the longer time period T_1 (lower frequency f_1) completes one full cycle in the range between $t = 0$ and $t = T_1$. While the signal with a shorter time period T_2 (higher frequency f_2) completes n full cycles in this range, and both signals return to the same position at $t = T_1$. Hence. T_1 is the time period of the combined signal $g = g_1 + g_2$. Therefore, the fundamental frequency and the fundamental time period will be $f_0 = f_1$ and $T_0 = T_1$, respectively. Note that f_0 is the lower frequency between the two signals.

For example, consider case (ii) earlier:

$$f_1 = 0.4 \text{ Hz} \Rightarrow T_1 = \frac{1}{0.4} = 2.5$$

$$f_2 = 1.2 \text{ Hz} \Rightarrow T_2 = \frac{1}{1.2} = 0.8333333.$$

For these two signals, we have

$$f_2 = 3 f_1 \Rightarrow$$

$$T_1 = 3T_2 = 3 \times \frac{1}{1.2} = 2.5.$$

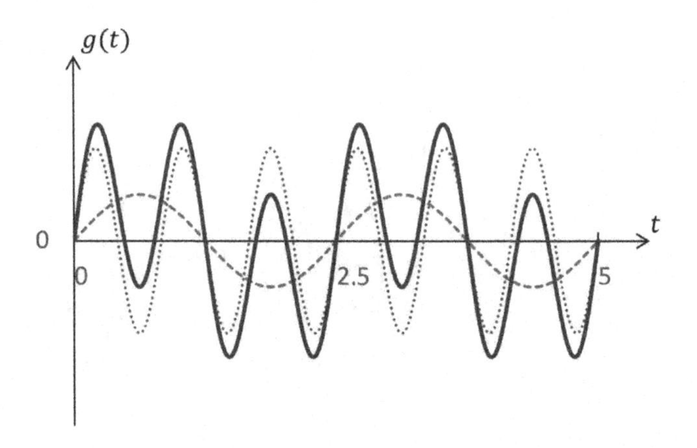

FIGURE 5.5 The signal with the lower frequency completes one full cycle at $t = 2.5$ but the signal with the higher frequency completes 3 cycles. The solid line is the graph of the two cycles of the combined signal; its frequency is the frequency of the lower frequency signal.

The lower frequency signal ($f_1 = 0.4$) completes one cycle in the range $t = 0$ to $t = 2.5$, while the higher frequency signal ($f_2 = 1.2$ Hz) completes 3 cycles in this range, and

$$\gcd(0.4, 1.2) = 0.4.$$

Hence, the combined signal $g = g_1 + g_2$ will have the frequency $f_0 = 0.4$ Hz (lower frequency) and a time period $T_0 = 2.5$. The graphical representation of sinusoidal signals $g_1(t) = \sin(0.8\pi t)$ and $g_2(t) = 2\sin(2.4\pi t)$ with frequencies of 0.4 Hz and 1.2 Hz, respectively, is given in Figure 5.5.

For the signals in part (i) earlier,

$$f_0 = \gcd(1, 100) = 1\,\text{Hz};$$

hence, the lower frequency $f_0 = 1$ Hz is the frequency of the combined signal.

For the signals, in part (iii) earlier the lower frequency1.5 Hz, is the frequency of the combined signal:

$$f_0 = \gcd(1.5, 3) = 1.5\,\text{Hz}.$$

5.5 THE BEAT PHENOMENON

When two signals with a small frequency difference are combined it creates a beat. The amplitude of the resulting signal alternates from maximum to zero, then recovers and returns to maximum, or the amplitude of the resulting signal alternates between destructive and constructive.

This process can be presented first graphically and then with a mathematical solution using trigonometric identities. In Figure 5.6a, at $t = 0$, two signals

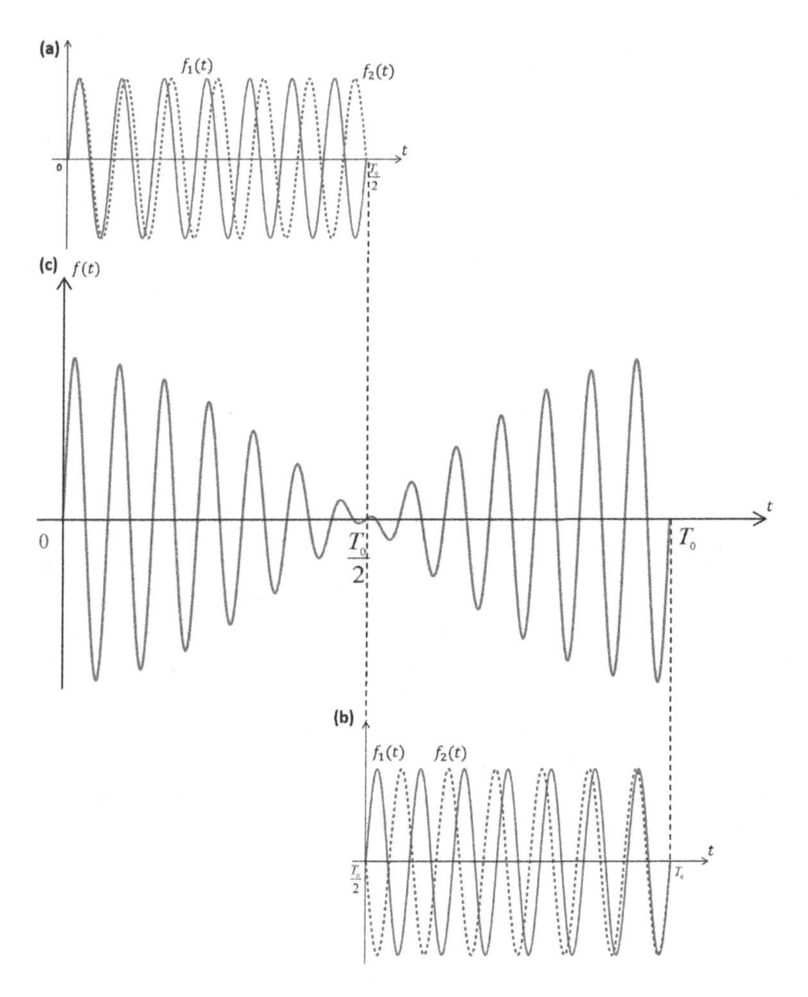

FIGURE 5.6 Visual explanation of the beat phenomenon. The amplitude of the oscillation drops from maximum to zero at $\frac{T_0}{2}$ (destructive interference) and then recovers and reaches a maximum amplitude at T_0 (constructive interference).

with frequencies f_1 and f_2 are initially in phase. But over time, the phase difference between these two waves increases and reaches to 180 degrees out of phase and hence causes destructive interference. When the waves are 180 degrees out of phase, then over time, the phase difference between the waves gradually decreases until they are back in phase and cause constructive interference as shown in Figure 5.6b.

Note that the composite signal, $f(t)$, is shown in Figure 5.6c over time period T_0. As shown in this figure, it can be said that the amplitude of the oscillations reaches zero at a certain time $\frac{T_0}{2}$ (which is associated with destructive interference) and then recovers and reaches the maximum amplitude at T_0 (which is associated with constructive interference). The process repeats itself every T_0 seconds.

The interference of two signals with slight frequency differences that undergo a repeated constructive and destructive process is called **beat**, and this process is shown in Figures 5.6a–c.

To understand the beat phenomenon mathematically, we consider the interference of two sine waves with the same amplitude and frequencies f_1 and f_2 with a small frequency difference:

$$\begin{aligned} f(t) &= \sin(2\pi f_1 t) + \sin(2\pi f_2 t) \\ &= 2\cos\left(2\pi \frac{f_1 - f_2}{2} t\right) \times \sin\left(2\pi \frac{f_1 + f_2}{2} t\right) \\ &= 2\cos(2\pi f_c t) \times \sin(2\pi f_s t) := A(t)\sin(2\pi f_s t), \end{aligned}$$

where $A(t) = 2\cos(2\pi f_c t)$, $f_c = \frac{f_1 - f_2}{2}$ and $f_s = \frac{f_1 + f_2}{2}$. This means that the interference of two sine waves (where their frequencies are close) can be described as a high-frequency sine wave whose amplitude is a slow-moving time-dependent cosine wave. Note that the frequency, f_s, of this oscillation is the average of frequencies f_1 and f_2. The modulated amplitude (with the frequency $f_c = \frac{f_1 - f_2}{2}$) is an **envelope** for the higher frequency oscillations $\sin(2\pi f_s t)$ as shown in Figure 5.7.

The interference of two signals as shown in Figure 5.7 consists of a higher frequency oscillation and a slow varying amplitude in the form of cosine wave described earlier. The beat frequency occurs when the interference of signals is constructive, destructive and constructive again. This process happens when the cosine wave drops from the maximum ±2 to 0 and then recovers and reaches back to the

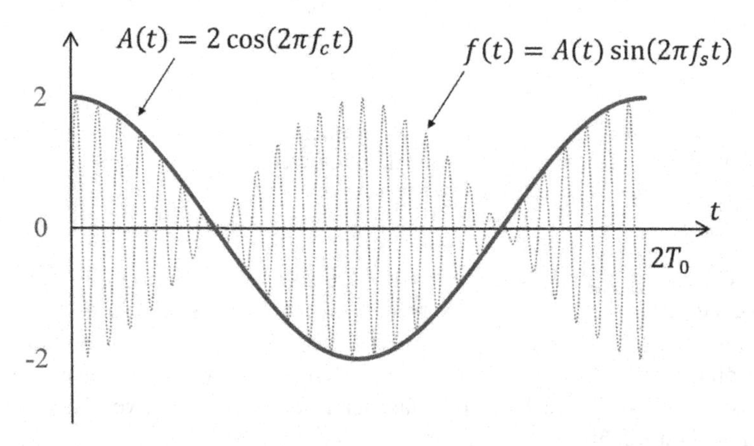

FIGURE 5.7 The amplitude of high-frequency oscillations with the frequency $f_s = \frac{f_1 + f_2}{2}$ decreases from maximum value ±2 to 0 then recovers and reaches back to the maximum value of ±2 at the time $t = T_0$. The low-frequency modulated amplitude with the frequency $f_c = \frac{f_1 - f_2}{2}$ and time period of $T_c = 2T_0$ is an **envelope** for the resultant signal with the frequency $f_0 = |f_1 - f_2|$ (twice the frequency f_c).

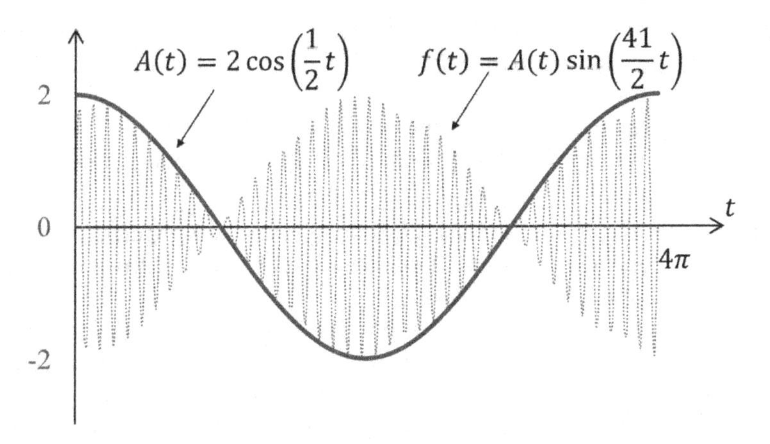

FIGURE 5.8 The amplitude of oscillations with the angular frequency $\frac{41}{2}$ decreases from maximum value ± 2 to 0 at $t=\pi$ then recovers and reaches back to the maximum value of ± 2 at $t=2\pi$. The modulated amplitude with the frequency $\frac{1}{4\pi}$ is an **envelope** for the resultant signal with the frequency $\frac{1}{2\pi}$.

maximum value ± 2. This means that the time period of the beat is $T_0 = \frac{1}{|f_1 - f_2|}$ (half period of the cosine wave), and its frequency is $f_0 = |f_1 - f_2|$.

For example, let us consider two periodic signals $g_1(t) = \sin(20t)$ and $g_2(t) = \sin(21t)$, the frequencies are $f_1 = \frac{20}{2\pi}$ and $f_2 = \frac{21}{2\pi}$, respectively. The combined signal can be written as below using the trigonometric identity:

$$f(t) = \sin(20t) + \sin(21t)$$
$$= 2\cos\left(\frac{1}{2}t\right) \times \sin\left(\frac{41}{2}t\right) := A(t) \times \sin\left(\frac{41}{2}t\right),$$

where $A(t) = 2\cos\left(\frac{1}{2}t\right)$ with the frequency $f_c = \frac{|f_1 - f_2|}{2} = \frac{1}{4\pi}$ and the "sine" function with the frequency $f_s = \frac{f_1 + f_2}{2} = \frac{41}{4\pi}$.

The frequency of the combined signal (beat frequency) occurs at

$$f_0 = |f_1 - f_2| = \left|\frac{20}{2\pi} - \frac{21}{2\pi}\right| = \frac{1}{2\pi}.$$

As shown in Figure 5.8, this happens when the cosine function drops from the maximum ± 2 to 0 at $t=\pi$ and then recovers and reaches back to the maximum value ± 2 at $t=2\pi$.

5.6 FREQUENCY DOMAIN (FREQUENCY SPECTRUM)

When signals are added together, it would be very difficult to identify that a combined signal is composed of several signals of different frequencies in time domain.

This can be achieved by drawing a graph where the amplitude (A) of each signal is presented in terms of their frequency (f). This graph is called frequency domain or frequency spectrum diagram.

To explain the frequency spectrum, we reconsider Example 5.1 where the two signals $g_1(t)=2\sin(2\pi t)$ and $g_2(t)=\sin(2\pi/1.25t)$ were added together to give $g(t)=2\sin(2\pi t)+\sin(2\pi/1.25t)$.

The graph of the signal g in the time domain is shown in Figure 5.2d. Using this graph, it is impossible to understand that the signal g consists of two signals of frequencies 0.8 Hz and 1 Hz.

The composite signal diagram can be appropriately represented by drawing a graph, where the vertical axis is the amplitude of each signal and the horizontal axis is their frequency (or angular frequency ω). In other words, the amplitude can be considered as a function of the frequency (angular frequency) and this is shown in Figure 5.9. Using this diagram, it is easy to recognise that the signal g consists of two signals with frequencies of 0.8 Hz and 1 Hz.

Drawing the amplitude of each signal against its angular frequency provides an easier representation of signals that are approximated using Fourier series, which is explained in Section 5.7.

Another example is to refer to Example 5.3 and draw the frequency spectrum of the following signal:

$$g(t)=2+\sin\left(\frac{1}{2}t\right)+3\sin\left(\frac{2}{3}t\right)+5\sin\left(\frac{7}{6}t\right)$$

$$=2+\sin(3\times\omega_0 t)+3\sin(4\times\omega_0 t)+5\sin(7\times\omega_0 t),\quad \omega_0=\frac{1}{6}.$$

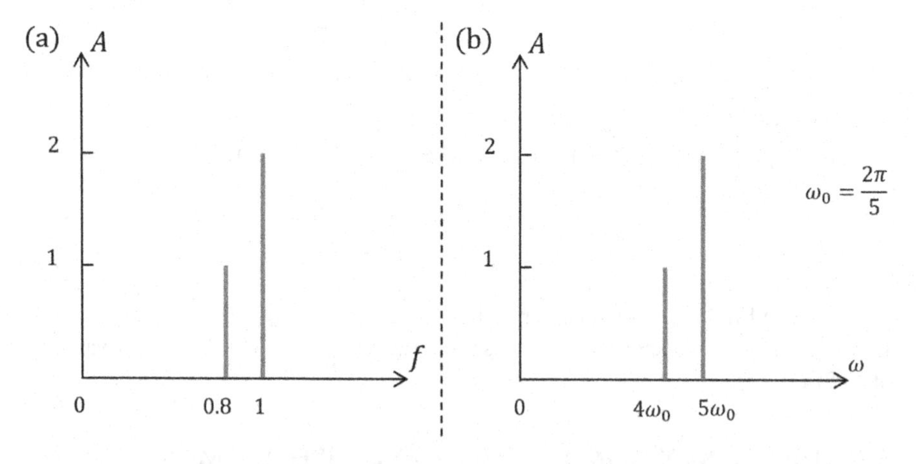

FIGURE 5.9 The frequency spectrum of the signal $g(t)=2\sin(2\pi t)+\sin(2\pi/1.25t)$. (a) The amplitude of each individual signal against its frequency. (b) The amplitude of each individual signal against its angular frequency.

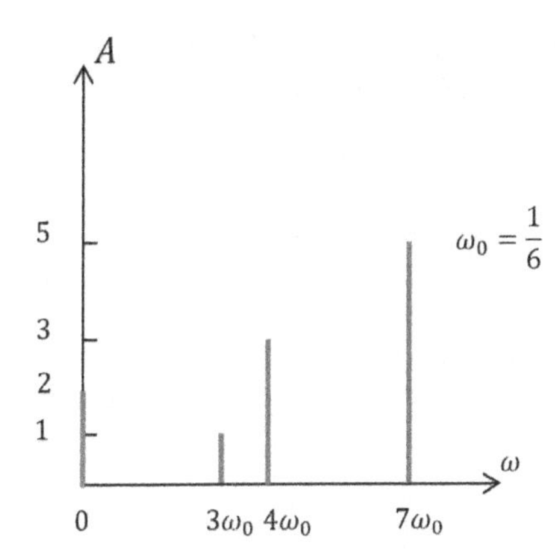

FIGURE 5.10 The frequency spectrum of the signal $g = 2 + \sin\left(\frac{1}{2}t\right) + 3\sin\left(\frac{2}{3}t\right) + 5\sin\left(\frac{7}{6}t\right)$.

The signal $g(t)$ consists of a constant term with zero frequency and three sine signals with the angular frequencies $3\omega_0, 4\omega_0$ and $7\omega_0$, respectively. The amplitudes are 2, 1, 3 and 5, respectively. Using this information, the frequency spectrum can be plotted as shown in Figure 5.10.

5.6.1 Exercises

1. Given the signal $g(t) = \sin(3t) + \sin(4t)$:
   (a) Show that it is a periodic signal.
   (b) Determine its time period and its frequency.
2. Given two sinusoidal periodic signals with the frequencies $f_1 = 1.4$ Hz and $f_2 = 1.618$ Hz:
   (a) Determine if the sum of these signals is periodic.
   (b) Determine its frequency.
3. Given two sinusoidal periodic signals with frequencies $f_1 = 0.8$ Hz and $f_2 = 1.6$ Hz with the amplitudes of 1 and 2, respectively:
   (a) Determine if the sum of these signals is periodic.
   (b) Determine its frequency.
4. Given two periodic signals $g_1 = \sin\left(\frac{2\pi}{0.75}t\right)$ and $g_2 = 2\sin\left(\frac{2\pi}{0.5}t\right)$:
   (a) Determine if the sum of these signals is periodic.
   (b) Determine its frequency.
5. Given two periodic signals $g_1 = \sin(5\pi t)$ and $g_2 = 2\cos(6\pi t)$:
   (a) Determine if the sum of these signals is periodic.
   (b) Determine its frequency.
6. Given two periodic signals $g_1 = \sin(2t)$ and $g_2 = 2\cos(6t)$:
   (a) Determine if the sum of these signals is periodic.
   (b) Determine its frequency.

7. Given two periodic signals $g_1 = \sin(0.5t)$ and $g_2 = 2\cos(0.8t)$:
   (a) Determine if the sum of these signals is periodic.
   (b) Determine its frequency.

8. Given $f = \sin(2\pi t) + \cos(4\pi t) + \sin(18\pi t)$:
   (a) Determine if the signal is periodic.
   (b) Determine the fundamental frequency and the fundamental angular frequency of the signal.
   (c) Write down f in terms of the fundamental angular frequency.
9. Sketch the graph of the frequency domain (spectrum) for Exercises 5.6.1 (1–8).

Answers

1.
   (a) Periodic
   (b) $f_0 = \frac{1}{2\pi} \Rightarrow T_0 = 2\pi$

2.
   (a) Periodic
   (b) $f_0 = \frac{1.4}{700} = \frac{1}{500} = 0.002 \text{ Hz or } 2 \text{ mHz}$
   $\Rightarrow T_0 = 500.$

3.
   (a) Periodic
   (b) $f_0 = \frac{0.8}{1} = 0.8 \text{ Hz (fundamental frequency)}$
   $\Rightarrow T_0 = 1.25.$

4.
   (a) Periodic
   (b) $f_0 = \frac{1}{1.5} \approx 0.6667 \text{ Hz (fundamental frequency)}$
   $\Rightarrow T_0 = 1.5.$

5.
   (a) Periodic
   (b) $f_0 = \frac{1}{2} \Rightarrow T_0 = 2$

6.
   (a) Periodic
   (b) $f_0 = \frac{1}{\pi} \Rightarrow T_0 = \pi$

7.
   (a) Periodic
   (b) $f_0 = \frac{1}{20\pi} \Rightarrow T_0 = 20\pi$

8.

(a) Periodic
(b) $f_0 = 1 \Rightarrow T_0 = 1 \Rightarrow \omega_0 = 2\pi \Rightarrow$
(c) $f = \sin(\omega_0 t) + \cos(2\omega_0 t) + \sin(9\omega_0 t)$

5.7 FOURIER SERIES OF THE SQUARE WAVE

In the previous sections, the sum of signals with unequal frequencies was explained. It was shown that for signals whose frequencies are an integer multiple of each other, the combined signal frequency is the signal frequency with the lowest frequency.

We now explain a simple case of Fourier series for periodic functions. A detailed explanation of the Fourier series is presented in Section 5.8. However, before giving a general definition of this series, let us consider a simple periodic signal known as a square wave with a frequency of $f = 1$ Hz (time period of $T = 1$ second). The graph of this waveform is given in Figure 5.11. Fourier claimed that this signal is composed of an infinite number of simple harmonics as given:

$$f(t) = b_1 \sin(2\pi t) + b_2 \sin(4\pi t) + b_3 \sin(6\pi t) + b_4 \sin(8\pi t) + \cdots$$
$$= \sum_{n=1}^{\infty} b_n \sin(2\pi n t). \tag{5.2}$$

So, what is the frequency of the series in Equation 5.2? To answer this question, we show that the lowest frequency of simple harmonics is the series frequency:

1. The frequency of the first harmonic, $b_1 \sin(2\pi t)$, is 1 Hz.
2. The frequency of the second harmonic is, $b_2 \sin(4\pi t)$, is 2 Hz.

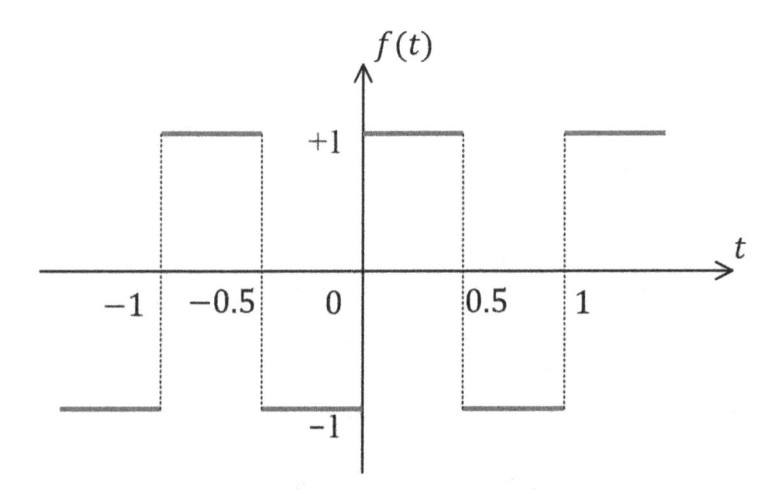

FIGURE 5.11 The graph of a square waveform with a frequency 1 Hz for 2 full cycles.

3. The combined signal $b_1\sin(2\pi t)+b_2\sin(4\pi t)$ will have the frequency of the signal with the lower frequency, that is 1 Hz.
4. Adding the third harmonic, $b_3\sin(6\pi t)$, with the frequency of 3 Hz, to $b_1\sin(2\pi t)+b_2\sin(4\pi t)$ will still imply a signal with the lower frequency of 1 Hz.
5. When we keep adding more harmonics, the frequency of sum will be kept at the lowest frequency 1 Hz.

The preceding series (Equation 5.2) can be better explained with a frequency spectrum diagram as shown in Figure 5.12a, which clearly shows that the frequency of each harmonic is an integer multiple of the fundamental frequency of 1 Hz.

The amplitudes of the sinusoidal signals (harmonics) as shown in Figure 5.12a are unknown yet and can be calculated as follows, which also applies to the Fourier series of any piecewise periodic function described in Section 5.8.

To obtain b_1 multiply both sides of equation (5.2) by $\sin(2\pi t)$ and integrate over the range 0 and 1. Except for the first term on the right-hand side the remaining terms will integrate to zero (try it!). To obtain b_2 multiply both sides of equation (5.2) by $\sin(4\pi t)$ and integrate over the range 0 and 1; except for the second term on the right-hand side, the remaining terms will integrate to zero (try it!), etc. We now generalise this process as described next.

In general, multiply both sides of $f(t)=\sum_{n=1}^{\infty} b_n\sin(2\pi nt)$ by $\sin(2\pi mt)$ and integrate it over 1 time period of the original signal:

$$\int_0^1 f(t)\sin(2\pi mt)\,dt=\sum_{n=1}^{\infty} b_n\int_0^1 \sin(2\pi nt)\sin(2\pi mt)\,dt,\ m=1,2,3,.. \tag{5.3}$$

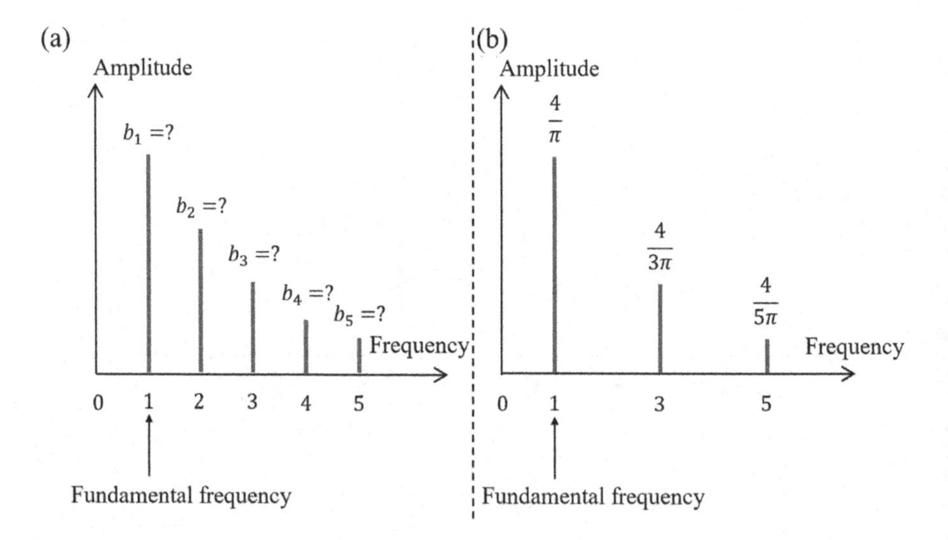

FIGURE 5.12 The frequency spectrum of the square wave.

Using the trigonometrical identity:

$$\sin(2\pi nt)\sin(2\pi mt) = \frac{1}{2}\left[\cos(2\pi nt - 2\pi mt) - \cos(2\pi nt + 2\pi mt)\right]$$
$$= \frac{1}{2}\left[\cos 2\pi(n-m)t - \cos 2\pi(n+m)t\right].$$

Hence,

$$\int_0^1 \sin(2\pi nt)\sin(2\pi mt)\,dt = \frac{1}{2}\int_0^1 \left[\cos 2\pi(n-m)t - \cos 2\pi(n+m)t\right]dt$$

$$= \frac{1}{2}\left[-\frac{1}{2\pi(n-m)}\sin 2\pi(n-m)t + \frac{1}{2\pi(n+m)}\sin 2\pi(n+m)t\right]_0^1$$

$$= \frac{1}{2}\left[-\frac{1}{2\pi(n-m)}\sin 2\pi(n-m) + \frac{1}{2\pi(n+m)}\sin 2\pi(n+m)\right].$$

For $n \neq m$, both $\sin 2\pi(n-m)$ and $\sin 2\pi(n+m)$ are zero, and hence, the integral will also be zero. For $n = m$ equation (5.3) can be written as

$$\int_0^1 \sin(2\pi nt)\sin(2\pi nt)\,dt = \frac{1}{2}\int_0^1 \left(-\cos(4\pi nt)\right)dt$$

$$= \frac{1}{2}\left[t + \frac{1}{4\pi n}\sin(4\pi nt)\right]_0^1$$

$$= \frac{1}{2}\left[\left(1 + \frac{1}{4\pi n}\sin(4\pi n \times 1)\right) - \left(0 + \frac{1}{4\pi n}\sin(4\pi n \times 0)\right)\right]$$

$$= \frac{1}{2}.$$

Thus, substituting in Equation 5.3 implies

$$\int_0^1 f(t)\sin(2\pi nt)\,dt = b_n \times \frac{1}{2} \Rightarrow$$

$$b_n = 2\int_0^1 f(t)\sin(2\pi nt)\,dt = 4\int_0^{0.5} \sin(2\pi nt)\,dt$$

$$= \frac{2}{n\pi}\left[1 - \cos(n\pi)\right] = \frac{2}{n\pi}\left[1 - (-1)^n\right].$$

Therefore, the square waveform (signal) can be expressed as a sine series as shown:

$$f(t) = \sum_{n=1}^{\infty} b_n \sin(2\pi nt)$$

$$= \sum_{n=1}^{\infty} \frac{2}{n\pi}\left[1-(-1)^n\right]\sin(2\pi nt))$$

$$= \frac{4}{\pi}\sin(2\pi t) + \frac{4}{3\pi}\sin(6\pi t) + \frac{4}{5\pi}\sin(10\pi t) + \cdots$$

Now that the amplitudes of the harmonics are known, we can complete the frequency spectrum diagram shown in Figure 5.12a and include the unknown amplitudes. The updated frequency spectrum diagram is given in Figure 5.12b.

5.8 FOURIER SERIES OF PERIODIC FUNCTIONS

In Section 5.7, we explained that a square wave with a frequency of 1 Hz can be expressed as a sine series, and we showed that the frequency of the series is also 1 Hz. In this section, this theory is generalised. The types of functions that we shall consider are periodic and piecewise continuous.

Let us consider a piecewise periodic graph like the one shown in Figure 5.13. The section of graph highlighted by an arrow represents one full cycle between $t = -L$ and $t = +L$, where L is the half of the time period T.

Suppose this cycle can be represented by a function $f(t)$ defined over a range between $t = -L$ and $t = L$, then the repetitive parts can be determined with the periodicity condition $f(t+T) = f(t)$.

Our aim is to construct an infinite series that is equal to the function $f(t)$ in the entire range between $-\infty$ and $+\infty$, and has the same frequency as the frequency of

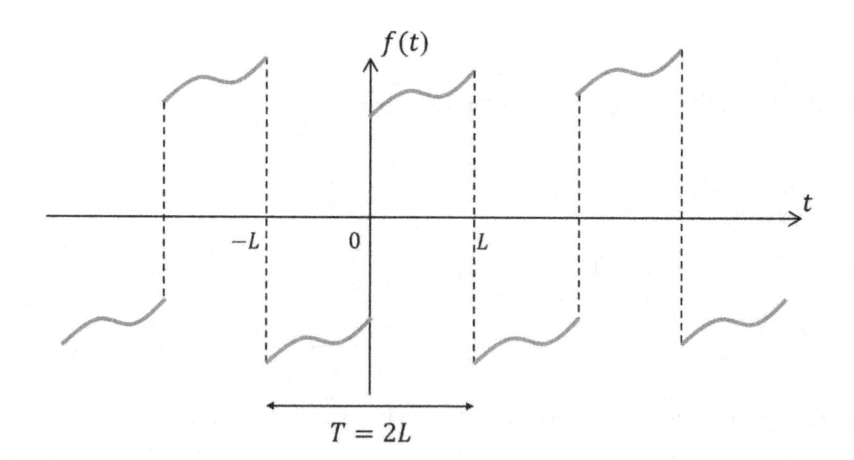

FIGURE 5.13 Graph of a piecewise periodic function with the time period $T = 2L$.

the function $f(t)$. Fourier claimed that $f(t)$ is composed of an infinite number of simple harmonics as shown:

$$f(t) = \frac{a_0}{2} + \sum_{n=1}^{\infty}\left[a_n \cos\left(\frac{n\pi t}{L}\right) + b_n \sin\left(\frac{n\pi t}{L}\right)\right]. \tag{5.4}$$

The coefficient a_n, for $n \neq 0$, can easily be obtained by multiplying both sides of Equation 5.4 by $\cos\left(\frac{m\pi t}{L}\right)$ and integrating it in the interval $[-L, L]$.

$$\int_{-L}^{L} f(t)\cos\left(\frac{m\pi t}{L}\right)dt = \sum_{n=1}^{\infty} a_n \int_{-L}^{L} \cos\left(\frac{n\pi t}{L}\right)\cos\left(\frac{m\pi t}{L}\right)dt +$$

$$\sum_{n=1}^{\infty} b_n \int_{-L}^{L} \sin\left(\frac{n\pi t}{L}\right)\cos\left(\frac{m\pi t}{L}\right)dt, \quad m = 1,2,3,\ldots \tag{5.5}$$

It can be shown that

$$\int_{-L}^{L} \sin\left(\frac{n\pi t}{L}\right)\cos\left(\frac{m\pi t}{L}\right)dt = 0$$

$$\int_{-L}^{L} \cos\left(\frac{n\pi t}{L}\right)\cos\left(\frac{m\pi t}{L}\right)dt = \begin{cases} 0 & n \neq m \\ L & n = m \end{cases}.$$

Hence, for the case $n = m$,

$$\int_{-L}^{L} f(t)\cos\left(\frac{n\pi t}{L}\right)dt = La_n$$

$$a_n = \frac{1}{L}\int_{-L}^{L} f(t)\cos\left(\frac{n\pi t}{L}\right)dt.$$

The coefficient a_0 can be determined by integrating both sides of Equation 5.4 over 1 period:

$$\int_{-L}^{L} f(t)dt = \frac{a_0}{2}\int_{-L}^{L} dt = L\,a_0$$

$$a_0 = \frac{1}{L}\int_{-L}^{L} f(t)dt.$$

The coefficient b_n can be obtained in the same way, but by multiplying both sides of Equation 5.4 by $\sin\left(\frac{m\pi t}{L}\right)$ and integrating both sides over the interval $[-L, L]$. We omit the calculations and present only the result:

$$b_n = \frac{1}{L}\int_{-L}^{L} f(t)\sin\left(\frac{n\pi t}{L}\right)dt, \quad n \neq 0.$$

Therefore, we obtain the Fourier series of the function $f(t)$ and is as follows:

$$f(t)=\frac{a_0}{2}+\sum_{n=1}^{\infty}a_n\cos\left(\frac{n\pi t}{L}\right)+\sum_{n=1}^{\infty}b_n\sin\left(\frac{n\pi t}{L}\right) \tag{5.6}$$

or

$$f(t)=\frac{a_0}{2}+\sum_{n=1}^{\infty}a_n\cos(n\omega t)+\sum_{n=1}^{\infty}b_n\sin(n\omega t),\quad \omega=\frac{2\pi}{T}.$$

This means that a periodic function $f(t)$ with a frequency $\frac{1}{T}$ can be expressed in terms of a series of simple harmonic functions. The series frequency is the same as the first harmonic frequency and is equal to $\frac{1}{T}$.

The constants a_0, a_n and b_n are known as Fourier coefficients for the series representation of the function $f(t)$, and as shown earlier, they are expressed as follows:

$$a_0=\frac{1}{L}\int_{-L}^{+L}f(t)\,dt \tag{5.7}$$

$$a_n=\frac{1}{L}\int_{-L}^{+L}f(t)\cos\left(\frac{n\pi t}{L}\right)dt \tag{5.8}$$

$$b_n=\frac{1}{L}\int_{-L}^{+L}f(t)\sin\left(\frac{n\pi t}{L}\right)dt. \tag{5.9}$$

These coefficients are the amplitude of the simple harmonics in Equation 5.4 and are used to construct the frequency spectrum that is explained in Section 5.9.

5.9 FREQUENCY SPECTRUM OF THE FOURIER SERIES

The frequency spectrum for combined signals is described in Section 5.5. In this section, we explain how it is implemented on the Fourier series.

The nth term of the Fourier series, that is

$$a_n\cos\left(\frac{n\pi t}{L}\right)+b_n\sin\left(\frac{n\pi t}{L}\right)=a_n\cos(n\omega_0 t)+b_n\sin(n\omega_0 t),$$

is called the nth harmonic, where $\omega_0=\frac{\pi}{L}$. The amplitude and the angular frequency of the nth harmonic are

$$A_n=\sqrt{a_n^2+b_n^2},\quad \omega_n=\frac{n\pi}{L},\quad n=1,2,3,\ldots$$

As stated in Section 5.7, it is often useful to plot the amplitude of the nth harmonic against the angular frequency, which is known as the frequency spectrum, and it is given in Figure 5.14. The frequency spectrum diagram has important applications in signal processing.

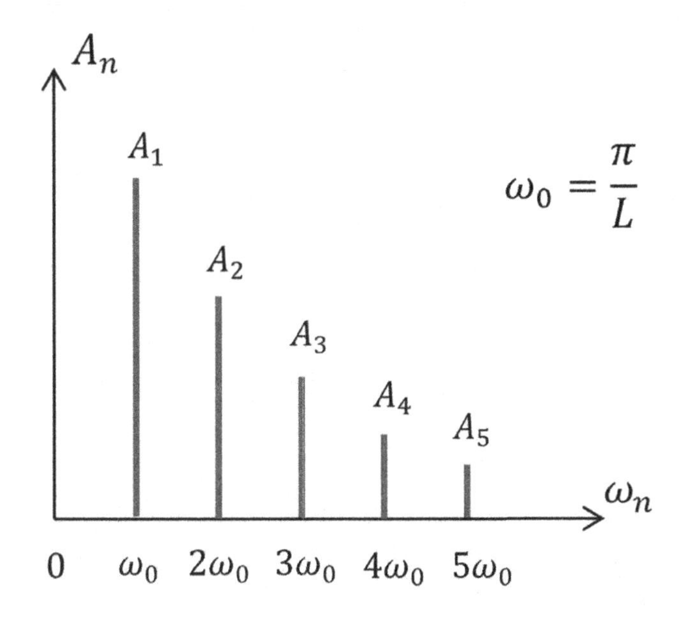

FIGURE 5.14 The frequency spectrum diagram.

Example 5.5 Given the following piecewise periodic function:

$$f(t)=\begin{cases}-1, & -\pi<t<0\\ +1, & 0<t<+\pi\end{cases}$$

(a) Sketch the graph of this function over three cycles (ranging from $t=-3\pi$ to $t=+3\pi$).
(b) Calculate the Fourier coefficients of this function.
(c) Write down the Fourier series of the function $f(t)$ up to $n=10$.
(d) Determine the fundamental frequency and the fundamental angular frequency of the Fourier series in part (c).
(e) Plot the frequency spectrum diagram up to $n = 10$.

Solution

(a) $T = 2\pi \Rightarrow L = \pi$ and $\omega = \dfrac{2\pi}{2\pi} = 1$ (See Figure 5.15).
(b) Since the function $f(t)$ is defined in two parts, to calculate the Fourier coefficients, we also divide the integrals into two parts:

$$a_0 = \frac{1}{L}\int_{-L}^{+L} f(t)dt = \frac{1}{\pi}\int_{-\pi}^{+\pi} f(t)dt = \frac{1}{\pi}\left[\int_{-\pi}^{0} f(t)dt + \int_{0}^{+\pi} f(t)dt\right]$$

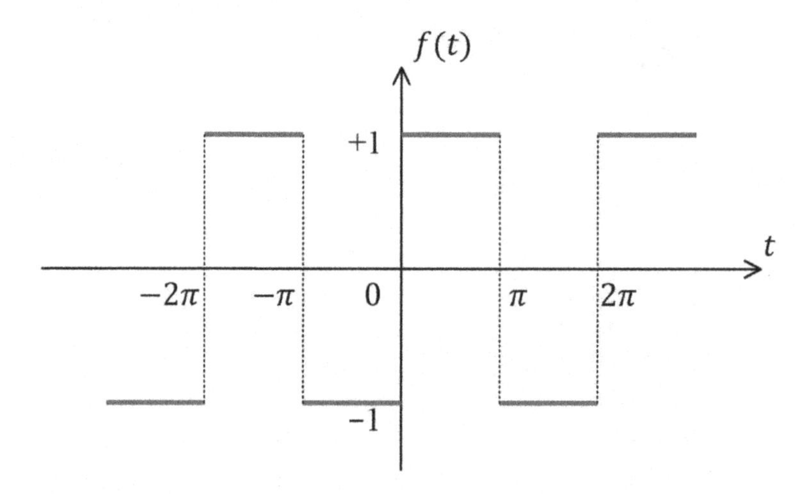

FIGURE 5.15 Graph of three cycles of the function $f(t)$ ranging from $t=-3\pi$ to $t=+3\pi$.

$$=\frac{1}{\pi}\left[\int_{-\pi}^{0}(-1)\,dt+\int_{0}^{+\pi}(+1)\,dt\right]=0$$

$$a_n=\frac{1}{L}\int_{-L}^{+L}f(t)\cos\left(\frac{n\pi t}{L}\right)dt=\frac{1}{\pi}\int_{-\pi}^{+\pi}f(t)\cos(nt)\,dt$$

$$=\frac{1}{\pi}\left[\int_{-\pi}^{0}f(t)\cos(nt)\,dt+\int_{0}^{+\pi}f(t)\cos(nt)\,dt\right]$$

$$=\frac{1}{\pi}\left[\int_{-\pi}^{0}(-1)\cos(nt)\,dt+\int_{0}^{+\pi}(+1)\cos(nt)\,dt\right]$$

$$=\frac{1}{\pi}\left[\left[-\frac{1}{n}\sin(nt)\right]_{t=-\pi}^{t=0}+\left[+\frac{1}{n}\sin(nt)\right]_{t=0}^{t=+\pi}\right]=0$$

$$b_n=\frac{1}{L}\int_{-L}^{+L}f(t)\sin\left(\frac{n\pi t}{L}\right)dt=\frac{1}{\pi}\int_{-\pi}^{+\pi}f(t)\sin(nt)\,dt$$

$$=\frac{1}{\pi}\left[\int_{-\pi}^{0}f(t)\sin(nt)\,dt+\int_{0}^{+\pi}f(t)\sin(nt)\,dt\right]$$

$$=\frac{1}{\pi}\left[\int_{-\pi}^{0}(-1)\sin(nt)\,dt+\int_{0}^{+\pi}(+1)\sin(nt)\,dt\right]$$

$$=\frac{1}{\pi}\left[\left[+\frac{1}{n}\cos(nt)\right]_{t=-\pi}^{t=0}+\left[-\frac{1}{n}\cos(nt)\right]_{t=0}^{t=+\pi}\right]=\frac{2}{n\pi}\left(1-(-1)^n\right).$$

(c) Now we construct the Fourier series by developing a table of Fourier coefficients first to minimise any errors (see Table 5.1).

TABLE 5.1
Fourier Coefficients

n	a_n	$b_n = \frac{2}{n\pi}\left(1-(-1)^n\right)$
1	$a_1 = 0$	$b_1 = \frac{4}{\pi}$
2	$a_2 = 0$	$b_2 = 0$
3	$a_3 = 0$	$b_3 = \frac{4}{3\pi}$
4	$a_4 = 0$	$b_4 = 0$
5	$a_5 = 0$	$b_5 = \frac{4}{5\pi}$
6	$a_6 = 0$	$b_6 = 0$
7	$a_7 = 0$	$b_7 = \frac{4}{7\pi}$
8	$a_8 = 0$	$b_8 = 0$
9	$a_9 = 0$	$b_9 = \frac{4}{9\pi}$
10	$a_{10} = 0$	$b_{10} = 0$

Hence,

$$\begin{aligned} f(t) &= b_1 \sin(t) + b_2 \sin(2t) + b_3 \sin(3t) + b_4 \sin(4t) + \cdots \\ &= \frac{4}{\pi}\sin(t) + \frac{4}{3\pi}\sin(3t) + \frac{4}{5\pi}\sin(5t) + \frac{4}{7\pi}\sin(7t) + \cdots \end{aligned}$$

In compact form, we can write

$$f(t) = \sum_{n=odd}^{\infty} \frac{4}{n\pi}\sin(nt) = \sum_{k=1}^{\infty} \frac{4}{(2k-1)\pi}\sin(2k-1)t.$$

In Figure 5.16, a graphical representation of this series is given for $n = 10$ and $n = 100$, which shows the convergence of the series and show that the series does indeed represent the function $f(t)$. Finite Fourier sums can closely approximate function $f(t)$ over the entire interval and the accuracy of the approximation depends on the number of harmonics used to construct the series. As mentioned earlier, in Figure 5.16, the numbers of harmonics are 10 and 100, respectively.

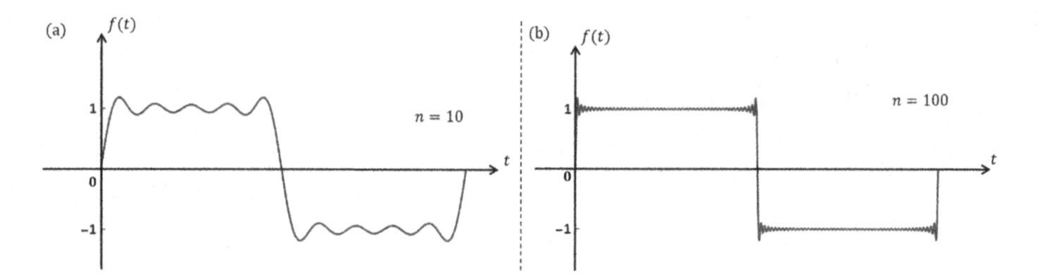

FIGURE 5.16 Graphs of the Fourier series for number of terms $n = 10$ and $n = 100$.

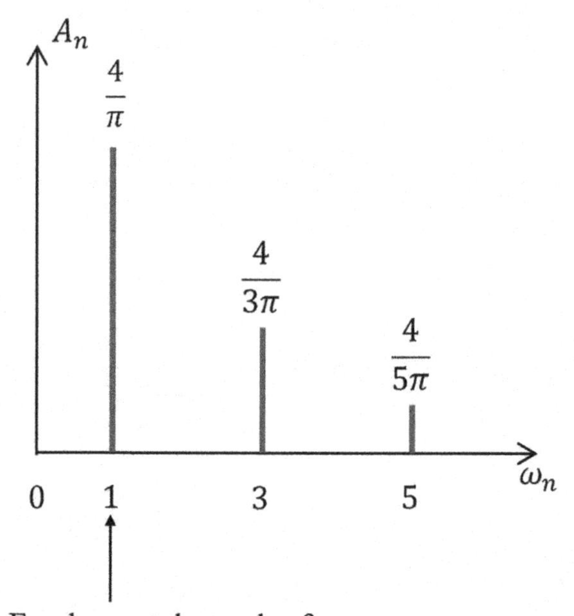

FIGURE 5.17 Frequency spectrum diagram.

(d) The fundamental frequency of the Fourier series is equal to the frequency of function $f(t)$. Hence $f_0 = \frac{1}{2\pi}$ and $\omega_0 = 2\pi \times f_0 = 1$.

(e) For plotting the frequency spectrum diagram, we must identify the amplitude and the angular frequency of each harmonic in the Fourier series. The amplitude of the nth harmonic is

$$A_n = \sqrt{a_n^2 + b_n^2} = \sqrt{0^2 + b_n^2} = b_n$$

$$\omega_n = n\omega_0 = n \times 1 = n, \quad n = 1,2,3,\ldots$$

Hence, using the amplitudes and the angular frequency we can plot the frequency spectrum – see Figure 5.17.

Example 5.6 Given the following function:

$$f(t) = 2(t+1), \quad -1 < t < 1, \text{ and } T = 2$$

(a) Sketch the graph of the (sawtooth waveform) function over five cycles.
(b) Calculate the Fourier coefficients of this function.
(c) Write down the Fourier series of the function $f(t)$ up to $n = 5$.
(d) Determine the fundamental frequency and the fundamental angular frequency of the Fourier series in part (c).
(e) Plot the frequency spectrum diagram up to $n = 5$.

Solution

(a) The graph of 5 cycles of $f(t)$ is shown in Figure 5.18.
(b) Using Equations 5.7, 5.8 and 5.9, we calculate the Fourier coefficients:

$$a_0 = \frac{1}{L}\int_{-L}^{+L} f(t)\,dt$$

$$= \frac{1}{1}\int_{-1}^{+1} 2(t+1)\,dt = 4$$

$$a_n = \frac{1}{L}\int_{-L}^{+L} f(t)\cos\left(\frac{n\pi t}{L}\right)dt$$

$$= \frac{1}{1}\int_{-1}^{+1} 2(t+1)\cos\left(\frac{n\pi t}{1}\right)dt = \int_{-1}^{+1} 2(t+1)\cos(n\pi t)\,dt = 0$$

$$b_n = \frac{1}{L}\int_{-L}^{+L} f(t)\sin\left(\frac{n\pi t}{L}\right)dt$$

$$= \frac{1}{1}\int_{-1}^{+1} 2(t+1)\sin\left(\frac{n\pi t}{1}\right)dt = \int_{-1}^{+1} 2(t+1)\sin(n\pi t)\,dt = \frac{4}{n\pi}(-1)^{n+1}.$$

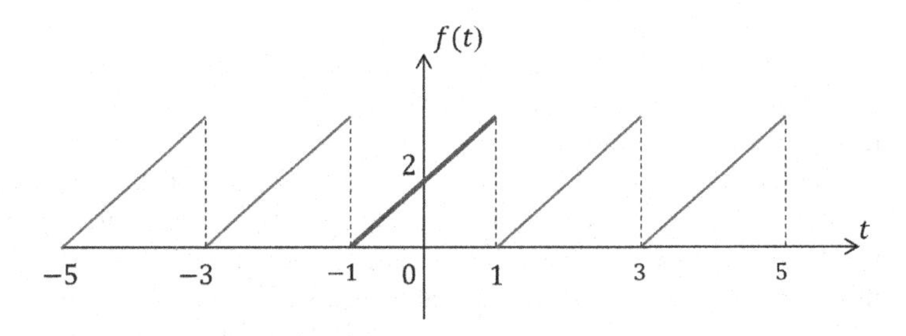

FIGURE 5.18 Five cycles of the function $f(t)$.

TABLE 5.2
Fourier Coefficients

n	a_n	$b_n = \frac{4}{n\pi}(-1)^{(n+1)}$
1	$a_1 = 0$	$b_1 = +\frac{4}{\pi}$
2	$a_2 = 0$	$b_2 = -\frac{4}{2\pi}$
3	$a_3 = 0$	$b_3 = +\frac{4}{3\pi}$
4	$a_4 = 0$	$b_4 = -\frac{4}{4\pi}$
5	$a_5 = 0$	$b_5 = +\frac{4}{5\pi}$

To calculate the Fourier coefficients, it requires integration, and in practice, this can be a bit complicated. In this example, we needed to determine the coefficients using integration by parts (differentiating $2(t+1)$ and integrating $\cos(n\pi t)$ or $\sin(n\pi t)$ that was omitted when constructing the Fourier series.

(c) Now we construct the Fourier series by developing a table of Fourier coefficients first to minimise any errors (see Table 5.2).

$$f(t) = \frac{a_0}{2} + \sum_{n=1}^{\infty} a_n \cos\left(\frac{n\pi}{L}t\right) + \sum_{n=1}^{\infty} b_n \sin\left(\frac{n\pi}{L}t\right)$$

$$= \frac{a_0}{2} + \sum_{n=1}^{\infty} b_n \sin\left(\frac{n\pi}{L}t\right), L = 1$$

$$= 2 + b_1 \sin(\pi t) + b_2 \sin(2\pi t) + b_3 \sin(3\pi t) + b_4 \sin(4\pi t) + \cdots$$

$$= 2 + \frac{4}{\pi}\sin(\pi t) - \frac{2}{\pi}\sin(2\pi t) + \frac{4}{3\pi}\sin(3\pi t) - \frac{1}{\pi}\sin(4\pi t) + \cdots$$

In Figure 5.19, a graphical representation of this series is given for $n = 10$ and $n = 100$, which shows the convergence of the series and show that the series does indeed represent the sawtooth waveform.

In this example, the graph of the original waveform shows instantaneous jumps; in mathematics, they are called jumps of discontinuities. In the Fourier series representation of the original function, each harmonic is continuous, and the jump of discontinuities is well preserved – a remarkable result. The small spikes in the corner of the Fourier series plot show the jumps of the discontinuities of the original function. They do not diminish with increasing n.

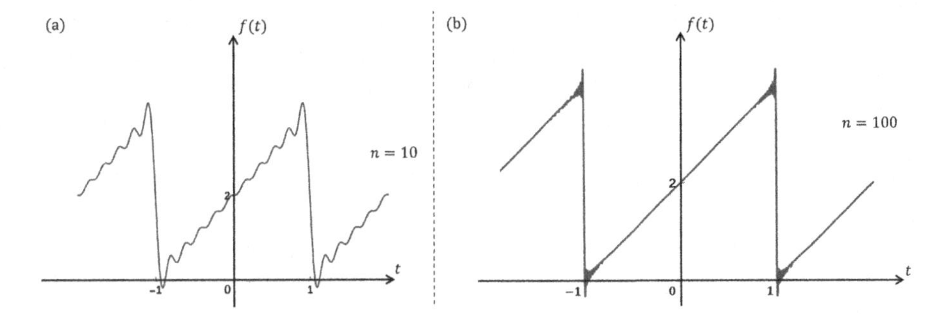

FIGURE 5.19 The graphs of the Fourier series for $n = 10$ and $n = 100$.

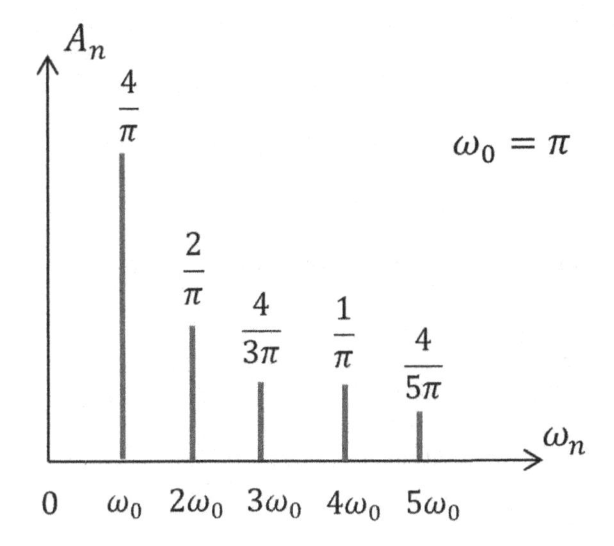

FIGURE 5.20 Frequency spectrum diagram.

(d) The fundamental frequency of the Fourier series is equal to the frequency of function $f(t)$. Hence $f_0 = \frac{1}{2}$ and $\omega_0 = 2\pi \times f_0 = \pi$.

(e) For plotting the frequency spectrum diagram, we must identify the amplitude and the angular frequency of each harmonic in the Fourier series.

$$A_n = \sqrt{a_n^2 + b_n^2} = \sqrt{0^2 + b_n^2} = |b_n|$$

$$\omega_n = n\omega_0 = n\pi, \quad n = 1, 2, 3, \ldots$$

Hence, using the amplitudes and the angular frequency, we can plot the frequency spectrum – see Figure 5.20.

Note: For convenience, sometimes a complete cycle of the piecewise periodic function $f(t)$ can be expressed in the interval $t = 0$ and $t = T$ instead of representing it in the interval $t = -L$ and $t = +L$. In this case, we can redefine the limits of integration and write it in the time intervals $t = 0$ and $t = T$. This will not affect the validity of the result and therefore the value of the Fourier coefficients:

$$a_0 = \frac{1}{L}\int_0^T f(t)\,dt$$

$$a_n = \frac{1}{L}\int_0^T f(t)\cos\left(\frac{n\pi t}{L}\right)dt$$

$$b_n = \frac{1}{L}\int_0^T f(t)\sin\left(\frac{n\pi t}{L}\right)dt.$$

Example 5.7 Given the following piecewise function, with one cycle defined from $t = 0$ to $t = 4$:

$$f(t) = \begin{cases} 0, & 0 < t < 1 \\ 10, & 1 < t < 2 \\ 2, & 2 < t < 4 \end{cases}$$

(a) Sketch the graph of the function $f(t)$ over two cycles (ranging from $t = 0$ to $t = 8$).
(b) Calculate the Fourier coefficients of $f(t)$.
(c) Write down the Fourier series the function $f(t)$ up to $n = 8$.
(d) Determine the fundamental frequency and the fundamental angular frequency of the Fourier series in part (c).
(e) Plot the frequency spectrum diagram up to $n = 8$.

Solution

(a) $T = 4$ (see Figure 5.21).

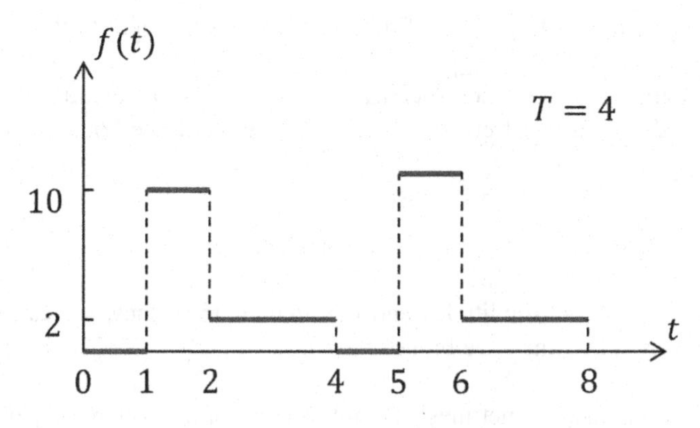

FIGURE 5.21 Graph of 2 cycles of the function $f(t)$.

(b) Since a full cycle of the function $f(t)$ is defined from $t=0$ to $t=4$, hence

$$a_0 = \frac{1}{L}\int_0^{2L} f(t)\,dt$$

$$= \frac{1}{4}\int_0^4 f(t)\,dt$$

$$= \frac{1}{4}\left[\int_0^1 0\,dt + \int_1^2 10\,dt + \int_2^4 2\,dt\right] = 7$$

$$a_n = \frac{1}{L}\int_0^{2L} f(t)\cos\left(\frac{n\pi t}{L}\right)dt$$

$$a_n = \frac{1}{2}\int_0^4 f(t)\cos\left(\frac{n\pi t}{2}\right)dt$$

$$= \frac{1}{2}\left[\int_0^1 0\times\cos\left(\frac{n\pi t}{2}\right)dt + \int_1^2 10\cos\left(\frac{n\pi t}{2}\right)dt + \int_2^4 2\cos\left(\frac{n\pi t}{2}\right)dt\right]$$

$$= \frac{1}{2}\left[\left[\frac{20}{n\pi}\sin\left(\frac{n\pi t}{2}\right)\right]_{t=1}^{t=2} + \left[\frac{4}{n\pi}\sin\left(\frac{n\pi t}{2}\right)\right]_{t=2}^{t=4}\right] = -\frac{10}{n\pi}\sin\left(\frac{n\pi}{2}\right)$$

$$b_n = \frac{1}{L}\int_0^{2L} f(t)\sin\left(\frac{n\pi t}{L}\right)dt$$

$$= \frac{1}{2}\int_0^4 f(t)\sin\left(\frac{n\pi t}{2}\right)dt$$

$$= \frac{1}{2}\left[-\frac{20}{n\pi}\cos\left(\frac{n\pi t}{2}\right)\right]_{t=1}^{t=2} + \frac{1}{2}\left[-\frac{4}{n\pi}\cos\left(\frac{n\pi t}{2}\right)\right]_{t=2}^{t=4}$$

$$= -\frac{10}{n\pi}\left[(-1)^n - \cos\left(\frac{n\pi}{2}\right)\right] - \frac{2}{n\pi}\left[1-(-1)^n\right]$$

$$= -\frac{10}{n\pi}\cos\left(\frac{n\pi}{2}\right) - \frac{8}{n\pi}(-1)^n - \frac{2}{n\pi}.$$

(c) Now we construct the Fourier series by developing a table of Fourier coefficients first to minimise any errors (see Table 5.3).

TABLE 5.3
Fourier Coefficients

n	$a_n = -\frac{10}{n\pi}\sin\left(\frac{n\pi}{2}\right)$	$b_n = \frac{10}{n\pi}\cos\left(\frac{n\pi}{2}\right) - \frac{8}{n\pi}(-1)^n - \frac{2}{n\pi}$
1	$a_1 = -\frac{10}{\pi}$	$b_1 = \frac{6}{\pi}$
2	$a_2 = 0$	$b_2 = -\frac{10}{\pi}$
3	$a_3 = \frac{10}{3\pi}$	$b_3 = \frac{2}{\pi}$
4	$a_4 = 0$	$b_4 = 0$
5	$a_5 = -\frac{2}{\pi}$	$b_5 = \frac{6}{5\pi}$
6	$a_6 = 0$	$b_6 = -\frac{10}{3\pi}$
7	$a_7 = \frac{10}{7\pi}$	$b_7 = \frac{6}{7\pi}$
8	$a_8 = 0$	$b_8 = 0$

$$f(t) = \frac{7}{2} - \frac{10}{\pi}\cos\left(\frac{\pi t}{2}\right) + \frac{10}{3\pi}\cos\left(\frac{3\pi t}{2}\right) - \frac{2}{\pi}\cos\left(\frac{5\pi t}{2}\right) + \frac{10}{7\pi}\cos\left(\frac{7\pi t}{2}\right) + \cdots$$
$$+ \frac{6}{\pi}\sin\left(\frac{\pi t}{2}\right) - \frac{10}{\pi}\sin(\pi t) + \frac{2}{\pi}\sin\left(\frac{3\pi t}{2}\right) + \frac{6}{5\pi}\sin\left(\frac{5\pi t}{2}\right) - \frac{20}{6\pi}\sin(3\pi t)$$
$$+ \frac{6}{7\pi}\sin\left(\frac{7\pi t}{2}\right) + \cdots$$

(d) The fundamental frequency of the Fourier series is equal to the frequency of function $f(t)$. Hence, $f_0 = \frac{1}{4}$ and $\omega_0 = 2\pi \times f_0 = \frac{\pi}{2}$.

(e) For plotting the frequency spectrum diagram, we must identify the amplitude and the angular frequency of each harmonic in the Fourier series. The amplitude and the angular frequency of the nth harmonic are

$$A_n = \sqrt{a_n^2 + b_n^2} = \sqrt{\left[-\frac{10}{n\pi}\sin\left(\frac{n\pi}{2}\right)\right]^2 + \left[\frac{10}{n\pi}\cos\left(\frac{n\pi}{2}\right) - \frac{8}{n\pi}(-1)^n - \frac{2}{n\pi}\right]^2}$$

$$\omega_n = n\omega_0 = \frac{n\pi}{2}, n = 1, 2, 3, \ldots$$

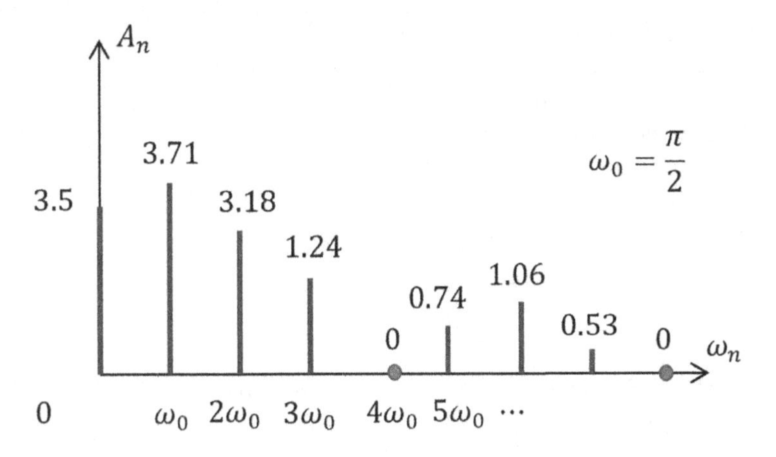

FIGURE 5.22 Frequency domain diagram.

$$A_1 \approx 3.71, A_2 \approx 3.18, A_3 \approx 1.24, A_4 = 0, A_5 \approx 0.74, A_6 \approx 1.06, A_7 \approx 0.53, A_8 = 0.$$

Hence, using the amplitudes and the angular frequencies, we can plot the frequency spectrum – see Figure 5.22.

5.9.1 Exercises

1. Given the following piecewise periodic function:

$$f(t) = \begin{cases} 0, & -1 < t < 0 \\ 1, & 0 < t < +1 \end{cases}, \quad T = 2$$

   (a) Sketch the graph of $f(t)$ over three cycles (from $t = -\frac{3T}{2}$ to $t = +\frac{3T}{2}$).
   (b) Calculate the Fourier coefficients of $f(t)$.
   (c) Write down the Fourier series of $f(t)$ up to $n = 8$.
   (d) Determine the fundamental frequency and angular frequency of the Fourier series in part (c).
   (e) Plot the frequency spectrum diagram up to $n = 8$.

2. Given the following piecewise periodic function:

$$f(t) = \begin{cases} 1, & -2 < t < 0 \\ 4, & 0 < t < +2 \end{cases}, \quad T = 4$$

   (a) Sketch the graph of the function $f(t)$ over three cycles (from $t = -\frac{3T}{2}$ to $t = +\frac{3T}{2}$).

(b) Calculate the Fourier coefficients $f(t)$.
(c) Determine the Fourier series of $f(t)$ up to $n=8$.
(d) Determine the fundamental frequency and angular frequency of the Fourier series in part (c).
(e) Plot the frequency spectrum diagram up to $n=8$.

3. Given the following piecewise periodic function:

$$f(t)=\begin{cases}-2, & -\pi<t<0\\ +2, & 0<t<+\pi\end{cases}, \quad T=2\pi$$

(a) Sketch the graph of the function $f(t)$ over three cycles (from $t=-\frac{3T}{2}$ to $t=+\frac{3T}{2}$).
(b) Calculate the Fourier coefficients of $f(t)$.
(c) Determine the Fourier series of $f(t)$ up to $n=8$.
(d) Determine the fundamental frequency and angular frequency of the Fourier series in part (c).
(e) Plot the frequency spectrum diagram up to $n=8$.

4. Given the following piecewise periodic function:

$$f(t)=\begin{cases}0, & 0<t<\pi\\ 4, & \pi<t<2\pi\\ 1, & 2\pi<t<4\pi\end{cases}, \quad T=4\pi$$

One cycle of $f(t)$ is ranging from $t=0$ to $t=4$.
(a) Sketch the graph of $f(t)$ over two cycles from $t=0$ to $t=8\pi$.
(b) Calculate the Fourier coefficients of $f(t)$.
(c) Determine the Fourier series of $f(t)$ up to $n=8$.
(d) Determine the fundamental frequency and angular frequency of the Fourier series in part (c).
(e) Plot the frequency spectrum diagram up to $n=8$.

Answers

1.

(a) $T=2\Rightarrow t=-3 \text{ to } t=3$.

(b) $a_0=1, a_n=0, b_n=\begin{cases}0, & n\text{ even}\\ \dfrac{2}{n\pi}, & n\text{ odd}\end{cases}$.

(c) Now we construct the Fourier series by developing a table of Fourier coefficients first to minimise any errors (see Table 5.4).

TABLE 5.4
Fourier Coefficients

n	a_n	b_n
1	$a_1 = 0$	$b_1 = \frac{2}{\pi}$
2	$a_2 = 0$	$b_2 = 0$
3	$a_3 = 0$	$b_3 = \frac{2}{3\pi}$
4	$a_4 = 0$	$b_4 = 0$
5	$a_5 = 0$	$b_5 = \frac{2}{5\pi}$
6	$a_6 = 0$	$b_6 = 0$
7	$a_7 = 0$	$b_7 = \frac{2}{7\pi}$
8	$a_8 = 0$	$b_8 = 0$

$$f(t) = \frac{1}{2} + \sum_{n=\text{odd}}^{\infty} \frac{2}{n\pi} \sin\left(\frac{n\pi}{1} t\right)$$

$$= \frac{1}{2} + \sum_{k=1}^{\infty} \frac{2}{(2k-1)\pi} \sin(2k-1)\pi t$$

$$= \frac{1}{2} + \frac{2}{\pi}\sin(\pi t) + \frac{2}{3\pi}\sin(3\pi t) + \frac{2}{5\pi}\sin(5\pi t) + \frac{2}{7\pi}\sin(7\pi t)$$

(d) The fundamental frequency of the Fourier series is equal to the frequency of function $f(t)$. Hence, $f_0 = \frac{1}{2}$ and $\omega_0 = 2\pi f_0 = \pi$, or it is the angular frequency of the first harmonic.

(e) For plotting the frequency spectrum diagram, we must identify the amplitude and the angular frequency of each harmonic in the Fourier series. The amplitude of the nth harmonic is

$$A_n = \sqrt{a_n^2 + b_n^2} = \sqrt{0^2 + b_n^2} = b_n$$

$$\omega_n = n\omega_0 = n\pi, n = 1, 2, 3, \ldots$$

Hence, using the amplitudes and the angular frequency, the frequency spectrum can easily be plotted.

TABLE 5.5
Fourier Coefficients

n	a_n	b_n
1	$a_1 = 0$	$b_1 = \frac{6}{\pi}$
2	$a_2 = 0$	$b_2 = 0$
3	$a_3 = 0$	$b_3 = \frac{6}{3\pi}$
4	$a_4 = 0$	$b_4 = 0$
5	$a_5 = 0$	$b_5 = \frac{6}{5\pi}$
6	$a_6 = 0$	$b_6 = 0$
7	$a_7 = 0$	$b_7 = \frac{6}{7\pi}$
8	$a_8 = 0$	$b_8 = 0$

2.
(a) $T = 4 \Rightarrow t = -6 \text{ to } t = 6$.
(b) $a_0 = 5, a_n = 0$

$$b_n = \begin{cases} 0, & n \text{ even} \\ \dfrac{6}{n\pi}, & n \text{ odd} \end{cases}.$$

(c) Now we construct the Fourier series by developing a table of Fourier coefficients first to minimise any errors (see Table 5.5).

$$f(t) = \frac{5}{2} + \sum_{n=\text{odd}}^{\infty} \frac{6}{n\pi} \sin\left(\frac{n\pi t}{2}\right) = \sum_{k=1}^{\infty} \frac{6}{(2k-1)\pi} \sin(2k-1)\frac{\pi t}{2}$$

$$= \frac{5}{2} + \frac{6}{\pi}\sin\left(\frac{\pi}{2}t\right) + \frac{6}{3\pi}\sin\left(\frac{3\pi}{2}t\right) + \frac{6}{5\pi}\sin\left(\frac{5\pi}{2}t\right) + \frac{6}{7\pi}\sin\left(\frac{7\pi}{2}t\right)$$

(d) The fundamental frequency of the Fourier series is equal to the frequency of function $f(t)$. Hence, $f_0 = \frac{1}{4}$ and $\omega_0 = 2\pi f_0 = \frac{\pi}{2}$.
(e) For plotting the frequency spectrum diagram, we must identify the amplitude and the angular frequency of each harmonic in the Fourier series. The amplitude of the nth harmonic is

$$A_n = \sqrt{a_n^2 + b_n^2} = \sqrt{0^2 + b_n^2} = b_n$$

$$\omega_n = n\omega_0 = \frac{n\pi}{2}, \quad n = 1,2,3,\dots$$

Hence, using the amplitudes and the angular frequency, the frequency spectrum can easily be plotted. Note the constant term $a_0 = \frac{5}{2}$ must also be included in the diagram.

3.

(a) $T = 2\pi \Rightarrow t = -3\pi$ to $t = 3\pi$.

(b) $a_0 = 0, \quad a_n = 0, \quad b_n = \begin{cases} 0, & n \text{ even} \\ \dfrac{8}{n\pi}, & n \text{ odd} \end{cases}.$

(c) Now we construct the Fourier series by developing a table of Fourier coefficients first to minimise any errors (see Table 5.6).

$$f(t) = \sum_{n=odd}^{\infty} \frac{8}{n\pi} \sin\left(\frac{n\pi t}{\pi}\right) = \sum_{k=1}^{\infty} \frac{8}{(2k-1)\pi} \sin(2k-1)t$$

$$= \frac{8}{\pi}\sin(t) + \frac{8}{3\pi}\sin(3t) + \frac{8}{5\pi}\sin(5t) + \frac{8}{7\pi}\sin(7t)$$

(d) The fundamental frequency of the Fourier series is equal to the frequency of function $f(t)$. Hence, $f_0 = 12\pi$ and $\omega_0 = 2\pi f_0 = 1$.

TABLE 5.6
Fourier Coefficients

n	a_n	b_n
1	$a_1 = 0$	$b_1 = \frac{8}{\pi}$
2	$a_2 = 0$	$b_2 = 0$
3	$a_3 = 0$	$b_3 = \frac{8}{3\pi}$
4	$a_4 = 0$	$b_4 = 0$
5	$a_5 = 0$	$b_5 = \frac{8}{5\pi}$
6	$a_6 = 0$	$b_6 = 0$
7	$a_7 = 0$	$b_7 = \frac{8}{7\pi}$
8	$a_8 = 0$	$b_8 = 0$

(e) For plotting the frequency spectrum diagram, we must identify the amplitude and the angular frequency of each harmonic in the Fourier series. The amplitude of the nth harmonic is

$$A_n = \sqrt{a_n^2 + b_n^2} = \sqrt{0^2 + b_n^2} = b_n$$

$$\omega_n = n\omega_0 = n, \quad n = 1,2,3,\ldots$$

Hence, using the amplitudes and the angular frequency, we plot the frequency spectrum can easily be plotted.

4.

(a) $T = 4\pi \Rightarrow t = 0 \text{ to } t = 8\pi$.

(b) $a_0 = 3, \quad a_n = -\frac{4}{n\pi}\sin\left(\frac{n\pi}{2}\right), \quad b_n = \frac{4}{n\pi}\cos\left(\frac{n\pi}{2}\right) - \frac{3}{n\pi}(-1)^n - \frac{1}{n\pi}.$

(c) Now we construct the Fourier series by developing a table of Fourier coefficients first to minimise any errors (see Table 5.7).

(d) The fundamental frequency of the Fourier series is equal to the frequency of function $f(t)$. Hence, $f_0 = \frac{1}{4\pi}$ and $\omega_0 = 2\pi f_0 = \frac{1}{2}$.

(e) For plotting the frequency spectrum diagram, we must identify the amplitude and the angular frequency of each harmonic in the Fourier series. The amplitude of the nth harmonic is

TABLE 5.7
Fourier Coefficients

n	a_n	b_n
1	$a_1 = +\frac{4}{\pi}$	$b_1 = +\frac{2}{\pi}$
2	$a_2 = 0$	$b_2 = -\frac{8}{2\pi}$
3	$a_3 = +\frac{4}{3\pi}$	$b_3 = +\frac{2}{3\pi}$
4	$a_4 = 0$	$b_4 = 0$
5	$a_5 = -\frac{2}{5\pi}$	$b_5 = +\frac{2}{5\pi}$
6	$a_6 = 0$	$b_6 = -\frac{8}{6\pi}$
7	$a_7 = +\frac{4}{7\pi}$	$b_7 = +\frac{2}{7\pi}$
8	$a_8 = 0$	$b_8 = 0$

$$A_n = \sqrt{a_n^2 + b_n^2}$$

$$\omega_n = n\omega_0 = \frac{n}{2}, \quad n = 1,2,3,\ldots$$

$$A_1 \approx 1.42, A_2 \approx 1.27, A_3 \approx 0.47, A_4 = 0,$$

$$A_5 \approx 0.28, A_6 \approx 0.426, A_7 \approx 0.20, A_8 = 0.$$

Hence, using the amplitudes and the angular frequency, the frequency spectrum can easily be plotted. Note that $\frac{a_0}{2} = \frac{3}{2}$ must also be included in the diagram.

5.10 ENGINEERING APPLICATIONS

Fourier series are used in a wide variety of physical and engineering applications, but in this chapter, we only present their applications on a simple mechanical system and an electrical system.

5.10.1 Vibration Analysis

Fourier series can be used in determining the response of mechanical systems subjected to vibrations. To explain this point, we use an example of valve vibration in hydraulic control systems.

In the study of valve vibration, the hydraulic control valve model consists of a valve of mass m attached (on one side) to an elastic spring with spring constant k, and a damper (dashpot), as shown in Figure 5.23a. Assume that in addition to the forces exerted by the spring and damper on one side, the other side of the valve is subjected to a periodic fluid pressure, $p(t)$ shown in Figure 5.23b, which changes as the valve opens (or closes) as follows:

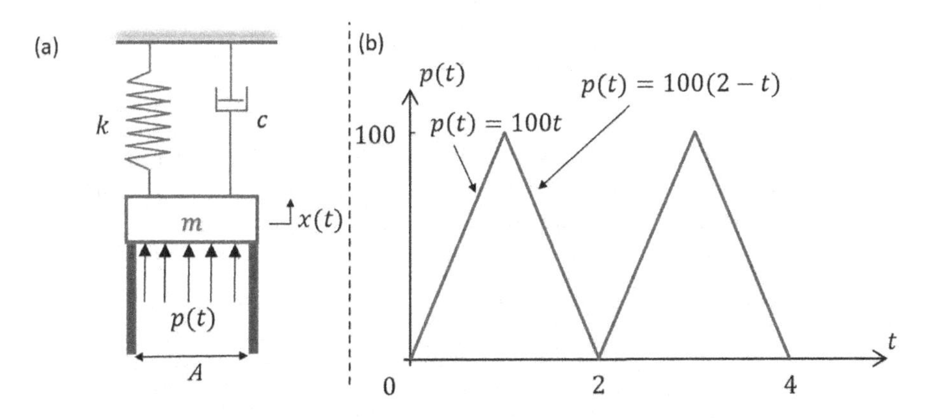

FIGURE 5.23 (a) Hydraulic control valve modelled as a spring-mass-damper system. (b) Pressure over time t.

$$p(t) = \begin{cases} 100\,t, & 0 < t < 1 \\ 100\,(2-t), & 1 < t < 2 \end{cases}.$$

The change in pressure over time causes the valve to vibrate. Assume the position of the valve at any instant of time is represented by $x(t)$.

The forces acting on the valve are as follows:

- Force resulting from the fluctuating pressure:

$$f(t) = A \times p(t) = \begin{cases} 100\,A\,t, & 0 < t < 1 \\ 100\,A(2-t), & 1 < t < 2 \end{cases}$$

  where A is the cross-sectional area of the fluid chamber shown in Figure 5.23a.
- The tension in the spring $= -kx$.
- The damping force applied by the dashpot is proportional to the vibration speed $= -c\dot{x}(t)$.

Newton's second law of motion states that Force = Mass × Acceleration, so

$$m\ddot{x} = -kx - c\dot{x} + f(t)$$
$$m\ddot{x} + c\dot{x} + kx = f(t).$$

This is a linear second-order differential equation with an unusual right-hand side and cannot be solved with the use of the traditional method of undetermined coefficients in its current form. However, since $p(t)$ is periodic with a time period $T = 2$, so $f(t)$ is also periodic with the same time period; hence, we can expand the right-hand side of the differential equation to its Fourier series and write the equation as

$$m\ddot{x} + c\dot{x} + kx = a_0 + \sum_{n=1}^{\infty}\left[a_n \cos\left(\frac{n\pi\,t}{L}\right) + b_n \sin\left(\frac{n\pi\,t}{L}\right)\right], \quad L = \frac{T}{2} = 1.$$

Now, this equation can be solved for $x(t)$ by using the principle of superposition, but it is not in the scope of this chapter to solve such an equation and we just calculate the Fourier coefficients here as follows:

$$a_0 = \frac{1}{1}\left[\int_0^1 100\,At\;dt + \int_1^2 100\,A(2-t)\,dt\right] = 100\,A$$

$$a_n = \frac{1}{1}\left[\int_0^1 100At\cos\left(\frac{n\pi t}{1}\right)dt + \int_1^2 100\,A(2-t)\cos\left(\frac{n\pi t}{1}\right)dt\right]$$

$$= \int_0^1 100At\cos(n\pi t)\,dt + \int_1^2 100\,A(2-t)\cos(n\pi t)\,dt$$

$$= \frac{200A\left[(-1)^n - 1\right]}{n^2\pi^2} == \begin{cases} -\frac{400A}{n^2\pi^2}, & \text{n odd} \\ 0, & \text{n even} \end{cases}$$

$$b_n = \frac{1}{1}\left[\int_0^1 100At\sin\left(\frac{n\pi t}{1}\right)dt + \int_1^2 100A(2-t)\sin\left(\frac{n\pi t}{1}\right)dt\right]$$

$$= \int_0^1 100At\sin(n\pi t)dt + \int_1^2 100A(2-t)\sin(n\pi t)dt$$

$$= -\frac{100A}{n\pi}\cos(n\pi) + 0 + \frac{100A}{n\pi}\cos(n\pi) - 0 = 0.$$

Hence, the Fourier series of $f(t)$ can be expressed as

$$F(t) = 50A - \frac{400A}{\pi^2}\cos(\pi t) - \frac{400A}{9\pi^2}\cos(3\pi t) - \cdots$$

As explained earlier, this series is used to solve the differential equation for the function $x(t)$ that represents the position of the valve at any instant of time during its vibration.

5.10.2 Voltage Output of a Rectifier

In most all electronic devices, the supply AC voltage is converted to DC voltage using a rectifier. The output voltage from the rectifier is not constant and still has ripples. Most applications will require a smooth DC voltage. To reduce the ripples, a filter is commonly used after the rectifier to remove the higher order harmonics. In order to design the filter, it is important to determine the magnitude of the harmonics where we can make use of Fourier analysis.

Let us consider a fully rectified waveform, which is given by

$$v(t) = |V\sin t| = \begin{cases} -V\sin t, & -\pi < t \le 0 \\ V\sin t, & 0 < t < \pi \end{cases}.$$

The graph of $v(t)$ is shown in Figure 5.24, and it is a 2π periodic function and its Fourier series can be obtained as follows:

$$a_0 = \frac{1}{\pi}\left[\int_{-\pi}^0 -V\sin t dt + \int_0^\pi V\sin t dt\right]$$

$$= \frac{2}{\pi}\int_0^\pi V\sin t\, dt = \frac{4V}{\pi}$$

$$a_n = \frac{1}{\pi}\int_{-\pi}^\pi |V\sin t|\cos\left(\frac{n\pi t}{\pi}\right)dt$$

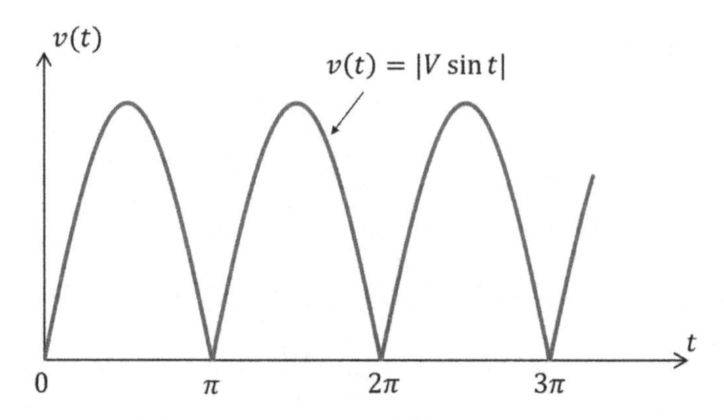

FIGURE 5.24 A fully rectified waveform

$$= \frac{2}{\pi}\int_0^{\pi} V \sin t \cos(nt)\,dt$$

$$= \frac{V}{\pi}\int_0^{\pi} \left\{\sin\left[(1+n)t\right] + \sin\left[(1-n)t\right]\right\} dt$$

$$= \begin{cases} \dfrac{-4V}{(n^2-1)\pi}, & n\,\text{even} \\ 0, & n\,\text{odd} \end{cases}$$

$$b_n = \frac{1}{\pi}\int_{-\pi}^{\pi} |V \sin t| \sin\left(\frac{n\pi t}{\pi}\right) dt$$

$$= \frac{2}{\pi}\int_0^{\pi} V \sin t \sin(nt)\,dt$$

$$= \frac{V}{\pi}\int_0^{\pi} \left[\cos(1-n)t - \cos(1+n)t\right] dt = 0.$$

Therefore, the Fourier series for the fully rectified waveform can be expressed as

$$v(t) = \frac{2V}{\pi} - \frac{4V}{3\pi}\cos(2t) - \frac{4V}{15\pi}\cos(4t) - \frac{4V}{35\pi}\cos(6t) - \cdots$$

5.10.3 Exercises

1. A hydraulic control valve shown in Figure 5.23 is subjected to a periodic fluid pressure fluctuation according to

$$p(t) = \begin{cases} 500t, & 0 < t < 1 \\ 500(2-t), & 1 < t < 2 \end{cases}.$$

Assuming the cross-sectional area of the valve chamber is 1,000 mm^2:

(a) Determine the force function, $f(t)$, for the hydraulic control valve.

(b) Write down the Fourier series for the force obtained in part (a) up to $n=4$.

2. A hydraulic control valve shown in Figure 5.23 is subjected to a periodic fluid pressure fluctuation according to

$$p(t) = \begin{cases} 1000t^2, & 0<t<1 \\ 1000(2-t)^2, & 1<t<2 \end{cases}.$$

Assuming the cross-sectional area of the valve chamber is 1,000 mm^2:

(a) Determine the force function, $f(t)$, for the hydraulic control valve.

(b) Write down the Fourier series for the force obtained in part (a) up to $n=3$.

3. Determine the Fourier series of the output voltage, $v(t)$, of a full wave rectifier up to $n=4$, given that

$$v(t) = |20\sin t|.$$

4. Determine the Fourier series of the output voltage, $v(t)$, of a half wave rectifier up to $n=4$, given that

$$v(t) = \begin{cases} 20\sin(\pi t), & 0<t<1 \\ 0, & 1<t<2 \end{cases}.$$

Answers

1. $f(t) = 0.25 - 2\pi^2\cos(\pi t) - \dfrac{2}{9\pi^2}\cos(3\pi t)$

2. $f(t) = \dfrac{1}{3} - \dfrac{4}{\pi^2}\cos(\pi t) + \dfrac{1}{\pi^2}\cos(2\pi t) - \dfrac{4}{9\pi^2}\cos(3\pi t)$

3. $v(t) = \dfrac{40}{\pi} - \dfrac{80}{3\pi}\cos(2t) - \dfrac{80}{15\pi}\cos(4t)$

4. $v(t) = \dfrac{20}{\pi} - \dfrac{40}{3\pi}\cos(2\pi t) - \dfrac{40}{15\pi}\cos(4\pi t) + 10\sin(\pi t)$

6 Statistics

6.1 INTRODUCTION

Statistics, in the sense of an academic discipline, is the study of methods of extracting and summarising information from collections, often large collections, of numerical facts. These collections of numerical facts are known as *data*, from the Latin for "things given".

Data can be collected in many different ways. For example, every 10 years, the UK government attempts to collect data on every individual resident in the UK on a certain date. This type of exercise is known as a *census*. Censuses are very expensive to mount, and researchers who want to find out something about a large population of people will usually collect data on only a part of the population, known as a *sample*. The subject of statistics includes the study of ways of selecting samples from populations.

Typically, data obtained from a sample are used to draw conclusions about the underlying population. The body of methods that ensures that such conclusions are as reliable as possible is known as *statistical inference*. The two branches of classical statistical inference are *statistical estimation* and *hypothesis testing*. Because samples are usually chosen from populations using a random mechanism, the mathematical theory of probability is fundamental to statistical inference.

In this chapter, some methods for describing data are presented. The process of data description is an essential part of statistical analysis and should always be carried out before any statistical tests or estimation methods are used.

Statistics can be divided into two branches, descriptive statistics and inferential statistics:

- Descriptive statistics involve organising data, presenting it graphically, and calculating summary statistics such as the mean and standard deviation. Descriptive statistics are essential for summarising and understanding data characteristics, serving as a preliminary step before proceeding to analyses that are more complex. Depending on the nature of the study, descriptive statistics may be sufficient as the sole method of analysing collected data.
- Inferential statistics, on the other hand, encompasses statistical techniques employed to make inferences or draw conclusions about a population based on a sample of data collected from that population. These methods enable researchers to extrapolate findings from a sample to the larger population, facilitating the generalisation of results. Inferential statistics play a crucial role in hypothesis testing, estimating parameters and making predictions, thereby providing valuable insights into populations beyond the scope of direct observation.

DOI: 10.1201/9781032630694-6

Before we discuss the descriptive and inferential statistics, we emphasise the terms used when dealing with *data*:

- Population
- Sample
- Random variable

6.2 POPULATION

A population encompasses the entirety of units, individuals, or objects, whether finite or infinite, from which data regarding specific characteristics are gathered or observed.

6.3 SAMPLE

Due to the large size of populations, direct observation of each member is usually impossible or impractical. Instead, researchers usually select a representative subset, known as a sample, to study, analyse and draw conclusions about the entire population. Thus, a sample is a representative subset of the entire population, collected at random, where each observation has an equal chance of being included in the collected sample.

6.4 RANDOM VARIABLE

Data can be collected from various methods such as surveys, questionnaires or experimental setups. The specific attribute being quantified or observed is termed a random variable due to its inherent variability. For instance, take the height of university students; although it can be measured, individual heights fluctuate and may differ with each observation.

6.5 QUALITATIVE VARIABLES

These variables are typically assessed categorically rather than numerically, as they pertain to qualitative characteristics. Examples include characteristics such as hair colour, blood type, marital status and gender.

6.6 QUANTITATIVE VARIABLES

These variables are quantifiable using numerical measurement, enabling comparisons based on magnitude. Examples include the height of university students, daily temperature readings, currency exchange rate and more. Unlike categorical variables, which are characterised by distinct categories, these quantitative variables are expressed as numerical values, allowing for precise measurement and analysis.

Measurements can be of two types:

- Discrete
- Continuous

TABLE 6.1
Number of Faulty Fire Extinguishers

0	0	2	0	0	3	2	0	0	6	0	0	0	2	1	1	0	0	0	3	0	6
1	0	0	1	0	0	1	0	0	4	2	1	1	0	1							

Discrete data consists of measurements that can only take clearly distinct values, for example the test scores out of 10. The continuous measurements are something like length, time and voltage, among others.

6.7 DESCRIPTIVE STATISTICS

Descriptive statistics is classified as follows:

- Organising data/summarising data
- Graphical representation
- Summary statistics

6.7.1 Summarising Data

Raw data collected from experiments or surveys can often be complex and challenging to interpret as a series of individual observations. To extract meaningful insights from these raw data, it is crucial to employ suitable methods to summarise and condense the information effectively. Two common approaches for summarising data are graphical and numerical summaries.

Graphical summaries utilise visual representations such as histograms and numerical summaries and, however, involve calculating summary statistics such as the mean, the median and the standard deviation. These numerical measures provide quantitative insights into central tendency, variability and distributional characteristics of the data. Both graphical and numerical summaries complement each other and are essential components of the data analysis process.

6.7.2 Frequency Distribution (Discrete Variable)

We now present examples of summarising the discrete data set.

Example 6.1 The numbers of faulty fire extinguishers in a survey of 37 small to medium-sized office premises are listed in Table 6.1.

Notes:

- The survey units are the premises.
- The variable is the number of faulty extinguishers.
- This type of variable is known as a discrete quantitative variable.
- Discrete quantitative variables usually take whole numbers (i.e. 0, 1, 2, 3, . . .) as possible values. Occasionally they can take fractional values also.

The frequency distribution is given in Table 6.2.

TABLE 6.2
Frequency Distribution and Relative Frequency Distribution of "Number of Faulty Fire Extinguishers"

Number of Faulty Extinguishers	Number of Premises (Frequency)	Relative Frequency (%)
0	20	54.05
1	8	21.62
2	4	10.81
3	2	5.41
4	1	2.70
5	0	0.00
6	2	5.41
Total	37	100%

Notes: The frequency distribution of a discrete quantitative variable should display all possible values, including any with zero frequency, such as 5 in Table 6.2. The relative frequencies are obtained by dividing frequencies into the total frequency 37 and multiplying the result by 100.

Graphical representation (bar chart) of Table 6.2 is given in Figure 6.1.

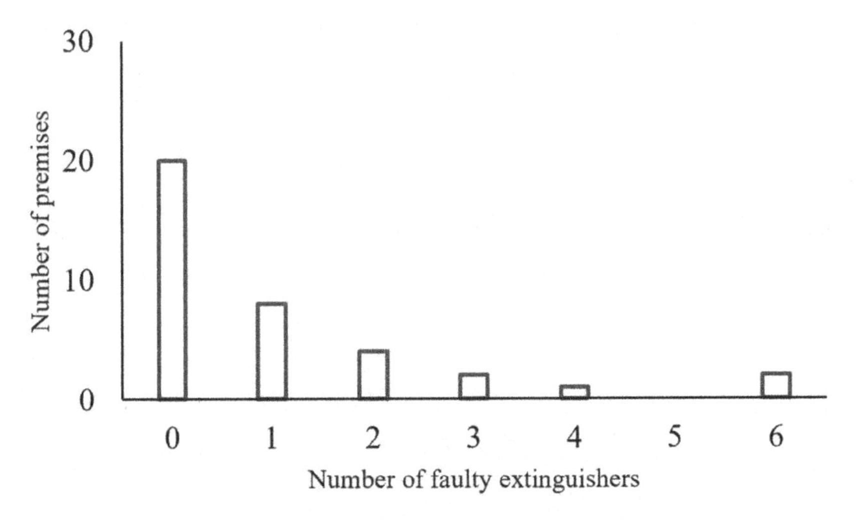

FIGURE 6.1 A bar chart to show the distribution of faulty fire extinguishers.

NOTES:

- This type of diagram is known as a bar chart.
- Bars should be drawn of equal width, with some spacing between them.
- The bar chart for a discrete quantitative variable should show all possible values on the horizontal axis, including any with zero frequency, such as 5 in Figure 6.1.

Example 6.2 Consider the set of data in Table 6.3, generated by an experiment where the occurrence of a certain type of subatomic particle was monitored. The particles passing through a detector in 10 minutes were counted 100 times with the following results.

If we look at the data closely, we see that there are many discrete values (2–74 inclusive) with low frequencies. A frequency distribution that considers the full discrete range of values would be long, "flat" and not very informative. For this type of data set, it is better to classify the data according to ranges of values, which is called **grouped frequency** distribution. We might choose the classification shown in Table 6.4 (other classifications could be used).

TABLE 6.3
Number of Particles Passing through the Detector in 10 Minutes

41	57	49	57	31	33	44	49	22	40
40	43	46	62	44	49	40	37	65	51
44	32	42	41	43	49	43	54	41	46
54	43	63	43	30	43	46	72	54	32
51	44	65	12	33	27	67	53	47	44
44	46	42	40	41	32	52	54	42	51
42	39	34	51	68	16	53	42	2	56
44	68	53	49	57	51	47	43	56	35
43	54	54	74	35	11	42	57	74	46
42	43	56	35	43	62	42	37	56	39

TABLE 6.4
Grouped Frequency Distribution of Discrete Data

Number of Particles	Number of Occasions	Percentage
0–10	1	1
10–20	3	3
20–30	3	3
30–40	18	18
40–50	41	41
50–60	23	23
60–70	8	8
70–80	3	3
	Total = 100	**100**

NOTES:

- Table 6.4 is known as a grouped frequency distribution table.
- The groups 0–10, 10–20 and so on are known as **class intervals**. The endpoints of class intervals are known as **class limits** (lower and upper class limits).
- The difference between the consecutive lower or upper class limits is called **class width**, which is 10.
- Table 6.4 does not make clear, for example, whether the data 30 should be allocated to the interval 20–30 or to 30–40. We shall adopt the rule used in the package Excel, which states that class intervals contain their *upper limits*, but not their *lower limits*. Thus 30 falls into the class interval 20–30
- The grouped frequency distribution is more informative than the ungrouped.

We use class intervals and their frequencies to draw the histogram. A histogram consists of a set of rectangles, adjacent to each other, on the horizontal axis as the base. The width of each rectangle is the class width of the particular class interval. These class widths are usually, but not necessarily, equal. If, however, the widths of the rectangles are equal, then the heights of the rectangles are proportional to the frequencies, so that the vertical scale usually represents the number of observations or frequency. Throughout this chapter, we have used Excel to draw the histogram. Figure 6.2 represents the histogram of the data presented in Table 6.4.

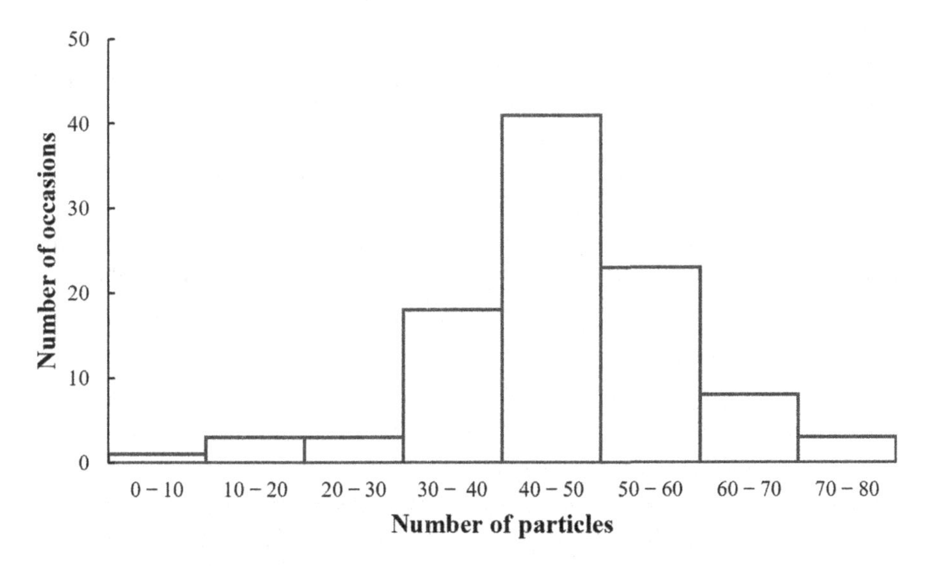

FIGURE 6.2 Histogram of particles passing through the detector.

6.8 THE MEAN, STANDARD DEVIATION AND MEDIAN

The arithmetic mean (or simply the mean) of a set of n observations $x_1, x_2, \ldots, x_n$ is denoted by $\bar{x}$ and is defined as

$$\bar{x} = \frac{x_1 + x_2 + \ldots + x_n}{n} = \frac{\Sigma x}{n}.$$

The standard deviation s is the square root of the variance (the variance is calculated using the expression under the square root):

$$s = \sqrt{\frac{\Sigma(x - \bar{x})^2}{n-1}}.$$

The mean is called a **measure of location** as it gives a representative value about which the data are spread. The standard deviation (*SD*) is a **measure of the spread** of the data about the mean; a large value for the *SD* means the data are well spread out, and a small value means that they are grouped quite tightly about the mean.

If a set of ungrouped data is arranged in ascending or descending order of magnitude, the *median* is defined as the value which divides the set into two numerically equal groups, that is with equal total frequencies. For example, 29, 34, 26, 37, 31 gives, in ascending order, 26, 29, 31, 34, 37; hence, the median is 31.

Before we calculate the mean and standard deviation for the given data, note that the formula for the standard deviation can be manipulated into a more convenient form:

$$s = \sqrt{\frac{1}{n(n-1)}\left[n\sum_{i=1}^{i=n} x_i^2 - \left(\sum_{i=1}^{i=n} x_i\right)^2\right]}.$$

For example, suppose an experiment that delivers some numerical value is run 6 times. The output values follow:

1.5 2.3 1.7 1.8 1.6 1.8

To work out the *SD*, we create a table like Table 6.5 in which the first column contains the individual data values and the second the squares of these values, we find the total value in each column and then use the formula:

$$\bar{x} = \frac{1}{n}\sum_{i=1}^{i=6} x_i = \frac{1}{6} \times 10.7 \approx 1.7833$$

$$s = \sqrt{\frac{n\sum_{i=1}^{i=n} x_i^2 - (\sum_{i=1}^{i=n} x_i)^2}{n(n-1)}} = \sqrt{\frac{6 \times 19.47 - 10.7^2}{5 \times 6}} = 0.2787.$$

TABLE 6.5
Table to Aid the Calculation of Standard Deviation

	x_i	x_i^2
	1.5	2.25
	2.3	5.29
	1.7	2.89
	1.8	3.24
	1.6	2.56
	1.8	3.24
Total	**10.7**	**19.47**

TABLE 6.6
Number of Mechanical Faults from 80 Audits

2	0	0	1	0	3	0	6	0	0	8
0	2	0	1	5	1	0	1	1	2	1
0	0	2	0	0	0	0	0	0	0	0
0	0	0	0	0	0	0	0	1	0	1
0	0	0	5	1	0	0	0	0	0	0
0	0	1	1	0	3	0	0	1	1	0
0	0	2	0	1	0	0	0	0	0	0
0	0	0								

6.8.1 Exercises

1. Determine the mean and standard deviation for each of the following sets of sample data:
   (i) Resistances in ohms

   152 146 149 153 152 147

   (ii) Output voltage

   2.4 2.6 2.4 2.5 2.7
   2.3 2.2 2.5 2.8 2.9

2. The number of mechanical faults reported for the survey of 80 audits in a certain engineering industry over 1 year is given in Table 6.6. Construct a frequency distribution table and draw a bar chart.
3. Table 6.7 shows the weights of 40 castings to the nearest kilogram. Construct a frequency distribution table and draw the corresponding histogram.

TABLE 6.7
Weights of 40 Castings; the Minimum Weight Is 19 and the Maximum is 76

38	64	50	32	44	25	49	57	46	58
36	48	52	44	27	38	**76**	63	**19**	54
73	43	47	35	53	40	35	61	45	35
56	45	28	68	40	47	65	46	42	50

TABLE 6.8
Chris's Score in a Sample of 100 Golf Games

90	96	105	94	84	83
94	94	88	96	86	102
76	99	95	100	94	92
96	91	87	89	115	92
97	102	92	84	98	94
100	102	88	80	94	93
94	103	90	101	90	94
94	102	88	85	88	86
84	101	94	106	96	80
100	91	93	95	95	92
94	99	82	98	98	83
84	93	95	102	95	83
94	91	94	99	80	93
90	99	104	99	108	86
95	85	102	86	115	86
104	95	87	93	80	108
98	112	95	102		

4. Chris Golfnut loves the game of golf. Chris also loves statistics. Combining both passions, Chris records a sample of 100 of his scores on his home course (par 72) and lists the data as shown in Table 6.8.
   (i) Draw the histogram and calculate summary statistics for the scores.
   (ii) Briefly describe what the statistics divulge.

Answers

(1) (i) $\bar{x} = 149.83, \quad s = 2.927$

(ii) $\bar{x} = 2.53, \quad s = 0.221$

(2) See Table 6.9 and Figure 6.3.

(3) See Table 6.10 and Figure 6.4.

TABLE 6.9
Frequency Distribution of Mechanical Faults

Mechanical Fault	Number of Audits
0	55
1	14
2	5
3	2
4	0
5	2
6	1
7	0
8	1

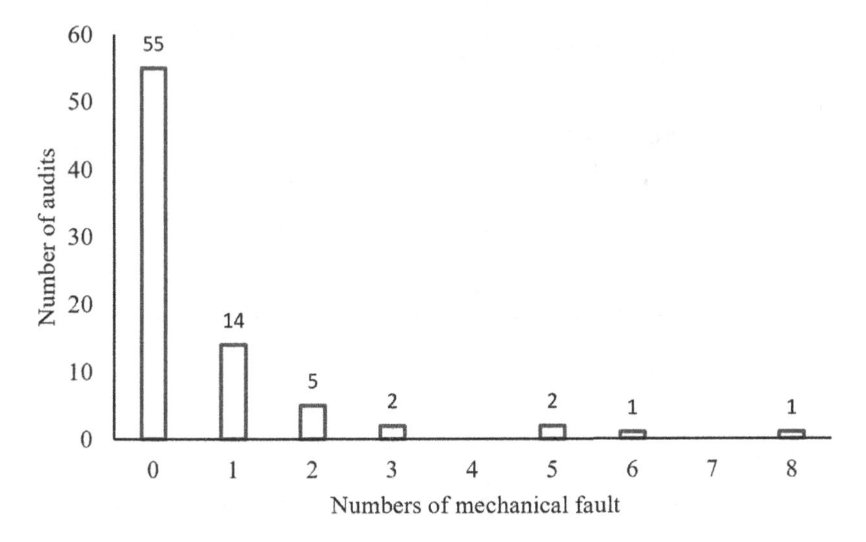

FIGURE 6.3 A bar chart to show the number of mechanical faults.

TABLE 6.10
Grouped Frequency Distribution of the Weight of Castings

Weight of Castings	Number of Castings
10–20	1
20–30	3
30–40	9
40–50	14
50–60	6
60–70	5
70–80	2

(4)

(i) See Table 6.11 and Figure 6.5. Mean = 93.5, Median = 94 and standard deviation = 7.7.

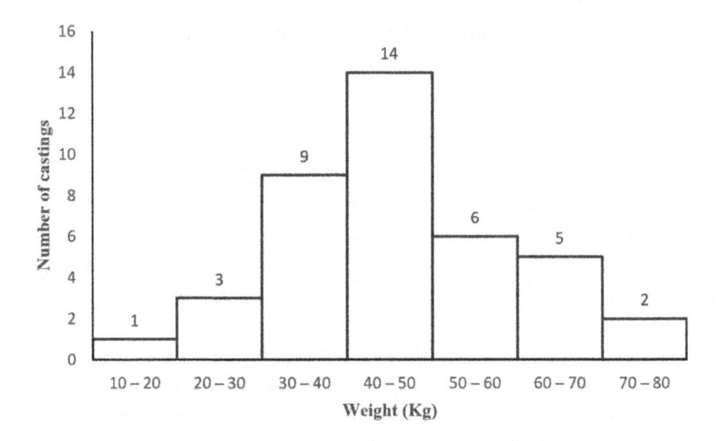

FIGURE 6.4 Histogram of the weights of 40 castings

TABLE 6.11
Grouped Frequency Distribution of Chris's Score

Scores	Number of Occasions
75–80	5
80–85	10
85–90	16
90–95	33
95–100	17
100–105	13
105–110	3
110–115	3

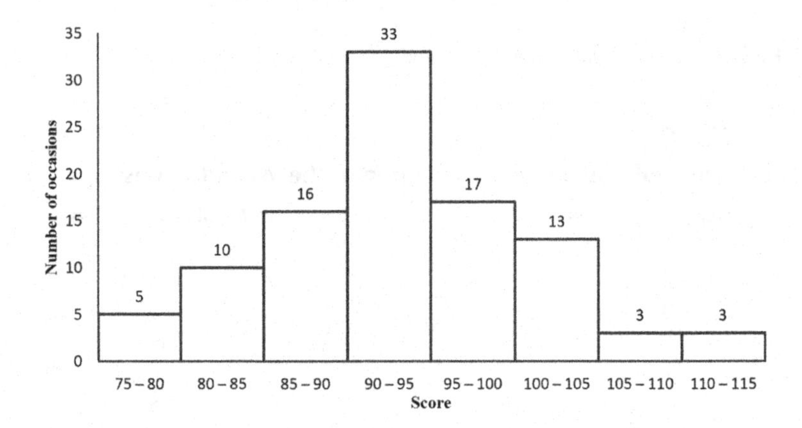

FIGURE 6.5 Histogram of Chris's score.

(ii) Chris is not a good golfer: his scores have a mean and a median of approximately 94, that is 22 strokes above par. The distribution of his scores is roughly symmetrical. Occasionally he has a good round (score below 80), but his scoring is erratic (maximum 115). About 2/3 of his scores are between 85 and 100.

6.9 CONTINUOUS QUANTITATIVE VARIABLES AND PROBABILITY DISTRIBUTION

When the measurement is something like length, time, voltage and so on, then, theoretically, we have a continuous measurement scale. That said, in practice, we can only measure to a certain degree of accuracy. For example, 2.4mm would be interpreted as 2.4mm to the nearest 0.1mm, which could actually be any value in the range 2.35–2.45mm; 2.40mm would be interpreted as 2.40mm to the nearest 0.01mm, meaning that the actual value lies somewhere between 2.395mm and 2.405mm. Therefore, all measurements are, in fact, discrete. For statistical purposes, however, it is sometimes convenient to specify class intervals so that they run together continuously.

Example 6.3 Travel times (minutes) to 32 fire incidents for an inner-city emergency fire service, during Tuesday, Wednesday and Thursday of 1 week are listed in Table 6.12.

NOTES:

- The survey units are the fires.
- The variable is the travel time; we represent it by X.
- This type of variable is known as a **continuous quantitative variable**. Continuous quantitative variables take real numbers as possible values. In the example shown earlier, times in minutes are recorded to one decimal place. It should be borne in mind that a recorded time of, for example, 5.9 minutes, has in reality an infinite sequence of decimal places, which are not shown. The grouped frequency distribution table is given in Table 6.13.

TABLE 6.12
Travel Times (minutes) to 32 Fire Incidents

5.9,	3.2,	3.2,	2.9,	2.1,	3.8,	0.7,	3.1,	5.3,	4.3,	3.0,
3.2,	4.7,	6.7,	5.8,	4.7,	4.1,	3.7,	4.6,	6.5,	4.1,	7.5,
4.1,	5.3,	2.1,	3.5,	4.0,	3.0,	2.6,	2.9,	1.3,	3.7	

TABLE 6.13
Grouped Frequency Distribution of the Continuous Data

Travel Time (minutes)	Frequency	Relative Frequency (%)
0.0–0.8	1	3.1
0.8–1.6	1	3.1
1.6–2.4	2	6.3
2.4–3.2	6	18.8
3.2–4.0	7	21.9
4.0–4.8	8	25.0
4.8–5.6	2	6.3
5.6–6.4	2	6.3
6.4–7.2	2	6.3
7.2–8.0	1	3.1
Total	**32**	**100**

TABLE 6.14
Probability Distribution Table

Travel time (minutes)	Frequency	Proportion (probability)
0.0–0.8	1	0.031
0.8–1.6	1	0.031
1.6–2.4	2	0.06
2.4–3.2	6	0.188
3.2–4.0	7	0.219
4.0–4.8	8	0.25
4.8–5.6	2	0.063
5.6–6.4	2	0.063
6.4–7.2	2	0.063
7.2–8.0	1	0.031
Total	**32**	**1**

Note: The grouped frequency distribution is more informative than the ungrouped. For example, it is clear that the most frequently occurring travel times are between 2.4 and 4.8 minutes.

Now we construct the **probability distribution.** To obtain the probability distribution of the random variable X (travel time), we divide the frequency of each class in Table 6.13 to the total number of frequencies, 32, as shown in Table 6.14.

TABLE 6.15
Probability Density Distribution

Travel Time (minutes)	Probability Density
0.0–0.8	0.03875
0.8–1.6	0.03875
1.6–2.4	0.075
2.4–3.2	0.235
3.2–4.0	0.27375
4.0–4.8	0.3125
4.8–5.6	0.07875
5.6–6.4	0.07875
6.4–7.2	0.07875
7.2–8.0	0.03875

For the continuous random variable X, in Example 6.3 "X = Travel time", we are usually interested in the **probability** that X falling within a certain range; for example, $P(1.6 < X < 2.4) = 0.06$ or $P(4.0 < X < 4.8) = 0.25$. However, using Table 6.14, we cannot determine $P(1.3 < X < 2.2)$. In order to determine probability of any random variable X in a range, say, (a, b), we must construct the probability density distribution and then define the probability density function. The probability density distribution for Example 6.3 is given in Table 6.15.

Note: The densities are obtained by dividing the probability of each class interval in Table 6.14 by the class width 0.8; for example, $0.03875 = \frac{0.031}{0.8}$.

We now construct the probability density histogram as shown in Figure 6.6.

NOTES:

- The diagram is known as a **probability density histogram**.
- Earlier, the class intervals are drawn rectangular bars whose *areas* are the probability of each class interval. For example, the bar above $4.0 - 4.8$ has the area $0.8 \times 0.3125 = 0.25$.
- The area under the entire histogram is 1.
- By increasing the observations while simultaneously decreasing the width of the class intervals, we would eventually obtain a limiting curve; that is, theoretically we can fit a curve to the histogram if we take a large sample and a very small class interval size! The continuous curve (which is superimposed on the histogram in Figure 6.6) is called **probability density function**.
- Generally speaking, the probability of the random variable X in a range is the area under the probability density function.

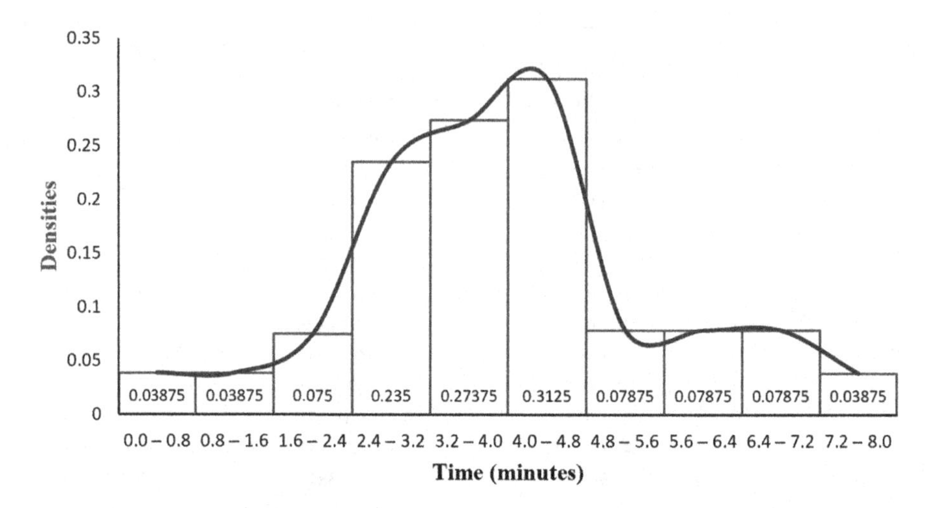

FIGURE 6.6 Probability density histogram.

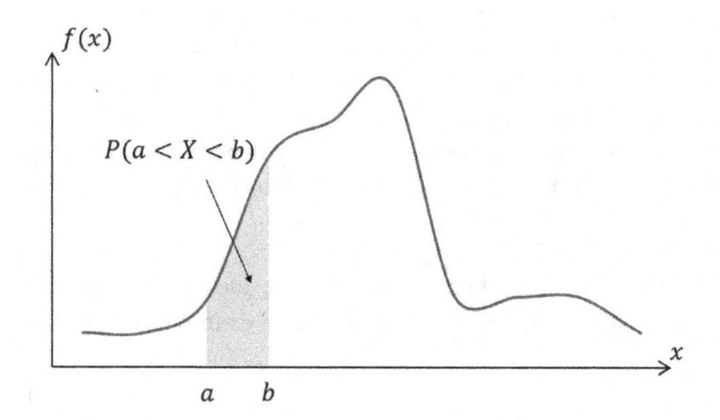

FIGURE 6.7 Probability density function.

Assume $f(x)$ is the probability density function of the random variable *X*, the likelihood (probability) of the random variable *X* falling within the interval (a,b) is represented by the area under the probability density curve $f(x)$ between the ordinates *a* and *b*. This concept is illustrated graphically by the shaded region in Figure 6.7.

6.10 NORMAL DISTRIBUTION

Among the probability distribution functions, we describe only the "normal distribution" shown in Figure 6.8. The term *normal distribution* only applies to populations. We have in mind a very large population of individuals, and a continuous quantitative variable, say, *X*, measured for each individual. If the density histogram

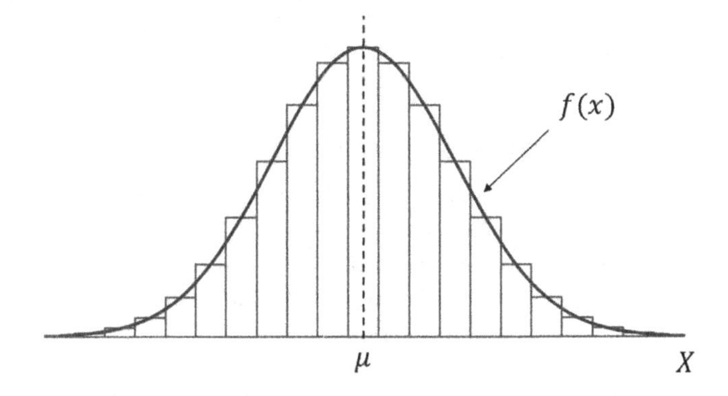

FIGURE 6.8 Normal distribution superimposed on a bell-shaped histogram.

of the variable in the population is closely approximated by the normal curve, then the variable is said to have a normal distribution in this population. The normal curve may be thought of as a symmetrical, bell-shaped curve superimposed on the density histogram, extending infinitely in both positive and negative directions; see Figure 6.8.

If an individual is chosen at random from the population, and the variable X is measured or observed on that individual, X is said to be a normally distributed random variable.

Undoubtedly, the normal distribution is the most important continuous probability distribution, which is of great significance in statistical inference and data analysis. Empirical investigations consistently validate the suitability of the normal distribution as an approximation for diverse physical variables.

The function that represents the Normal distribution can be expressed as follows:

$$f(x) = \frac{1}{\sigma\sqrt{2\pi}} e^{-\frac{1}{2}\left(\frac{x-\mu}{\sigma}\right)^2},$$

where μ is the mean of the random variable X and σ is its standard deviation.

NOTES:

- The variable x takes values between $\pm\infty$.
- The function $f(x)$ is always positive.
- The area under the normal curve is 1.

A shorthand notation for the normal random variable X with the mean μ and the standard deviation σ is $X = N(\mu, \sigma)$.

6.10.1 Standard Normal Distribution

Let us consider a population whose distribution of values is characterised by a normal distribution with a mean value of $\mu = 0$ and a standard deviation of $\sigma = 1$; this is called the standardised normal distribution **(Z-distribution)**. The graph of Z-distribution is shown in Figure 6.9, and the bell-shaped curve of the standardised Normal distribution is given mathematically by

$$f(z) = \frac{1}{\sqrt{2\pi}} e^{-\frac{z^2}{2}}.$$

The variable Z can (theoretically) take any population value between $-\infty$ and $+\infty$. Suppose we now intend to randomly select a value from the Normal population. The probability that the value will be greater than some specified value u is given by the integral

$$P(Z > u) = \int_{z=u}^{\infty} f(z)\,dz$$

Graphically, this is represented by the area under the bell curve to the right of $Z = u$ in Figure 6.9.

Note: The total area under the bell curve (in Figure 6.9) is equal to 1. This will prove extremely useful for determining areas under the bell curve that are not in tail areas.

The standard normal distribution does not occur in nature but is a useful mathematical tool. Areas under the standard normal curve have been calculated and are published in statistical tables. We shall use tables compiled by Murdoch and Barnes (Table 6.16).

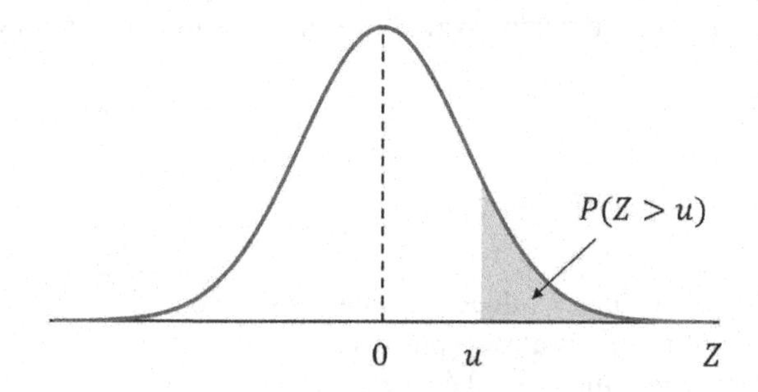

FIGURE 6.9 Z-distribution.

TABLE 6.16
Statistical Table of Murdoch and Barnes

u	.00	.01	.02	.03	.04	.05	.06	.07	.08	.09
0.0	.5000	.4960	.4920	.4880	.4840	.4801	.4761	.4721	.4681	.4641
0.1	.4602	.4562	.4522	.4483	.4443	.4404	.4364	.4325	.4286	.4247
0.2	.4207	.4168	.4129	.4090	.4052	.4013	.3974	.3936	.3897	.3859
0.3	.3821	.3783	.3745	.3707	.3669	.3632	.3594	.3557	.3520	.3483
0.4	.3446	.3409	.3372	.3336	.3300	.3264	.3228	.3192	.3156	.3121
0.5	.3085	.3050	.3015	.2981	.2946	.2912	.2877	.2843	.2810	.2776
0.6	.2743	.2709	.2676	.2643	.2611	.2578	.2546	.2514	.2483	.2451
0.7	.2420	.2389	.2358	.2327	.2296	.2266	.2236	.2206	.2177	.2148
0.8	.2119	.2090	.2061	.2033	.2005	.1977	.1949	.1922	.1894	.1867
0.9	.1841	.1814	.1788	.1762	.1736	.1711	.1685	.1660	.1635	.1611
1.0	.1587	.1562	.1539	.1515	.1492	.1469	.1446	.1423	.1401	.1379
1.1	.1357	.1335	.1314	.1292	.1271	.1251	.1230	.1210	.1190	.1170
1.2	.1151	.1131	.1112	.1093	.1075	.1056	.1038	.1020	.1003	.0985
1.3	.0968	.0951	.0934	.0918	.0901	.0885	.0869	.0853	.0838	.0823
1.4	.0808	.0793	.0778	.0764	.0749	.0735	.0721	.0708	.0694	.0681
1.5	.0668	.0655	.0643	.0630	.0618	.0606	.0594	.0582	.0571	.0559
1.6	.0548	.0537	.0526	.0516	.0505	.0495	.0485	.0475	.0465	.0455
1.7	.0446	.0436	.0427	.0418	.0409	.0401	.0392	.0384	.0375	.0367
1.8	.0359	.0351	.0344	.0336	.0329	.0322	.0314	.0307	.0301	.0294
1.9	.0287	.0281	.0274	.0268	.0262	.0256	.0250	.0244	.0239	.0233
2.0	.02275	.0222	.02169	.02118	.02068	.02018	.01970	.01923	.01876	.01831
2.1	.01786	.0174	.01700	.01659	.01618	.01578	.01539	.01500	.01463	.01426
2.2	.01390	.0136	.01321	.01287	.01255	.01222	.01191	.01160	.01130	.01101
2.3	.01072	.0104	.01017	.00990	.00964	.00939	.00914	.00889	.00866	.00842
2.4	.00820	.0080	.00776	.00755	.00734	.00714	.00695	.00676	.00657	.00639
2.5	.00621	.0060	.00587	.00570	.00554	.00539	.00523	.00508	.00494	.00480
2.6	.00466	.0045	.00440	.00427	.00415	.00402	.00391	.00379	.00368	.00357
2.7	.00347	.0034	.00326	.00317	.00307	.00298	.00289	.00280	.00272	.00264
2.8	.00256	.0025	.00240	.00233	.00226	.00219	.00212	.00205	.00199	.00193
2.9	.00187	.0018	.00175	.00169	.00164	.00159	.00154	.00149	.00144	.00139
3.0	.00135									
3.1	.00097									
3.2	.00069									
3.3	.00048									
3.4	.00034									
3.5	.00023									
3.6	.00016									
3.7	.00011									
3.8	.00007									
3.9	.00005									
4.0	.00003									

The probability $P(Z > u)$, which is referred to as right-tail areas in Figure 6.9, is difficult to compute directly from the integral, but it is tabulated in the statistical table of Murdoch and Barnes (shown by Table 6.16). In this table, each of the four-/five-decimal-point-tabulated values corresponds to the probability $P(Z > u)$, where u falls in the range 0–4.

Example 6.4 Determine the following probabilities using the statistical Table 6.16.

(i) $P(Z > 1.5) = ?$
A graphical representation of this probability is shown in Figure 6.10a, the tail area is determined directly from the statistical Table 6.16: $P(Z > 1.5) = 0.0668$.

(ii) $P(Z < -1) = ?$
A graphical representation is given in Figure 6.10b. Although the statistical Table 6.16 does not deal with tail areas to the left, the symmetry of the graph means that we look up the tabulated value for +1.0: $P(Z < -1) = P(Z > +1) = 0.1587$.

(iii) $P(Z < 1.25) = ?$
We want the shaded area in Figure 6.10c. If we look up 1.25 in Table 6.16, we get the unshaded tail area to the right of 1.25. Because the total area under the curve is equal to 1, the shaded area is given by $P(Z < 1.25) = 1 - 0.1056 = 0.8944$.

(iv) $P(-2 < Z < 1.5) = ?$
The shaded area in Figure 6.10d is clearly the entire area below the curve minus the two tail areas: $P(-2 < Z < 1.5) = 1 - \left(P(Z > 1.5) + P(Z > +2)\right)$

$$= 1 - (0.0668 + 0.02275) = 0.91045.$$

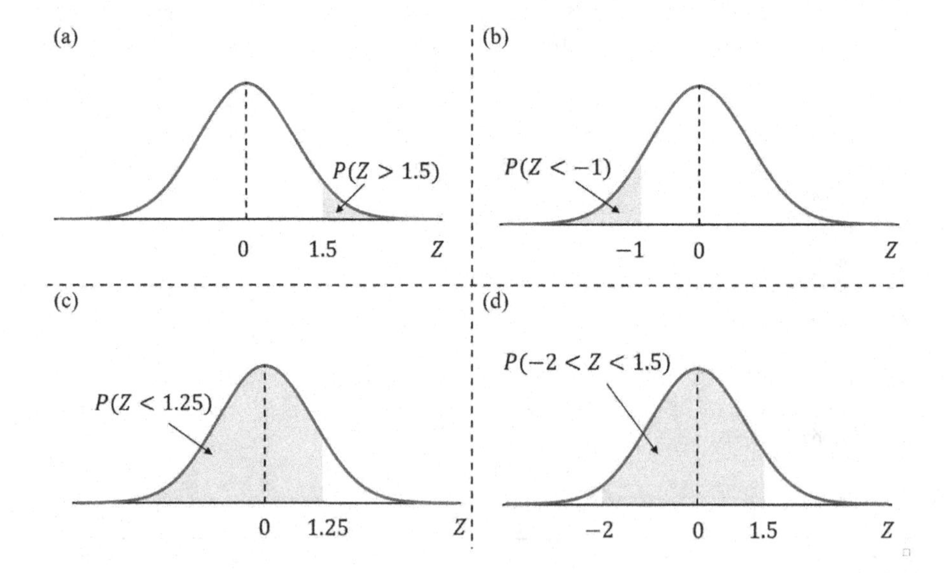

FIGURE 6.10 Shaded areas are associated with the given probabilities.

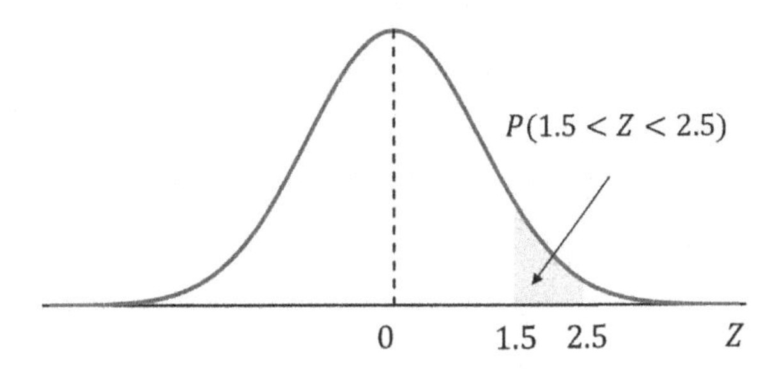

FIGURE 6.11 $P(1.5 < Z < 2.5)$.

(v) $P(1.5 < Z < 2.5) = ?$
This time, the shaded area, in Figure 6.11, is given by the tail area to the right of 1.5 minus the tail area to the right of 2.5:

$$P(1.5 < Z < 2.5) = P(Z > 1.5) - P(Z > 2.5)$$
$$= 0.0668 - 0.00621 = 0.06059.$$

6.10.2 Exercises

Let Z be a random variable with the standard normal distribution. Use the table of upper tail areas of the normal distribution to obtain the probabilities that follow:

(i) $Z > 1.00$ (ii) $Z > 1.79$ (iii) $Z < -0.54$
(iv) $Z < 1.83$ (v) $Z > -1.411$ (vi) $0.52 < Z < 1.06$
(vii) $-2.04 < Z < -1.62$ (viii) $-0.82 < Z < 1.39$

Answers

(i) 0.1587 (ii) 0.0367 (iii) 0.2946

(iv) 0.9664 (v) 0.9207 (vi) 0.1569

(vii) 0.03192 (viii) 0.7116

6.10.3 Standardising the Random Variable X

To compute normal probabilities for a given range of values of the random variable X, we use the following transformation:

$$Z = \frac{X - \mu}{\sigma}.$$

This transformation shifts the basis of probability calculation from X to Z, resulting in what is known as the standard normal distribution. The graphical representation of the change of bases of probability calculation is shown in Figure 6.12.

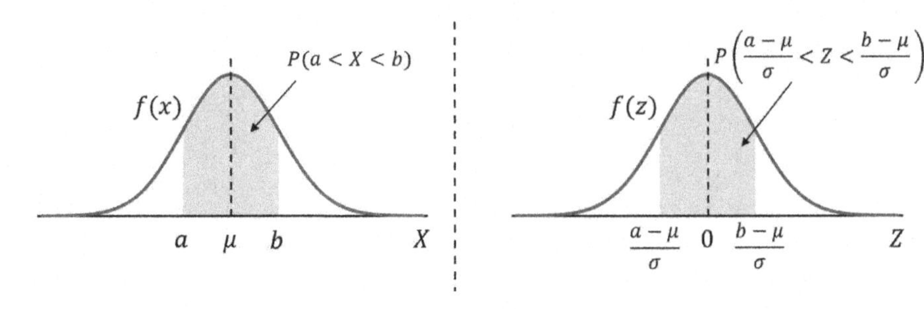

FIGURE 6.12 Standardising the random variable X.

Example 6.5 Suppose we have a population with a mean $\mu = 15$ and a standard deviation $\sigma = 2$. Determine the following probabilities:

(i) $P(X > 18)$
(ii) $P(X < 13)$
(iii) $P(X < 17.5)$
(iv) $P(11 < X < 18)$
(v) $P(18 < X < 20)$

Solution

(i) $P(X > 18) = P\left(Z > \frac{18-15}{2}\right) = P(Z > 1.5)$

(See Example 6.4i.)

(ii) $P(X < 13) = P\left(Z < \frac{13-15}{2}\right) = P(Z < -1)$

(See Example 6.4ii.)

(iii) $P(X < 17.5) = P\left(Z < \frac{17.5-15}{2}\right) = P(Z < 1.25)$

(See Example 6.4iii.)

(iv) $P(11 < X < 18) = P\left(\frac{11-15}{2} < Z < \frac{18-15}{2}\right) = P(-2 < Z < 1.5)$

(See Example 6.4iv.)

(v) $P(18 < X < 20) = P\left(\frac{18-15}{2} < Z < \frac{20-15}{2}\right) = P(1.5 < Z < 2.5)$

(See Example 6.4v.)

Example 6.6 LEDs (light-emitting diodes) produced in a certain factory have normally distributed lifetimes with a mean lifetime of $\mu = 25{,}000$ hours and a standard deviation of $\sigma = 500$ hours. Determine the percentage of LED lamps that will have the following lifetimes:

(i) Greater than 25,750 hours
(ii) Between 24,000 and 25,750 hours
(iii) Between 25,750 and 26,000 hours.

Solution

$$X = N(25000, 500) = \text{lifetime of LEDs}$$

(i) $P(X > 25750) = P\left(Z > \frac{25750-25000}{500}\right) = P(Z > 1.5) = 0.0668 \sim 6.68\%.$

(ii) $P(24000 < X < 25750) = P\left(\frac{24000-25000}{500} < Z < \frac{25750-25000}{500}\right)$

$$= P(-2 < Z < 1.5) = 1 - \left(P(Z > 2) + P(Z > 1.5)\right)$$

$$= 1 - (0.0668 + 0.02275) = 0.91045 \sim 91.045\%$$

(See Example 6.4iv.)

(iii) $P(25750 < X < 26{,}000) = P\left(\frac{257500-25000}{500} < Z < \frac{26000-25000}{500}\right)$

$$= P(1.25 < Z < 2) = P(Z > 1.25) - P(Z > 2)$$

$$= 0.1056 - 0.02275 \sim 8.285\%$$

(See Example 6.4(v) and adjust the graph by replacing 1.25 with 1.5 and 2 with 2.5).

6.10.4 Exercises

1. A random variable X is normally distributed with a mean of 25 and a standard deviation of 5. Determine the probabilities that a value of X will fall in the following ranges:
   (i) $X > 26$ (ii) $X > 35$ (iii) $X < 23$
   (iv) $22 < X < 30$ (v) $27 < X < 32$ (vi) $17 < X < 19$
2. A machine in a factory produces components whose lengths are approximately normally distributed with a mean 102mm and a standard deviation 1mm.
   (a) Find the probability that if a component is selected at random and measured, its length will be
       (i) less than 100 mm. (ii) greater than 104 mm.
   (b) If an output component is only accepted when its length lies in the range 100–104 mm, find the expected proportion of components that are accepted.

   Note: Proportion is equivalent to probability; percentage is proportion × 100.

3. Tests on incandescent electric light bulbs of a certain type indicated that their lifetimes were normally distributed with a mean of 1,860 hours and a standard deviation of 70 hours. Determine the percentage of bulbs that can be expected to last
   (i) more than 2,000 hours. (ii) less than 1,790 hours.
4. The fire load of a large population of hotels is normally distributed with a mean of 500 MJ/m^2 and a standard deviation of 70 MJ/m^2. Find the proportion of hotels with the following fire loads:
   (i) Greater than 600 MJ/m^2
   (ii) Less than 350 MJ/m^2
   (iii) Between 420 and 570 MJ/m^2
5. The diameters of a production run of ball bearings are normally distributed with a mean of 0.6140 cm and a standard deviation of 0.0025 cm. Assuming that measurements are made to the nearest thousandth of a centimetre, determine the percentage of bearings with the following diameter measurements:
   (i) Equal to 0.615 cm
   (ii) Between 0.610 cm and 0.618 cm inclusive
   (iii) Greater than 0.617 cm
   (iv) Less than 0.608 cm.

Note: A value measured as 0.615 cm to the nearest thousandth of a centimetre could have an actual value anywhere between 0.6145 cm and 0.6155 cm. So, for part (i), you are looking to calculate $P(0.6145 < X < 0.6155) \times 100$. Similar adjustments have to be made for the other parts.

Answers

1. (i) 0.4207 (ii) 0.02275 (iii) 0.3446
   (iv) 0.5670 (v) 0.2638 (vi) 0.0603
2. (a) (i) 0.02275; (ii) 0.02275 (b) 0.9545
3. (i) 2.275% (ii) 15.87%
4. (i) $Z = (600 - 500)/70 = 1.43$. Hence, the required probability = 0.0764.

   (ii) $Z = (350 - 500)/70 = -2.14$. Hence, the required probability = 0.01618.

   (iii) $Z = (420 - 500)/70 = -1.14$. $Z = (570 - 500)/70 = 1$.

   Hence, required probability = $1 - (0.1587 + 0.1271) = 1 - 0.2858 = 0.7142$.
5. (i) 14.64% (ii) 92.82%

   (iii) 8.08% (iv) 0.466%

6.10.5 Calculating the X-Value by Knowing the Probability

In Section 6.10.1, we knew the $Z = z_\alpha$ value, and we found the probabilities $P(Z > z_\alpha)$ using the statistical Table 6.16. Sometimes, we may need to do a "reverse" calculation

TABLE 6.17
The Statistical Table of Murdoch and Barnes to Determine the Standard Normal Variable Z_α for Which $P(Z > z_\alpha) = \alpha$

α	z_α	α	z_α	α	z_α	α	z_α	α	z_α	α	z_α
.50	0.0000	.050	1.6449	.030	1.8808	.020	2.0537	.010	2.3263	.05	1.6449
.45	0.1257	.048	1.6646	.029	1.8957	.019	2.0748	.009	2.3656	.01	2.3263
.40	0.2533	.046	1.6849	.028	1.9110	.018	2.0969	.008	2.4089	.001	3.0902
.35	0.3853	.044	1.7060	.027	1.9268	.017	2.1201	.007	2.4573	.0001	3.7195
.30	0.5244	.042	1.7279	.026	1.9431	.016	2.1444	.006	2.5121	.00001	4.2655
.25	0.6745	.040	1.7507	.025	1.9600	.015	2.1701	.005	2.5758	.025	1.9600
.20	0.8416	.038	1.7744	.024	1.9774	.014	2.1973	.004	2.6521	.005	2.5758
.15	1.0364	.036	1.7991	.023	1.9954	.013	2.2262	.003	2.7478	.0005	3.2905
.10	1.2816	.034	1.8250	.022	2.0141	.012	2.2571	.002	2.8782	.00005	3.8906
.05	1.6449	.032	1.8522	.021	2.0335	.011	2.2904	.001	3.0902	.000005	4.4172

in which we have a probability $P(Z > z_\alpha) = \alpha$ and want to determine $Z = z_\alpha$. The statistical method of Murdoch and Barnes (see Table 6.17) is provided to do the reverse calculation and find $Z = z_\alpha$.

For any normal distribution X, with a mean μ and a standard deviation σ, we use the transformation $Z = \dfrac{X - \mu}{\sigma}$ and Table 6.17 to obtain $X = x_0$ with a given probability $P(X > x_0) = \alpha$. We elaborate this with the following example.

Example 6.7 For the same distribution as in Example 6.5 ($\mu = 15$ and $\sigma = 2$) determine x_0 for the following:

(i) $P(X > x_0) = 0.05$
(ii) $P(X > x_0) = 0.75$

Solution

(i) Consider the equivalent standardised normal calculation:

$$P(X > x_0) = 0.05 \Rightarrow P\left(Z > \frac{x_0 - 15}{2}\right) = 0.05.$$

The probability of 0.05 is associated either with the left or right tail of the Z-distribution. Given that we have a "greater than" inequality, 0.05 must be a right-tail probability (area) as shown in Figure 6.13a.
Using Table 6.17, the probability $\alpha = 0.05$ corresponds to $z_\alpha = 1.6449$, and hence,

$$\frac{x_0-15}{2}=1.6449 \Rightarrow x_0=15+2\times 1.6449=18.3298.$$

(ii) The process for $P(X > x_0)=0.75$ is similar. Consider the equivalent standardised normal calculation:

$$108.\ P(X > x_0)=0.75 \Rightarrow P\left(Z > \frac{x_0-15}{2}\right)=0.75.$$

This time the probability of 0.75 is greater than a half, given the inequality is greater than; hence, $\frac{x_0-15}{2}$ must be on the negative side of zero as shown in Figure 6.13b.

The tail area to the left of $\frac{x_0-15}{2}$ must be 0.25 (i.e. $1-0.75$). Using the statistical Table 6.17, the right tail probability $\alpha=0.25$ corresponds to $z_\alpha=0.6745$, since $\alpha=0.25$ is the left-tail area; hence, it must be associated to $-z_\alpha=-0.6745$. Therefore,

$$\frac{x_0-15}{2}=-0.6745 \Rightarrow x_0=15+2\times(-0.6745)=13.651.$$

Example 6.8 A machine makes electronic resistors with a mean resistance of 100 Ω and a standard deviation of 5 Ω. 10% of resistors are rejected because their resistances are not within the firm's tolerance limits (5% too great, 5% too low). Assuming that the values are normally distributed, determine the tolerance limit of the firm.

Solution

$$\mu=100, \sigma=5$$

$$X=N(100,5)=\text{resistance of the resistors}$$

$$B=\text{Tolerance limit}$$

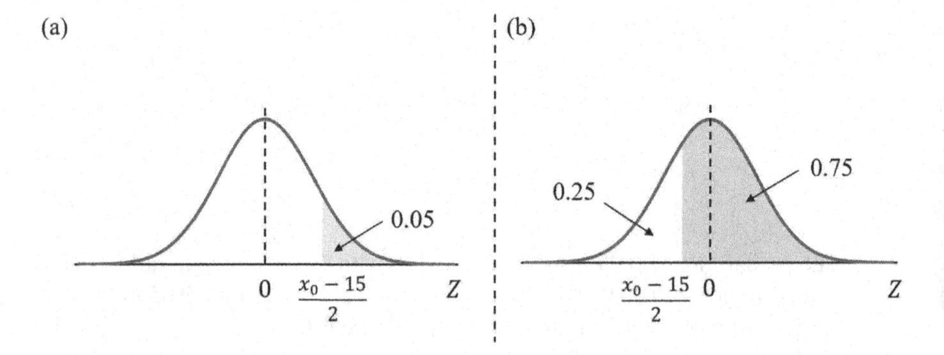

FIGURE 6.13 Finding the z-value associated with given probability.

$$P(X < B) = 0.05$$

$$P(X > B) = 0.05$$

$$\Rightarrow$$

$$P\left(X < \frac{B-100}{5}\right) = 0.05 \quad \text{or}$$

$$P\left(X > \frac{B-100}{5}\right) = 0.05$$

The probability 0.05 is associated with the left tail and right tail; hence,

$$\frac{B-100}{5} = -1.6449 \text{ or } \frac{B-100}{5} = +1.6449$$

$$\Rightarrow B = 100 \pm 5 \times 1.6449 \approx 100 \pm 8.2.$$

6.10.6 Exercises

1. The lifetimes of TV tubes fitted to television sets manufactured by the Hishobi company are approximately normally distributed with a mean of 75 months and a standard deviation of 8 months. The tubes are guaranteed for a period of 5 years.
   (a) What proportion of tubes may be expected to fail within the guarantee period?
   (b) The manufacturer wishes to reduce the proportion failing within 5 years to 1% by reducing the standard deviation of the lifetime. What must the new standard deviation be to achieve this target (assuming the mean lifetime is still 75 months)?
2. A random variable X is normally distributed with a mean of 10 and a standard deviation of 4. Determine the value of the variable whose chance of being exceeded is 0.05. Find another value greater than 12 such that the probability associated with the interval from 12 to that value is 0.2.
3. The fire load of a large population of hotels is Normally distributed with a mean 500 MJ/m^2 and a standard deviation 70 MJ/m^2. Find the value of fire load, x MJ/m^2, say, such that 80% of hotels have a fire load less than x.
4. Fire protective clothing and associated respiratory equipment impose significant heat stress on firefighters. From experimental data on young (18–23) male firefighters, wearing standard-issue clothing, it is estimated that during a maximal physical work training of about 1.5 hours, which includes typical firefighter tasks, sweat loss is approximately normally distributed with a mean of 1.9 litres and a standard deviation of 0.8 litres.
   (a) Determine the probability that, during such a training session, a randomly chosen firefighter's sweat loss would be the following:
   (i) Greater than 3 litres (ii) Less than 2.5 litres

(b) In such an exercise, what sweat loss in litres would be exceeded by 10% of the population from which the experimental participants were drawn?

(c) By redesigning certain features of the clothing, manufacturers aim to ensure that 90% of the firefighting population have a sweat loss less than 2.75 litres. What would be the **mean** sweat loss in the maximum work training using the redesigned clothing? You may assume that the standard deviation remains unaltered at 0.8 litres.

Answers

1. (a) 3.04% (b) Approximately 6.5 months.
2. 16.58 ; 14.94.
3. Upper 80% point of the standard normal distribution = 0.8416. Hence, $\frac{x-500}{70} = 0.8416$. Solving for x yields $x = 558.9$ MJ/m^2.
4. (a) (i) 0.0838 (ii) 0.7734
   (b) $x = 2.924$ litres
   (c) Let the unknown mean = μ; $\mu = 1.72$ litres.

6.11 ESTIMATION AND CONFIDENCE INTERVALS

Many studies or experiments involve taking measurements on a *sample* of devices, businesses or products. It is usual to take a sample that is random and representative of the *population* under study. *Statistical inference* deals with the question of what can be inferred about the population from a sample (i.e., what can we say about the population from data collected on a sample). Statistical inference will only be valid if the experiment has been properly designed to ensure that the sample is random and representative of the population.

Important topics to be considered within statistical inference include the following:

- Estimation
- Distribution of the sample mean
- Confidence interval
- Hypothesis testing

In the statistical inference in this book, we will almost always assume that variables are "normally" distributed.

6.11.1 Point Estimation

Table 6.18 shows formulae for sample statistics and population parameters. The sample statistics are *point estimates* of the corresponding population parameters.

(Note the sample size is denoted by n, the population size by N and the data values by x_i.)

TABLE 6.18
Formulae for Sample Statistics and Population Parameters

	Sample Statistic	Population Parameter
Mean	$\bar{x} = \frac{1}{n}\sum_{i=1}^{n} x_i$	$\mu = \frac{1}{N}\sum_{i=1}^{N} x_i$
Standard deviation	$s = \sqrt{\frac{1}{n-1}\sum_{i=1}^{n}(x_i - \bar{x})^2}$	$\sigma = \sqrt{\frac{1}{N}\sum_{i=1}^{N}(x_i - \mu)^2}$

Example 6.9 Fire loads (MJ/m^2) are obtained for a random sample of 10 commercial properties. The results are as follows:

273, 421, 3260, 1235, 496, 397, 672, 548, 523, 701

We estimate the mean fire load of commercial properties in the underlying population by the sample mean $\bar{x} = 853$ MJ/m^2. This is an example of a *point estimate* and lacks precision.

6.11.2 Distribution of the Sample Mean

In Example 6.9, if we were to have measured the fire loads for a different sample of 10 commercial properties, then our point estimate would be slightly different. We require knowing how much the sample mean varies from sample to sample.

Suppose that in this example the experiment was repeated for 5 different samples, each sample consisting of the fire loads for 10 commercial properties. The sample means were calculated for each sample giving the following results:

Sample means 775 879 801 755 872.

The variability from sample mean to sample mean is much less than the variability from the data presented for the fire loads of individual commercial properties.

Now if we draw all possible samples of size 10 fire load of commercial properties and determine their means, then we obtain a different distribution. The distribution of an indefinitely large number of sample means, obtained by repeatedly sampling from the same population, is known as the sampling distribution of the sample mean and it is represented by $\bar{X}$.

The sampling distribution of the sample mean

(i) has the **same mean** as the population;
(ii) has **less** variability than the population;
(iii) has a normal distribution when the population has a normal distribution.

We can be precise about the extent of (ii). In fact, the standard deviation of the distribution of the sample mean (known as the *standard error of the mean* to distinguish it from the standard deviation of the population of individual values) is given by the formula

$$\sigma_{\bar{x}} = \frac{\sigma}{\sqrt{n}}.$$

Note: For a sample size greater than 30 (large sample) the population can have any distribution however the distribution of the sample mean will be normal. Figure 6.14 shows the theoretical distribution of

- the fire load of the individual commercial properties;
- the mean of the fire loads for 10 commercial properties.

Note: To rescale to a standardised normal distribution, we use

$$Z = \frac{\bar{X} - \mu}{\frac{\sigma}{\sqrt{n}}}.$$

Example 6.10 A population of specimens of southern pine, exposed to an input flux of 40 kW m^{-2}, has a mean heat release rate of 96.12 kW m^{-2} and a standard deviation of 6.40 kW m^{-2}. You may assume that the population of heat release rate is normal. A random sample of 16 specimens is obtained. Find the probability that the sample mean is less than 100 kW m^{-2}

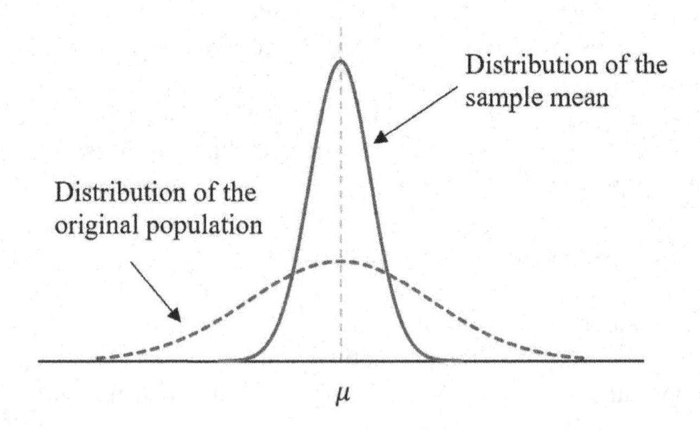

FIGURE 6.14 Sampling distribution of the sample mean.

Solution

$$\bar{X} = \text{mean Heat Release Rate of 16 specimens}$$

$$P(\bar{X} < 100) = P\left(Z < \frac{100 - 96.12}{\frac{6.4}{\sqrt{16}}}\right)$$

$$= P(Z < 2.425) = 1 - \frac{0.00755 + 0.00776}{2} \approx 0.99.$$

6.11.3 Exercises

1. Small electric motors produced by a manufacturer have a mean life of 800 hours with a standard deviation of 60 hours. Determine the probability that a random sample of 16 motors will have the following mean lives:
   (i) More than 820 hours
   (ii) Less than 785 hours
   (iii) Between 790 and 810 hours
   (iv) Between 770 and 785 hours
2. The weights of parcels received at a warehouse are normally distributed with a mean of 300 kg and a standard deviation of 50 kg. Determine the probability that 25 packages arriving at random and loaded on a lift will exceed the lift's safety limit of 8,200 kg.

Answer

1. (i) 0.0918 (ii) 0.1587 (iii) 0.4972 (iv) 0.13595
2. 0.00256

6.11.4 Interval Estimates

In Example 6.8, we estimated the mean fire load in a population of commercial properties, by the mean of a sample of 10 such properties, $\bar{x} = 853$ MJ/m^2. By good luck, this estimate may be exactly equal to the mean fire load in the population. However, it is more likely that there will be a discrepancy between the estimate and the population mean. We pointed out that the point estimates vary from sample to sample and hence lack precision.

It is good practice to summarise data by presenting *interval estimates* for the important parameters (e.g. μ). The interval should be accompanied by a probability that indicates how likely the interval is to include the parameter.

The probability is called the ***level of confidence***, and the resulting interval is called a ***confidence interval***. Common choices for the level of confidence are 90%, 95% and 99%.

The basic idea underlying the construction of a confidence interval relates to the **distribution of the sample mean**. Let us first elaborate the specific case of finding

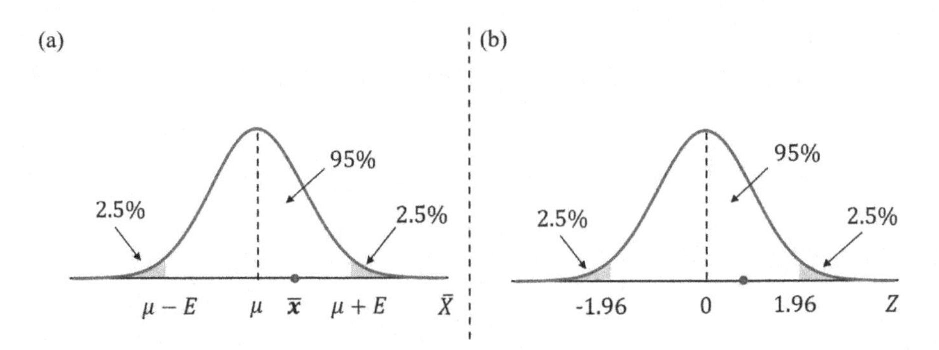

FIGURE 6.15 Standardising the distribution of the sample mean.

$95\% := (100-5)\%$ confidence interval around the sample mean, $\overline{x}$. Consider the distribution of the sample mean with 0.95 (95%) probability in the middle area as shown in Figure 6.15a.

There is a 95% chance that the any sample mean $\overline{x}$ lies between $\mu - E$ and $\mu + E$:

$$\mu - E < \overline{x} < \mu + E \Rightarrow$$

$$\overline{x} - E < \mu < \overline{x} + E.$$

What is the value of E? Standardising the distribution of the sample mean using $Z = \frac{\overline{X}-\mu}{\frac{\sigma}{\sqrt{n}}}$ gives (see Figure 6.15b)

$$1.96 = \frac{\mu + E - \mu}{\frac{\sigma}{\sqrt{n}}} \Rightarrow$$

$$E = 1.96 \times \frac{\sigma}{\sqrt{n}}.$$

Hence, 95% confidence limits can be obtained by adding and subtracting E to the sample mean $\overline{x}$; that is

$$95\%\,\text{confidence limits} = \overline{x} \pm E = \overline{x} \pm 1.96 \times \frac{\sigma}{\sqrt{n}}.$$

We now generalise the above explanation for the $(100-5)\%$ confidence limits to cover $(100-\alpha)\%$ confidence limits (the common values for α are 1, 5 and 10). In general, if the level of confidence is $(100-\alpha)\%$ then Figure 6.15 can be generalised as Figure 6.16.

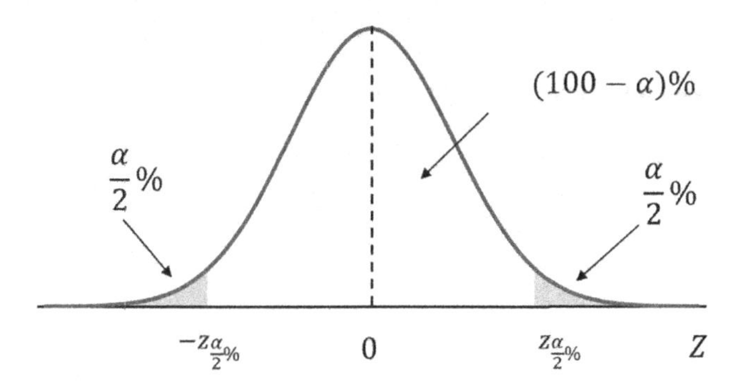

FIGURE 6.16 Generalised graph for finding the confidence interval.

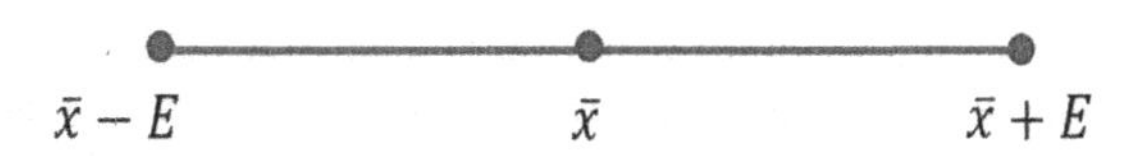

FIGURE 6.17 Confidence interval.

Therefore, referring to the description of the 95% confidence limits, we can write

$$(100-\alpha)\%\,\text{confidence limits} = \bar{x} \pm E = \bar{x} \pm z_{\frac{\alpha}{2}\%} \times \frac{\sigma}{\sqrt{n}}.$$

Note: $z_{\frac{\alpha}{2}\%}$ is the z-value associated with $\frac{\alpha}{2}\%$ area (probability).

Graphically, we have results such as in Figure 6.17.

Interpretation of confidence interval: there is a $(100-\alpha)\%$ chance that the interval contains the unknown parameter u. A simpler interpretation is that the confidence interval provides a range of *plausible values* of the unknown parameter μ.

As mentioned earlier, common choices for the level of confidence are 90%, 95% and 99% (i.e., **α = 10, 5, 1**). The confidence limits for these values are given in Table 6.19.

TABLE 6.19
90%, 95% and 99% Confidence Limits

90% confidence interval	$\bar{x} \pm z_{5\%} \times \frac{\sigma}{\sqrt{n}}$	$\bar{x} \pm 1.6449 \times \frac{\sigma}{\sqrt{n}}$
95% confidence interval	$\bar{x} \pm z_{2.5\%} \times \frac{\sigma}{\sqrt{n}}$	$\bar{x} \pm 1.9600 \times \frac{\sigma}{\sqrt{n}}$
99% confidence interval	$\bar{x} \pm z_{0.5\%} \times \frac{\sigma}{\sqrt{n}}$	$\bar{x} \pm 2.5758 \times \frac{\sigma}{\sqrt{n}}$

Example 6.11 A firm making light bulbs wanted to check the quality of its production. The quality control manager selected a random sample of 60 bulbs and determined their lifetimes with the following results: $\bar{x} = 11602$ hours and $s = 592$ hours.

Assuming the lifetime of bulbs is normally distributed, determine 90% confidence limits for the mean lifetime of the full production (i.e., the population mean μ).

Solution

A 90% level of confidence implies $\alpha = 10\%$; hence, $z_{\frac{\alpha}{2}\%} = z_{5\%} = 1.6449$. The confidence limits for 90% are

$$\bar{x} \pm z_{\frac{\alpha}{2}\%} \times \frac{\sigma}{\sqrt{n}} = 11602 \pm z_{5\%} \frac{592}{\sqrt{60}}$$

$$= 11602 \pm 1.6449 \times \frac{592}{\sqrt{60}} = 11602 \pm 125.7146.$$

Thus, the left limit $\approx$ 11,476 hours, and the right limit $\approx$ 11,728 hours.

PARTICULAR RULES COMMONLY USED

If $n > 30$, we can replace σ with a sample standard deviation s when σ is unknown.

If $n \leq 30$, the population must have a normal distribution, and we must know σ.

If $n \leq 30$ and σ is unknown, we use the t-distribution, which is explained in Section 6.1.9.

6.11.5 Exercises

1. A wholesaler receives 1,000 sacks of a commodity each with a nominal weight of 100 kg. A random sample of 50 of these sacks has a mean weight of 103.6 kg and a standard deviation of 2.50 kg. Calculate 95% and 99%

confidence limits for the mean weight of all the sacks received and the total weight of all the sacks received.

2. A random sample of 400 bottles was drawn from a large consignment. The capacities of the bottles have a mean of 507.3 ml and a standard deviation of 8.2 ml. Calculate 98% confidence limits for the mean capacity of the bottles in the consignment.
3. A quality control engineer wants to estimate the proportion of defective parts being manufactured by his company. A sample of 74 components showed that 43.24% were defective, with a standard deviation of 11%. Determine 99% confidence limits for the mean proportion of defective parts.
4. A car manufacturer wishes to estimate, in the form of a 95% confidence interval, the mean number of miles per gallon (mpg) achieved in city driving for a specific model of car.
   The city miles-per-gallon results were recorded for a random sample of 41 cars of this model driven under similar conditions. The calculated mean and standard deviation of these recordings were 32 mpg and 6 mpg, respectively. Calculate the required confidence limits.
5. There is a complaint that the rods manufactured by a company are not made up to specification. The mean strength of steel rods manufactured is required to be estimated. The breaking strength of a random sample of 107 rods was found to have a mean of 19,570 kg with a standard deviation of 366 kg. Evaluate 95% confidence limits for the mean strength of the rods manufactured by the company. Is the complaint justified?

Answers

1. 103.6 ± 0.6930; 103.6 ± 0.9107; $103,600 \pm 693$; $103,600 \pm 911$
2. 507.3 ± 0.95
3. 43.24 ± 3.29
4. 32 ± 1.83
5. $19,570 \pm 69.35$
   $\Rightarrow \left[19,501,\ 19,639\right]$-No

6.11.6 The *t*-Distribution

If the random variable $X = N(\mu, \sigma)$, then

$$Z = \frac{X - \mu}{\sigma} = N(0,1) \tag{6.1}$$

$$Z = \frac{\bar{X} - \mu}{\frac{\sigma}{\sqrt{n}}} = N(0,1). \tag{6.2}$$

Equation 6.2 is valid because the standard error of the sample mean is the population standard deviation divided by the square root of the sample size. We use Equation (6.2) to construct confidence intervals, as we have already seen.

Now, in practice, we rarely know the value of σ. What we do when σ is unknown (or we have small sample size) is to estimate it using the sample standard deviation s.

If we substitute s in place of σ in Equation (6.2) above, the Z-statistic becomes a "t-statistic with $n-1$ degrees of freedom" (often the Greek letter v is used to denote degrees of freedom). Recall that when s was calculated, a divisor of $n-1$ was used in the calculation, instead of n.

Hence, we obtain

$$t=\frac{\overline{X}-\mu}{\frac{s}{\sqrt{n}}} \quad \text{with } v=(n-1)\,\text{degree of freedom}.$$

The t-distribution (sometimes called Student's t-distribution after the pseudonym of its inventor) is actually a family of distributions since there is a different distribution for each value of v. Figure 6.18a shows the t-distribution with 10 degrees of freedom (*df* for short).

As we see from the plot, the t-distribution is very similar to the Normal distribution. The percentage points of the t-distribution are shown in Table 6.20. As the degrees of freedom $v=n-1$ increase, the percentage points become closer to the equivalent points from the normal distribution. Graphically, the symbol $t_{\alpha,v}$ is the value of the variable t for which $P\left(t>t_{\alpha,v}\right)=\alpha$ (see Figure 6.18b).

If we recall that a $(100-\alpha)\%$ Confidence Interval for μ when σ is known is $\overline{x}\pm z_{\frac{\alpha}{2}\%}\times\frac{\sigma}{\sqrt{n}}$, we see the close similarity to the following result:

A $(100-\alpha)\%$ confidence interval for μ when σ is unknown and sample size is small, is $\overline{x}\pm t_{\frac{\alpha}{2}\%,v}\times\frac{s}{\sqrt{n}}$, where $t_{\frac{\alpha}{2}\%,v}$ denotes the t-value with v *df* and with upper tail probability $\frac{\alpha}{2}\left(\frac{\alpha}{2}\%\right)$.

Example 6.12 A punctuality performance measure of a train-operating company is the percentage of trains arriving "on time". This phrase is defined as "arriving by, or within 5 minutes, of the scheduled arrival time".

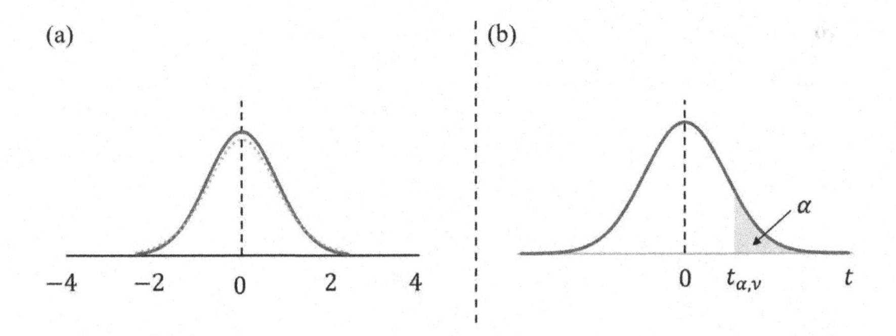

FIGURE 6.18 (a) Comparison of the normal Z-distribution (solid line) with the t-distribution with 10 *df* (dashed line). (b) $P\left(t>t_{\alpha,v}\right)=\alpha$.

On 7 randomly selected days, the following measurements were recorded for the company. The data refer to the percentages of trains that failed to arrive on time.

14.14 14.23 14.20 14.20 14.17 14.21 14.19

Construct a 90% confidence interval for the company's overall daily mean percentage of trains that fail to arrive on time, and state any assumptions made.

TABLE 6.20
Percentage Points of the *t*-Distribution

α =	0.10	0.05	0.025	0.01	0.005	0.001	0.0005
ν = 1	3.078	6.314	12.706	31.821	63.656	318.31	636.62
2	1.886	2.920	4.303	6.965	9.925	22.326	31.598
3	1.638	2.353	3.182	4.541	5.841	10.213	12.924
4	1.533	2.132	2.776	3.747	4.604	7.173	8.610
5	1.476	2.015	2.571	3.365	4.032	5.893	6.869
6	1.440	1.943	2.447	3.143	3.707	5.208	5.959
7	1.415	1.895	2.365	2.998	3.499	4.785	5.408
8	1.397	1.860	2.306	2.896	3.355	4.501	5.041
9	1.383	1.833	2.262	2.821	3.250	4.297	4.781
10	1.372	1.812	2.228	2.764	3.169	4.144	4.587
11	1.363	1.796	2.201	2.718	3.106	4.025	4.437
12	1.356	1.782	2.179	2.681	3.055	3.930	4.318
13	1.350	1.771	2.160	2.650	3.012	3.852	4.221
14	1.345	1.761	2.145	2.624	2.977	3.787	4.140
15	1.341	1.753	2.131	2.602	2.947	3.733	4.073
16	1.337	1.746	2.120	2.583	2.921	3.686	4.015
17	1.333	1.740	2.110	2.567	2.898	3.646	3.965
18	1.330	1.734	2.101	2.552	2.878	3.610	3.922
19	1.328	1.729	2.093	2.539	2.861	3.579	3.883
20	1.325	1.725	2.086	2.528	2.845	3.552	3.850
21	1.323	1.721	2.080	2.518	2.831	3.527	3.819
22	1.321	1.717	2.074	2.508	2.819	3.505	3.792
23	1.319	1.714	2.069	2.500	2.807	3.485	3.768
24	1.318	1.711	2.064	2.492	2.797	3.467	3.745
25	1.316	1.708	2.060	2.485	2.787	3.450	3.725
26	1.315	1.706	2.056	2.479	2.779	3.435	3.707
27	1.314	1.703	2.052	2.473	2.771	3.421	3.689
28	1.313	1.701	2.048	2.467	2.763	3.408	3.674
29	1.311	1.699	2.045	2.462	2.756	3.396	3.660
30	1.310	1.697	2.042	2.457	2.750	3.385	3.646
40	1.303	1.684	2.021	2.423	2.704	3.307	3.551
60	1.296	1.671	2.000	2.390	2.660	3.232	3.460
120	1.289	1.658	1.980	2.358	2.617	3.160	3.373
∞	1.282	1.645	1.960	2.326	2.576	3.090	3.291

Solution

We assume that the daily percentage of trains failing to arrive on time is approximately normally distributed.

We have the following summary statistics: $n=7$; $\overline{x}=14.191$; $s=0.029$. A 90% confidence interval for the mean daily percentage of trains failing to arrive on time is given by

$$\overline{x} \pm t_{5\%,6} \times \frac{s}{\sqrt{n}} = 14.191 \pm 1.943 \times \frac{0.029}{\sqrt{7}} = 14.191 \pm 0.021.$$

The lower 90% confidence limit is 14.170; the upper 90% confidence limit is 14.212.

So, a range of plausible values for the mean daily percentage of trains failing to arrive on time is given by (14.170, 14.212).

6.11.7 Exercises

1. The 2.5% point of the standard normal distribution is 1.96, that is the point above which 2.5% of the area under the curve lies. What are the corresponding points for the t-distributions with degrees of freedom equal to the following:
   (i) 9 (ii) 20 (iii) 30 (iv) 60 (v) 200
2. A metallurgist carried out four tests to determine the melting point of manganese. The results were $1259^o\ C, 1262^o\ C, 1253^o\ C$ and $1265^o\ C$. Calculate 95% confidence limits for the melting point of manganese.
3. In 8 test runs, an experimental engine consumed, respectively, 14, 12, 11, 13, 15, 12, 16 and 13 litres of petrol per second. Construct a 90% confidence interval for the mean petrol consumption.
4. The breaking strength of a random sample of 12 specimens of cotton thread of a particular type had a mean of 7.38 oz and a standard deviation of 1.24 oz. Assuming that the breaking strength of cotton thread of this type is approximately normally distributed, construct and interpret 90% confidence limits for the mean breaking strength of this type of thread.
5. The variation in capacitance in a large batch of capacitors is normally distributed. Assume we do not know the mean capacitance and we want to approximate it. Hence, we have to take a sample of size 7 from the large batch of capacitors to find an approximation for the mean capacitance. We found that the mean capacitance is 20 μF with a standard deviation of 6.19 μF. Determine the 90% confidence limits for the mean capacitance.

Answers

1. 2.262, 2.086, 2.042, 2.00, 1.96
2. 1259.75 ± 8.1514
3. 13.25 ± 1.118
4. 7.38 ± 0.643
5. 20 ± 4.55

6.12 HYPOTHESIS TESTING

Life is full of theories. At home and at work, we constantly need to make decisions by, often, relying on our beliefs or theories. We make theories about the weather, about people, about industrial processes and measurements. However, making decisions based on unsubstantiated theories is, to say the least, unwise. Data are collected in order to test a theory. Nevertheless, how can we use these data objectively to check our theory?

Our theories will relate to one or more populations. The data to test our theories will have been collected from samples drawn randomly from the study population(s). If we were to repeat our data collection, we would obtain a different set of data.

Hypothesis testing or significance testing offers us an objective method of checking theories or testing one theory against another.

In this book, we restrict our consideration of hypothesis testing to *only one* scenario: Comparing **one sample** mean against a "standard" or "target" value μ_0.

In the theory of hypothesis testing, we set up a hypothesis concerning the parameter of a population and use the information in a sample drawn from the population to test the hypothesis.

Suppose we want to test whether there is a change in the population parameter (μ_0) or not. We set up two hypotheses.

A null hypothesis: It is a theory about the population, which represents a "status quo" in which there is no change from a specified value (μ_0). It takes the form of $\mu = \mu_0$. This is often written

$$H_0 : \mu = \mu_0.$$

It is interpreted as "there has been no change in the mean".

An alternative hypothesis: It represents the alternative theory and specifies the direction of change, which is of interest. It takes the form of $\mu \neq \mu_0$ or $\mu < \mu_0$ or $\mu > \mu_0$. This is written as

$$H_1 : \mu \neq \mu_0 \ \text{(two sided)}$$

$$H_1 : \mu < \mu_0 \ \text{(one sided)}$$

$$H_1 : \mu > \mu_0 \ \text{(one sided)}.$$

We either reject H_0 or do not reject H_0. We make the decision essentially on the basis of the sample mean $\bar{x}$. If the sample mean is far below or above the hypothesised mean μ_0, we reject H_0. However, if the sample mean is close to the hypothesised mean, we do *not* reject H_0. But how far $\bar{x}$ should be away from μ_0 to warrant rejection of H_0? The sampling distribution of the sample mean provides the answer.

We draw the distribution of the sample mean and introduce the level of significance α (say, $\alpha = 0.05$, which is usually represented as 5%). We associate $\frac{\alpha}{2}\%$ (2.5%) to two tails (right and left) of the distribution for the two-sided alternative (see Figure 6.19) and $\alpha\%$ (5%) to the right or the left tail for the one-sided alternative.

The significance level α is the small probability that we will reject the null hypothesis when, in fact, the null hypothesis is true.

Note: The standard significance level to choose is 0.05 ~ 5%, but we may require a smaller chance of rejecting a true null hypothesis.

In Figure 6.19, the shaded areas on the tails are called the rejection regions as the α% (very small percentage of the sample mean) which are most unlikely if H_0 were true, will fall in these regions. So, if the sample mean lies in the lower or upper tail of sampling distribution (i.e. in the rejection region for the two-side alternative), we reject the null hypothesis. In Figure 6.19 the sample mean $\bar{x}$ does not lie in the rejection region.

In order to identify the rejection region easily, we standardise the distribution of the sample mean using the following transformations and evaluate the standardised value of the sample mean $\bar{x}$, which is called the test statistic value:

$$\text{(i) } Z = \frac{\bar{X} - \mu_0}{\frac{\sigma}{\sqrt{n}}} \text{ if } \sigma \text{ is known,} \quad \textbf{or} \quad \text{(ii) } t = \frac{\bar{X} - \mu_0}{\frac{s}{\sqrt{n}}} \text{ if } \sigma \text{ is unknown.}$$

We reject the null hypothesis based on the value of the test statistic and the rejection region on the standardised scale. The rejection region will depend upon the size of α (the significance level), the form of the alternative hypothesis and the type of distribution that is used to find the critical values. For $\alpha = 0.05 \sim (5\%)$ and using the Z-distribution, the rejection region along with critical values are shown in Figure 6.20.

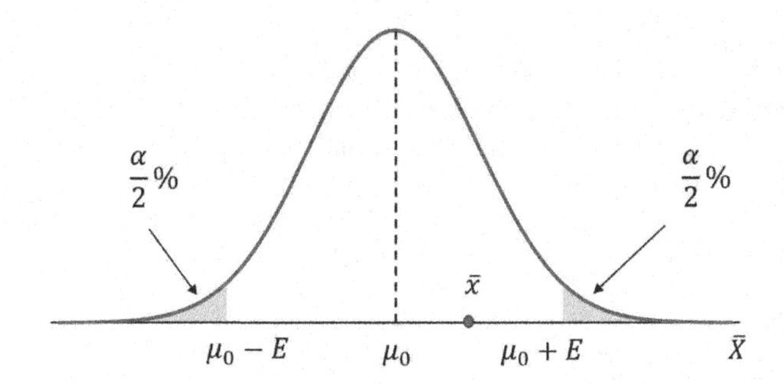

FIGURE 6.19 $\frac{\alpha}{2}\%$ of the significance level is associated with each tail for 2-sided alternative.

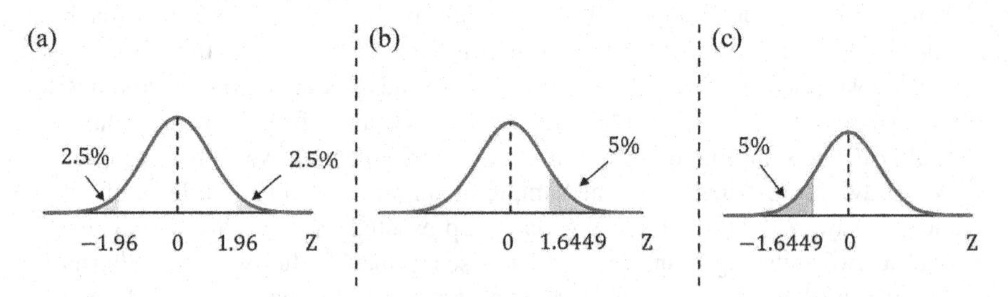

FIGURE 6.20 (a) Rejection region for the two-sided alternative (H_1: $\mu \neq \mu_0$). (b) Rejection region for the one-sided alternative hypothesis (H_1: $\mu > \mu_0$). (c) Rejection region for the one-sided alternative hypothesis (H_1: $\mu < \mu_0$).

Example 6.13 A factory production line produces 15-cm nails. The production output is to be tested to see if the nail lengths are as they should be (i.e., mean of 15 cm). A sample of 100 nails is taken and their lengths measured. The mean length turns out to be 14.72 cm with a standard deviation of 2.1 cm. Taking a significance level of 10% (i.e., $\alpha = 0.1$), test whether the population mean is significantly different from 15.

Solution

(a) Null hypothesis: H_0: $\mu = 15$
Alternative hypothesis: H_1: $\mu \neq 15$ (two-sided alternative).
(b) Significance level: $\alpha = 0.1 \sim (10\%)$.
(c) $\bar{x} = 14.72$, $s = 2.1$, $n = 100$; since the sample size is greater than 30, we use *Z*-test:

$$Z = \frac{\bar{X} - \mu_0}{\frac{s}{\sqrt{n}}} = \frac{14.72 - 15}{\frac{21}{\sqrt{100}}} \approx -1.333.$$

(d) Determine the critical regions (see Figure 6.21).
(e) **Decision**: $z = -1.333$ does not lie in either of the critical tail regions. Accept H_0. Mean length is not significantly different from 15 cm. Production standards are fine.

Example 6.14 A particular brand of 100-W light bulbs should have a lifetime of 10,000 hours. A sample of 80 bulbs is taken and tested to destruction. The sample is found to have a mean lifetime of 9,954 hours with a standard deviation of 225. Using a significance level of 10%, test whether the mean lifetime is significantly less than 10,000.

Solution

(a) Null hypothesis: H_0: $\mu = 10{,}000$
Alternative hypothesis: H_1: $\mu < 10{,}000$ (one-sided alternative)
(b) Significance level: $\alpha = 0.1 \sim (10\%)$.

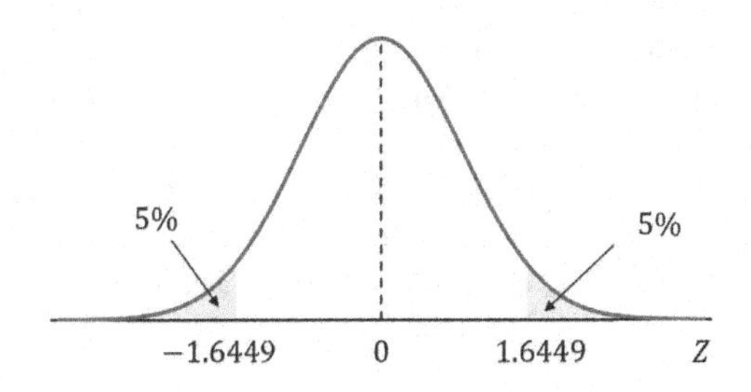

FIGURE 6.21 Critical region for two-sided alternative.

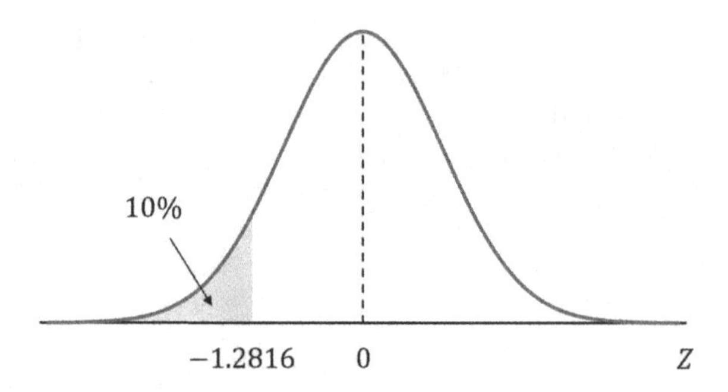

FIGURE 6.22 Critical region for one-sided alternative.

(c) $\bar{x} = 9{,}954, s = 225, n = 80$; since sample size is greater than 30, we use Z-test:

$$Z = \frac{\bar{X} - \mu_0}{\frac{s}{\sqrt{n}}} = \frac{9954 - 10000}{\frac{225}{\sqrt{81}}} \approx -1.83.$$

(d) Determine critical region (see Figure 6.22).

(e) **Decision:** $z = -1.83$ lies in the critical tail region. Reject H_0 and accept H_1; the mean lifetime is significantly less than 10,000 hours. Production standards are below the requirements.

6.12.1 Exercises

1. A random sample of 36 observations is drawn from a normal population with a known standard deviation of 4.2. The mean of the sample is found to be 62.3. Test whether the mean of the population is equal to 64.4. [Use a 5% significance level.]
2. Is it likely that a sample of 300 items with a mean value of 16.0 is a random sample from a large population whose mean is 16.8 and standard deviation is 5.2? [Use a 10% significance level.]
3. The mean breaking strength of steel rods is specified at 20,000 kg with a standard deviation of 457 kg. There is a complaint that the breaking strength of the rods is less than specification. A sample of 100 rods was found to have a mean breaking strength of 19,920 kg. Is the complaint justified? [Use a 5% significance level.]
4. A buyer of bricks from a wholesaler claims that the quality has deteriorated. From past experience, he knows the crushing strength to have an average of 400 kg and a standard deviation of 20 kg. He tests a sample of 100 and finds a mean of 395.8 kg. Does the buyer have a justifiable claim? [Use a 1% significance level.]

5. A manufacturer produces steel cables with a breaking strength of 20,000 N and a standard deviation of 1000 N. It is claimed that by using a new manufacturing process, the breaking strength can be increased. To test this claim, a sample of 50 cables made under the new process is tested and found to have an average breaking strength of 20,500 N. Can the claim be justified? [Use a 1% significance level.]
6. A random sample of 40 sacks of coal has a mean weight of 110 kg. Test at the 5% level of significance whether the population from which they are drawn has a mean weight of 112 kg and a standard deviation of 5 kg.
7. The resistors produced by a certain manufacturing process should have a resistance of 11.5 ohms as stated in the product specification. As part of a quality control scheme, 10 resistors were chosen at random from the production. Their resistances had a mean of 10.48 ohms and a standard deviation of 1.29 ohms. On the basis of the information given, conduct a suitable test at a 5% level of significance to decide whether the process is satisfying the product specification.
8. The producer of a certain make of flashlight dry-cell battery claims that the batteries have a mean life of ***at least*** 750 minutes. A random sample of 15 such batteries was taken and tested. Their mean lifetimes were 745 minutes, and the standard deviation was 24 minutes.
   Is there sufficient evidence, at the 1% level of significance, to ***refute*** the manufacturer's claim?

Answers

1. Critical Z = ±1.96	Test statistic: Z = −3.0	Decision: No (H_1)
2. Critical Z = ±1.6449	Test statistic: Z = −2.66	Decision: No (H_1)
3. Critical Z = −1.6449	Test statistic: Z = −1.75	Decision: Yes (H_1)
4. Critical Z = −2.3263	Test statistic: Z = −2.1	Decision: No (H_0)
5. Critical Z = +2.3263	Test statistic: Z = +3.54	Decision: Yes (H_1)
6. Critical Z = ±1.96	Test statistic: Z = 2.53	Decision: No (H_1)
7. Critical Z = ±2.262	Test statistic: t = −2.5	Decision: No (H_1)
8. Critical Z = ±2.624	Test statistic: t = −0.81	Decision: No (H_0)

6.13 CORRELATION AND REGRESSION

6.13.1 INTRODUCTION

One of the most interesting and useful applications of statistical methods involves assessing whether there is a relationship between two variables.

Examples:

1. The defective number of electrical components and weeks of experience of engineering staff.
2. The mileage per gallon that a particular make of car obtains and the speed of the car.

3. Energy consumption and average daily temperature.
4. A customer's waiting time at a ticket office and the number of ticket clerks.
5. The market value and the age of a model of car.

Correlation quantifies how much association there is between the two variables. **Regression** goes further: it gives us an equation which uses one variable to explain the variation in the other variable.

Independent and dependent variables: We can distinguish two types of variables, *independent* variable and *dependent* variable.

An independent variable is one which is either under the experimenter's control (e.g. the number of ticket clerks) or can be observed but not controlled (e.g. the daily temperature). As a result of changes that are deliberately made, or simply take place, in the independent variable, an effect is transmitted to the dependent variable.

In general, we are interested in assessing how changes in the independent variable affect the dependent variable.

The distinction between independent and dependent variables is not always clear-cut and the choice is sometimes made by the experimenter to suit his objectives (e.g. which would *you* choose as the independent variable in Question 4 earlier?).

Scatterplots: It is customary to label the independent variable X and the dependent variable Y. Suppose we have available n pairs $(X_1,Y_1),(X_2,Y_2),\ldots,(X_n,Y_n)$ of corresponding values of X and Y. The first step in any investigation of the relationship between X and Y is to represent the data graphically using a *scatterplot* (or *scatter diagram* or *scatter graph*). In this graph, each pair is represented by a point, where the X-axis is always horizontal and the Y-axis is vertical.

Example 6.15 A random sample of 8 employees is taken from the production department of a light engineering factory. The data in Table 6.21 show, for each employee, the number of weeks of experience in the wiring of components and the corresponding number of components that were rejected as unsatisfactory.

The scatterplot looks like Figure 6.23.

TABLE 6.21
Data for Number of Weeks of Experience and Number of Components Rejected for a Sample of 8 Employees

Employee	A	B	C	D	E	F	G	H
Weeks of experience (X)	4	5	7	9	10	11	12	14
Number of rejects (Y)	21	22	15	18	14	14	11	13

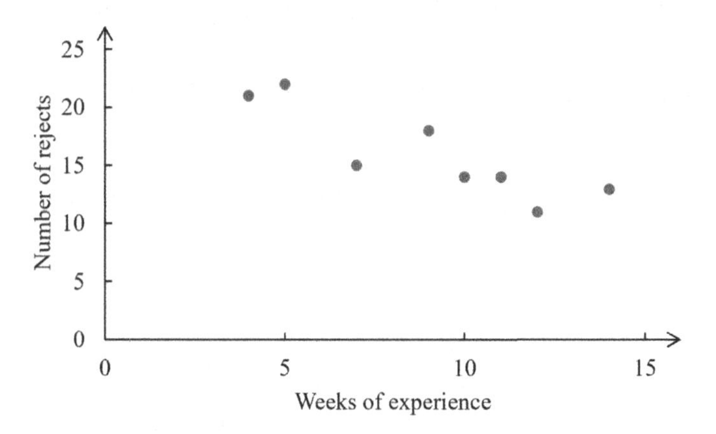

FIGURE 6.23 Scatterplot of the number of rejects against weeks of experience.

The underlying pattern in Figure 6.23 appears to be that lower experience (X) are associated with higher number of rejects (Y): X and Y have a **negative (inverse) relationship**. A **positive (direct) relationship** occurs when Y increases as X increases.

6.13.2 Correlation

There are various ways of measuring (i.e., quantifying) the association between two variables.

Pearson's Coefficient of Correlation: The most common measure of *linear* association is called *Pearson's product moment coefficient of correlation* or simply the *correlation.* This is computed from the sample data by combining, using a formula, the pairs of values into a single number, usually denoted by r_p. There are several alternative formulae for r_p: the most convenient for hand calculation using a calculator is

$$r_p = \frac{n\sum XY - \sum X \times \sum Y}{\sqrt{\left[n\sum X^2 - \left(\sum X\right)^2\right] \times \left[n\sum Y^2 - \left(\sum Y\right)^2\right]}}.$$

The correlation coefficient is designed to give a value between –1 and +1 (inclusive) and is a measure of the linear nature of the data:

$r_p = -1$: perfect linear relationship with a negative gradient;
$r_p = +1$: perfect linear relationship with a positive gradient;
$r_p = 0$: no linear relationship indicated by the data.

Usually, we will get a value somewhere between these values, but hopefully "close to" ±1 ± 1. For example, we calculate the correlation coefficient r_p for the data given in Example 6.15 as follows. First, we construct a table like Table 6.22.

TABLE 6.22
Table to Aid the Calculation of the Correlation Coefficient

X	Y	X^2	Y^2	$X \times Y$
4	21	16	441	84
5	22	25	484	110
7	15	49	225	105
9	18	81	324	162
10	14	100	196	140
11	14	121	196	154
12	11	144	121	132
14	13	196	169	182
$\sum X = \mathbf{72}$	$\sum Y = \mathbf{128}$	$\sum X^2 = \mathbf{732}$	$\sum Y^2 = \mathbf{2156}$	$\sum X \times Y = \mathbf{1069}$

Now we can easily compute the correlation coefficient by substitution of the numbers at the bottom row of Table 6.22 in the given formula:

$$r_p = \frac{8 \times 1069 - 72 \times 128}{\sqrt{\left[8 \times 732 - (72)^2\right] \times \left[8 \times 2156 - (128)^2\right]}} \approx -0.87.$$

The calculated value of $r_p = -0.87$ shows high inverse linear association that can be interpreted as the more experience an employee has in wiring components, the fewer number of rejects they may be expected to produce.

6.13.3 Regression

Regression is concerned with establishing the form of the relationship between a dependent variable Y and one or more independent variables X. However, we will consider only a linear model where there is only a single independent variable in our model, and it is called the **simple linear regression model**. The model takes the following form:

$$Y = \beta_0 X + \beta_1.$$

To define the relationship between X and Y, we need to know the values of β_0 and β_1. However, these values are population parameters, which are always unknown. We need to find a way to estimate these parameters. The method of estimating these parameters, which is known as the *least squares approximation*, is beyond the scope of this book, and we only provide formulas to calculate the approximate value of these parameters as follows:

$$\hat{\beta}_0 = \frac{n \sum XY - \sum X \times \sum Y}{n \sum X^2 - \left(\sum X\right)^2}$$

$$\hat{\beta}_1 = \frac{\Sigma Y - \beta_0 \Sigma X}{n}.$$

The approximated model $Y = \hat{\beta}_0 X + \hat{\beta}_1$ is known as the **best-fitted line**.

For example, we revisit the data given in Example 6.15 and determine the equation of the best-fitted line using the data in Table 6.22:

$$\hat{\beta}_0 = \frac{n\Sigma XY - \Sigma X \times \Sigma Y}{n\Sigma X^2 - (\Sigma X)^2} = \frac{8 \times 1069 - 72 \times 128}{8 \times 732 - 72^2} \approx -0.9881$$

$$\hat{\beta}_1 = \frac{\Sigma Y - \hat{\beta}_0 \Sigma X}{n} = \frac{128 + 0.9881 \times 72}{8} \approx 24.893.$$

Hence, the equation of the best-fitted line is

$$Y = -0.9891X + 24.893.$$

The best-fitted line plot should now look like the line presented in Figure 6.24.

Interpretation of the $\hat{\beta}_0$ and β_1: The equation can be demonstrated as

$$\text{Number of reject} = -0.99(\text{weeks of experience}) + 24.9.$$

The slope is $\hat{\beta}_0 = -1$, which means that for each additional week of experience, the number of rejects produced in the week considered decreases by an average of 1.

The intercept is $\hat{\beta}_1 = 24.9$. Technically, the intercept is the point at which the sample regression line and the Y-axis intersect. This means that when $X = 0$ (i.e., the employee has no experience at all) the number of rejects produced is about 25. We might be tempted to interpret this as the number of rejects produced by a new

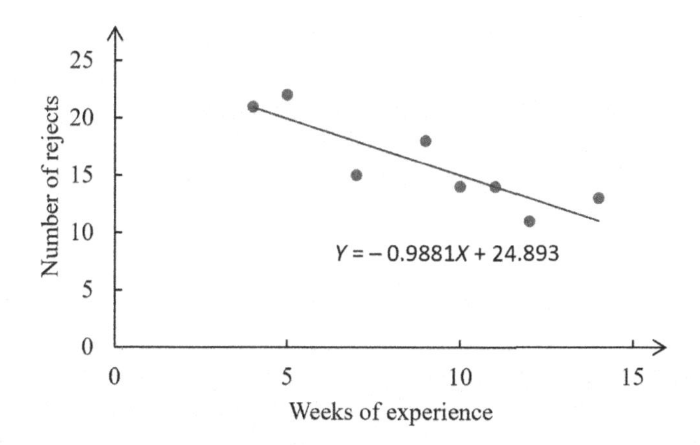

FIGURE 6.24 Graph of the best-fitted line superimposed on the scatterplot.

employee. However, this is a dubious interpretation. Because our sample did not include any employees with no experience at all, we have no basis for interpreting $\hat{\beta}_1$. As a general rule, we cannot validly determine the value of Y for a value of X that is well outside the range of sample values of X. In this example, the smallest and largest values of X are 4 and 14, respectively. Because $X = 0$ is not in this interval, we cannot safely interpret the value of Y when $X = 0$.

6.13.4 Exercises

1.

(i)

X	2	4	3	8	6	5	6	7	5	4
Y	15	10	16	6	14	12	12	9	11	15

$X = 6$

(ii)

X	5	7	9	11	13	15
Y	1.7	2.4	2.8	3.4	3.7	4.4

$X = 6$

For each of the tables of data given in part (i) and part (ii):

(a) Plot a scatter diagram
(b) Use your judgement to draw a straight line that attempts to fit the data.
(c) Determine the regression line Y on X, superimpose it on your scatter diagram and compare it with your estimated line.
(d) Determine the correlation coefficient and comment.
(e) Use the regression equation to estimate Y for the stated value of X.

2. A young apprentice once remarked, in a smug tone, "Younger carpenters are much more productive than older carpenters". Table 6.23 shows, for a random sample of 10 carpenters; the carpenters' ages, in years; and their productivity, measured by the number of days required to complete a standard set of kitchen units.
   (i) Construct a scatter graph of the data and estimate (without calculation) the value of Pearson's correlation coefficient between age and productivity.
   (ii) Calculate r_p and interpret your result in the data context.
3. The mileage per gallon (mpg) that a particular make of car obtains varies with speed (mph). The following data in Table 6.24 were obtained under test conditions.

TABLE 6.23
Data from a Sample of Carpenters

Age (X)	54	49	28	26	59	27	33	57	48	30
Productivity (Y)	12	15	21	14	12	26	20	12	14	21

(i) Plot a scatter diagram and determine the regression line Y on X, superimpose it on your scatter diagram. Use MPG as the dependent variable (Y) and MPH as the independent variable (X).

(ii) State explicitly the equation of the best-fitting line of MPG on MPH in the form
$\text{MPG} = \hat{\beta}_1 + \hat{\beta}_0 (\text{MPH})$. Provide (if possible) interpretations of $\hat{\beta}_0$ and $\hat{\beta}_1$.

(iii) Comment on the usefulness of the fitted regression equation in estimating fuel consumption for any particular speed.

(iv) Estimate the MPG that could be expected at 80 mph under the test conditions.

4. A typical residential house was selected by a local power company to develop an empirical model for energy consumption (in kilowatts per day) as a function of average daily temperature during the winter months. Over a 15-day period, the following information (Table 6.25) was obtained.

(i) Obtain a scatter diagram and fitted line plot of the data. Is a linear relationship suggested?

TABLE 6.24
MPG and MPH Data from a Sample of Cars

MPH (X)	30.0	35.0	40.0	45.0	50.0	55.0	60.0	65.0	70.0
MPG (Y)	45.2	46.2	47.3	47.9	46.3	41.3	36.2	33.5	28.3

TABLE 6.25
Average Temperature and Energy Consumption over a 15-Day Period

Average Temperature (°C)	Energy Consumption (kW)
0	70
8	57
7.5	60
13.5	63
14	57
8.5	66
4.5	67
−11	107
−7.5	96
−8.5	88
1.5	80
0.5	64
2	79
−6	82
−4	97

(ii) Interpret the estimated regression coefficients.
(iii) Estimate the energy consumption for this house over the period if the average temperature had been

(a) 6°C (b) –15°C

Comment on the reliability of these estimates.

Answers

1. (i)
   - (a) See Figure 6.25.
   - (c) $Y = 18.667 - 1.333X$.
   - (d) $r_p = -0.778$ [reasonable negative correlation].
   - (e) When $X = 6, Y = 10.667$.

   (ii)
   - (a) See Figure 6.26.
   - (c) $Y = 0.495 + 0.257X$
   - (d) $r_p = +0.995$ [very good positive correlation]
   - (e) When $X = 10, Y = 3.067$.

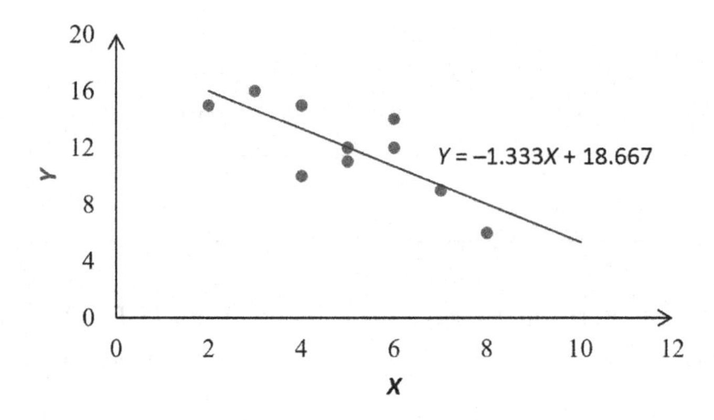

FIGURE 6.25 Graph of the best-fitted line superimposed on the scatterplot.

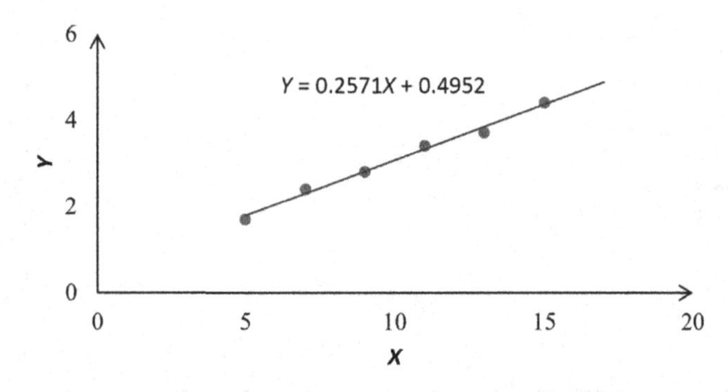

FIGURE 6.26 Graph of the best-fitted line superimposed on the scatterplot.

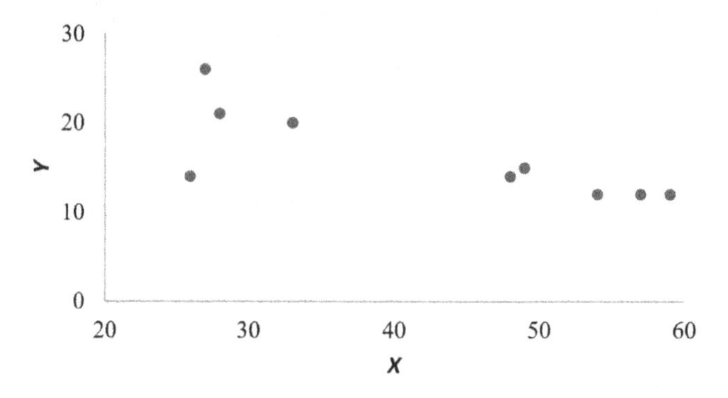

FIGURE 6.27 Scatterplot of productivity and age.

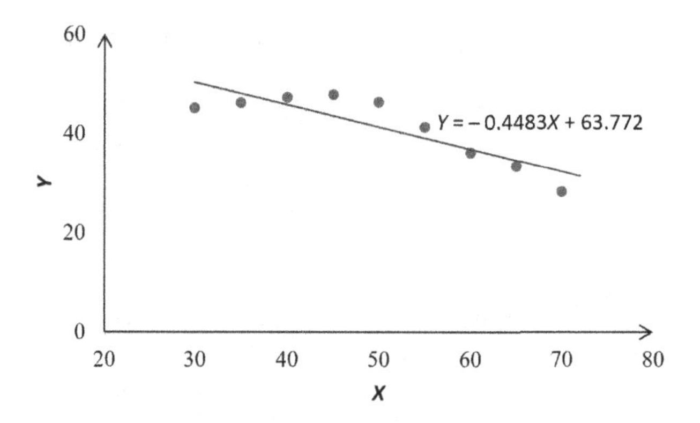

FIGURE 6.28 Graph of the best-fitted line superimposed on the scatterplot. The scatterplot shows non-linearity.

2.
  (i) See Figure 6.27. Estimated r_p is –0.8.
  (ii) The calculated value of $r_p \approx 0.791$, which represents a moderately strong inverse linear relationship between age and productivity: older carpenters tend to have a lower productivity score than younger carpenters, bearing out the smug apprentice's remark.

3.
  (i) See Figure 6.28.
  (ii) $\text{MPG} = 63.77 - 0.448(\text{MPH})$ or $Y = 63.77 - 0.448X$.
  Interpretation of $\hat{\beta}_0 = -0.448$: For each increase of 1 mph in the speed, the mpg on average drops by 0.45 miles.
  Interpretation of $\hat{\beta}_1 = 63.77$: No sensible interpretation, since when a car is stationary (mph = 0), the question of fuel consumption does not arise.
  (iii) The *linear* relationship between mpg and mph is weak, suggesting that the fitted equation will *not* be particularly useful for estimating fuel consumption from knowledge of speed.

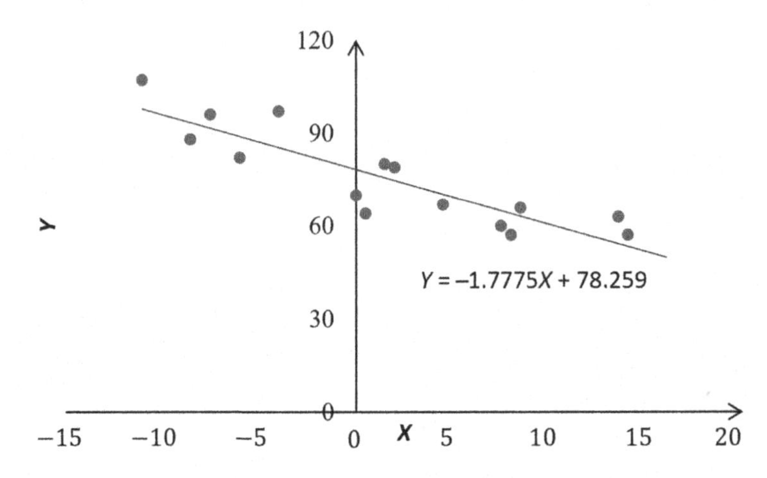

FIGURE 6.29 Graph of the best-fitted line superimposed on the scatterplot.

(iv) When mph = 80, mpg = 63.77 – 0.448(80) = 27.9, but note that, since 80 mph is outside the range of sample data and the plot is not linear anyway, this prediction is not reliable.

4.

(i) See Figure 6.29.

(ii) A fairly strong inverse linear relationship is suggested.

(iii) The best-fitting line has the equation

Energy consumption $=78.3-1.78$ (Average temperature).

The average energy usage per day is 78.3 kW when the temperature is at freezing point (0°C).

Each 1°C *decrease* in average temperature results, on average, in an *increase* of 1.78 kW in daily energy consumption.

(iii) (a) When average temperature = 6 °C, energy consumption = 67.6 kW.

(b) When average temperature = −15 °C, energy consumption = 105.0 kW.

The estimate in (a) is quite reliable, since 6 °C is within the range of sample values.

The estimate in (b) is not reliable, since −15 °C is well outside the range of sample values.

Index

A

acceleration, 41, 82, 256
air resistance, 82, 98
algebraic equations, 41, 42, 101, 112
algebraic multiplicity, 161
alternative hypothesis, 297–299
amplitude, 90–92, 226–231
analytical solution, 44, 75, 79, 81, 214, 217
angular frequency, 230–231, 238
approximate solution, 44, 75, 77–79, 81, 212, 214
approximation, 44, 75–77, 241, 275, 304
area under the curve, 1, 4, 20, 278

B

bar chart, 263–264
beat, 90, 92, 226–229
best fitted line, 305
block diagram, 150

C

capacitor, 95–97
census, 260
centre of mass, 30–36
characteristic equation
  complex eigenvalues, 163, 203
  differential equation, 57, 61, 68
  matrix, 154, 157
  repeated eigenvalues, 161
  roots of, 87
charge, 96
circular disc, 23, 26
class interval, 265, 271, 273–273
  limit, 265
  width, 265, 273
combined signal frequency, 219–220, 233
commutative property, 144
complementary function, 67–73, 92, 97
completing the square, 108–109, 113
complex conjugate eigenvalues, 163, 166, 205–207
complex eigenvalues, 162–168, 203–208
complex variable, 102
composite signal, 227, 230
confidence interval, 289–292
  limit, 289–292, 296
constant of integration, 43, 45, 51, 93
constructive interference, 227, *227*
continuous data, **272**
continuous quantitative variable, 271, 274
control
  engineering, 148
  system, 137, 148–150, 255
convolution, 137–145
  integral, 130, 138–145, 149–150, **152**
  properties, 144
correlation, 301–304
cosine wave, 228–229
critical damping, 87–89
critical value of step function, 115
current, 56, 96–97, 134, 137, 153
cylinder intersecting a paraboloid, 26–28

D

damped oscillation, 66
damper, 87, 91–92, 215, 255, 256
damping, 87–92, 99, 149, 215, 256
damping coefficient, 92, 215
dashpot *see* damper
data, 260–262
defective matrix, 177, 188, 190, 196
density, 31–32
  distribution, 30–32
dependent variable, 44, 302, 304
derivatives, 39, 42, 49, 51, 60, 68–70
  and the Laplace transform, 110–114
descriptive statistics, 260, 262
destructive interference, 227, *227*
discrete data, 262, 264
discrete quantitative variable, 262–264
determinant, 154, 157
diagonal matrix, 171
diagonalisation, 169–172, 174, 176, 181
differential equations, 39, 42
  first-order, 50–55
  order of, 43
  second-order, 56–73
  separable, 44–48
Dirac delta function, 129–134, **152**
Dirac function, 111, 112, 129–130, 143–144
direct integration, 43–44, 103, 174, 176
displacement, 40, 65, 133, 153
distributive property, 144
double integration
  general polar region, 25–28
  non-rectangular regions, 8–15
  polar rectangular region, 19–22
  rectangular regions, 3–8
  reversing the order, 15–18

type, 1 region, 9
type, 2 region, 10
double summation, 5

E

eigenvalues, 154–155
eigenvectors, 155
electrical circuit, 49, 95–96, 101, 115, 153
electrical systems, 114, 255
equation
circle, 26
motion, 64, 67
paraboloid, 26
straight line, 119
Euler
formula, 59, 103
method, 75–81
exact solution, 75–77
exponential function, **69**, 71, 103, 143

F

falling object, 82
filter, 257
fluid mechanics, 93
force, 40–41, 82–83, 87, 89–92, 115, 129, 131, 133, 153, 255–256
Fourier coefficients, 238
Fourier series
periodic functions, 236–238
sawtooth waveform, 243–246
square wave, 233–236
Fourier's law of heat conduction, 93–94
frequency, 218–219, 222, 229–230, 238, 263–265
frequency distribution, 263–265, **269**
frequency domain, 104, 229–231
frequency spectrum, 229–231, 238–239
friction, 40–41, 153
frictional force, 125
frictionless surface, 153
fundamental angular frequency, 222, 224
fundamental frequency, 222, 224–225, 234

G

general polar region, 25–28
generalised eigenvector, 178–179, 187, 191
geometric multiplicity, 162
gravitational force, 40, 82–83
grouped frequency distribution, 264–265, **269**, 271–272

H

harmonic motion, 39, 41, 49, 87
harmonic oscillator, 149
harmonics, 218–219, 224, 233–234, 236–237, 241
heat flux, 94
heat transfer, 93–94
Heaviside function, 114
histogram, 262, 265, 273–276
Hooke's law, 40, 153
hydraulic control valve, 255–256
hypothesis testing, 260, 297–300

I

identity matrix, 154
improper integral, 102
impulse function *see* Dirac delta function
impulse response, 130, 137, 148–149
independent variable, 42, 44, 102, 302, 304
inductor, 95–96
inferential statistics, 260
inherent variability, 261
initial value problems, 63–66
input signal, 148
integral
inner, 7
outer, 7
single, 1
integrating factor, 50–54, 83–84, 86, 96
integration by parts, 54, 103, 110, 143, 145, 244
interval estimate, 289
inverse Laplace transform, 104–106, 108, 121

K

Kirchhoff's voltage law, 96

L

lamina
centre of mass, 33–35
density, 31–32
density distribution, 30–32
mass, 31–32
moment of inertia, 36–37
Laplace domain, 149–150
Laplace transform, 101–102
first shifting theorem, 106–108, 120
linear property of, 102–103
table of, **150–152**
least squares approximation, 304
linear combination, 56, 58–59, 160, 187, 190–191, 193, 195, 199
linear differential equation, 39, 49, 50, 56, 82, 111, 216
linear operator, 105
linear systems of differential equations, 153–154
linear time invariant systems, 137, 145, 148, 150
linearly dependent vectors, 160
linearly independent, 57–60, 178–179, 187, 204
eigenvectors, 160–162, 171–172, 175–178, 182–186, 190

real eigenvectors, 169
solution, 178, 180, 187–188, 191–192, 194, 196, 198
vectors, 160, 180, 189, 192, 196, 198

M

mass balance equation, 85
mass per unit area, 31
McLaren series, 217
mean, 260, 262, 266, 275–276, 280, 287–288, 297–300
median, 262, 266
method of undetermined coefficients, 67–69, 71–73, 89, 112, 256
mixing problem, 209–215
mixture of solutions, 85–87
moment, 33–35
moment of inertia, 36–37

N

natural frequency, 90–92
Newton's laws of motion, 82
Newton's second law, 41, 82–83, 153, 215, 256
non-homogeneous equations, 67–73
non-linear differential equation, 39
normal distribution, 274–276, 279, 283, 287–288, 292, 294
null hypothesis, 297–299
numerical solution, 75, 77, **78**, 81, **81**

O

oscillation, 41, 90–91, 125, 215, 228, *227*
damped, 66
forced, 90
speed of, 40, 125, 133
overcritical damping, 87–89

P

paraboloid, 26–28
partial differential equation, 39, 42
partial fraction decomposition, 105, 108–109, 113, 145
Pearson's coefficient of correlation, 303
periodic function, 217–218, 233–234, 236–239, 245, 257
piecewise periodic function, 234, 237, 245
point estimate, 286–287, 289
polar coordinates, 19, 25–26
polar rectangular region, 19–22
population, 260–261, 274–276, 285–289, 292–293, 297, 299
position, 40, 65, 84, 125, 131, 256
pressure, 255–256
principle of superposition, 58–59, 256
probability, 271–278, 279–280, 282–285, 288–291, 294, 297
distribution, 271–273, 274–275
probability density, 272–274
function, 273, 274, *274*
histogram, 273
projection of the surface, 8

Q

qualitative variable, 261
quantitative variable, 262–264, 271, 274

R

random variable, 261, 272–273, 274–276, 279–280, 293
raw data, 262
real eigenvalues, 169, 174–177, 181, 184
rectifier
full wave, 257–259
half wave, 259
reduction of order, 60
regression, 301–302, 304, 305–306
rejection region, 298–299
relative frequency, **263, 272**
repeated eigenvalues, 161–162, 177, 179, 186, 190, 192, 194, 196
resistor, 95–97, 284
resonance, 91–92
RLC circuit, 95, 97
roots
of the characteristic equation, 61, 87, 154, 163, 182, 203
complex, 61
complex conjugate, 58, 163, 165–166
imaginary, 71, 73
real distinct, 58, 61–62, 68–69
repeated, 59, 61, 180
Runge-Kutta method, 77

S

sample, 260–261, 273
mean, 287–290, 293, 297–299
sawtooth waveform, 243–246
scanning line
horizontal, 10, 17
radial, 23, 26, 30
vertical, 9
scatterplot, 302–303, 305
second order coupled system, 177, 181
second order linear ODEs
homogeneous, 57–66
non-homogeneous, 67–73
second moment of mass, 36
second shifting theorem, 120–122, 134
separating variables, 39

shifted function, 115
simple harmonic function, 218, 238
significance level, 297–299
significance testing, 297
simple linear regression model, 304
sine wave, 219, 228–229
sinusoidal wave, 217, 219, 222
speed, 40–41, 65, 82, 133, 256, 301
sphere, 25
spring, 39–41, 87, 89–90, 91, 130, 255–256
spring constant, 40, 99, 130, 153, 255
spring stiffness, 40, 124
spring-mass system, 40, 87, 92, 132, 149, 153–154
  coupled, 173
spring-mass-damper, 87, 209, 256
  2 degrees-of-freedom, 215–216
streamlines, 93
square matrix, 154, 157
square wave, 219, 233–236
standard deviation, 260, 262, 266, 275–276
standard normal distribution, 276, 279
statistical
  estimation, 260
  inference, 260, 275, 286
  table, 276, **277**, 278, **283**
summary statistics, 260, 262, 296
summation, 5
superposition principle, 56
system of coupled equations, 174
system response, 130

T

t-distribution, 292–296
Taylor series, 217
temperature gradient, 94
thermal conductivity, 94
time dependent differential equation, 41
time dependent function, 102, 104, 111
time domain, 101, 149, 229–230
time period, 217
transfer function, 148–150
trigonometric function, 103, 217
trigonometric identities, 218, 226

U

uncoupled systems of equations, 175
under-critical damping, 87–89
unit step function, 114–119
  Laplace transform, 119–120

V

variance, 266
vibrating spring, 39–41, 56, 67, 124
vibration, 41, 56, 87, 89–92, 255–256
vibration analysis, 41, 82, 255–256
voltage, 56, 95–97, 114–115, 124, 129, 131, 257, 262, 271
voltage drop, 95–97
volume
  hemisphere, 23–25
  solid, 2–6, 20

Z

Z-distribution, 276, *276*, 283, 294, 298

Printed in the United States
by Baker & Taylor Publisher Services